Plant Life
in the
Devonian

Plant Life in the Devonian

Patricia G. Gensel
Henry N. Andrews

PRAEGER SPECIAL STUDIES • PRAEGER SCIENTIFIC

New York • Philadelphia • Eastbourne, UK
Toronto • Hong Kong • Tokyo • Sydney

Library of Congress Cataloging in Publication Data

Gensel, Patricia G., 1944–
  Plant life in the Devonian.

  Bibliography: p.
  Includes index.
  1. Paleobotany—Devonian. I. Andrews, Henry Nathaniel,
1910–   . II. Title.
QE918.G46 1984   560'.1'726   83-22964
ISBN 0-03-062002-3

Published in 1984 by Praeger Publishers
CBS Educational and Professional Publishing
A Division of CBS, Inc.
521 Fifth Avenue, New York, New York 10175 U.S.A.

© 1984 by Praeger Publishers

3456789 052 987654321
Printed in the United States of America

To

E. A. Newell Arber 1870–1918

for his book DEVONIAN FLORAS *in which he presented
the early land plants as they were known 60 years ago*

and to

Sir J. William Dawson 1820–1899

*for his pioneering studies of Devonian plants
in northeastern America*

Contents

1. Introduction   1

    Some Important Summary Accounts in Devonian Plant
Literature   2

    Some Historical Reflections   3

2. A Timetable and Classification   7

    Timetable   7

    Classification   11

3. Invasion of the Land: Pre-Devonian Plants   15

    Problematic Microfossil Remains   16

    Cooksonia: The Earliest Land Vascular Plant   19

    Other Possibly Vascular Plants   23

    Silurian–Devonian Nonvascular Plants   26

    Concluding Remarks   31

**4. Early Land Vascular Plants and the Origins of
Diversity**   33

    Part I. The Trimerophytes and the Rhyniophytes   36

    Part II. The Zosterophyllophytes   76

    The Question of Gametophytes   106

    Summary   112

5. The Lycopods   115

    I. Prelycopods   117

II. Lower and Middle Devonian Fossils that Clearly Establish the Lycopod Line   129

III. Chiefly Upper Devonian Fossils that Reveal the Origins of Heterospory and the Arborescent Habit in Lycopods   144

Summary   156

6. **The Sphenopsids (Horsetails)**   161

Devonian Plants Allied to the Sphenophyllales   164

Possible Equisetalean Devonian Plants   169

Summary   170

7. **The Cladoxylopsids, Coenopterids, and Some Possibly Related Forms**   173

The Coenopterids and *Rhacophyton*   175

Some Notes on Cambial Origins and Evolution   181

The Cladoxyls   184

The Iridopterids and *Reimannia*   203

Summary   211

8. **The Progymnosperms**   213

The Concept of Progymnosperms and *Archaeopteris* Dawson, 1871   214

Other Progymnosperms   227

Archaeopteridales, continued   228

Aneurophytales   237

Incertae sedis—*Triradioxylon*   254

Periderm formation in early land vascular plants   256

Protopityales   256

Summary   260

9. **The Early Seed Plants**   263

The Earliest Fossil Seeds   264

Some Apparently Primitive Seeds from the Lower Carboniferous   270

Some Platyspermic Seeds 279

Evolutionary Trends in Early Seeds 282

Great Forest Trees of the Early Carboniferous 285

Summary 292

10. Heterospory in the Devonian 295

Introductory Comments 295

Some Heterosporous Devonian Plants 298

Evidence from Dispersed Spores 312

Summary 312

11. Palynological Considerations 315

General Surveys of Palynological Data Pertaining
to Vascular Plant Evolution 316

Biostratigraphy 318

In Situ Spore Studies 319

Other Considerations 326

Ultrastructure 326

12. Devonian Floras and Concluding Remarks 329

Biogeography 329

An Early Devonian Landscape 332

A Middle Devonian Landscape 335

An Upper Devonian Landscape 335

Evolutionary Considerations 340

Future Possibilities 340

Literature Cited 347

Index 371

About the Authors 381

Acknowledgments

We have drawn our information for this account from many sources, other than our own experience and thoughts, as is indicated through the text and in the literature-cited section. Therein lies our greatest obligation and we acknowledge with thanks these efforts of the many investigators of the past century and a half and the publishing houses that have allowed us to reproduce portions of their journals and books.

There are some who have been especially generous with their services whom we wish to mention individually. Patricia Bonamo and Dianne Edwards have read two of what we regard as the more important sections and their comments have significantly improved our text. Among others who have supplied special items of information or illustrations, we especially wish to acknowledge: Charles B. Beck, David F. Brauer, William G. Chaloner, Jeffrey B. Doran, David S. Edwards, Muriel Fairon-Demaret, James D. Grierson, Christine M. Hartman, Ove Abro Høeg, William S. Lacey, Xing-xue Li, Albert G. Long, A. G. Lyon, Sergius H. Mamay, Lawrence C. Matten, Stephen E. Scheckler, Bruce S. Serlin, and Danuta Zdebska. Additionally, Mitchell Sewell and Susan Sizemore, Biology Department, University of North Carolina, have provided considerable technical assistance with the photographs and drawings. James Belcher, Herb Harwell, Linda Raubeson and Charles Trant greatly assisted in the preparation of plates and in proofreading.

Plant Life
in the
Devonian

1 / Introduction

It is appropriate first to state our reasons for preparing the following account and to outline our aims and objectives. During the past three decades there has been a strongly increased interest in the plants of Devonian age along with those that immediately preceded them in the late Silurian and those that followed in the basal part of the Carboniferous.

It seems reasonably clear now that land vascular plants originated in the late Silurian and early Devonian; we find in this time range some well-preserved and remarkably small, simple and primitive plants that represent the beginnings of a land flora. None of these have survived to the present day but once established they began to evolve into several diverse lines which become evident by the late Early Devonian and early Middle Devonian. As we progress through the Middle Devonian and into the Late Devonian, major groups of pteridophytes appear; and by Late Devonian times the first seed plants are recognized. One cannot, of course, draw sharp lines of plant classification across the formal geologic time boundaries and since we now have a considerable knowledge of early gymnosperms in the basal Carboniferous we have extended our account to that point. In essence we have some of the most important phases in the evolution of the plant world within this time period of about 50 million years (Late Silurian to basal Carboniferous).

It is especially important to state the limits of our coverage and the audience that our account is designed to serve. As might be expected in dealing with plants that date back in the vicinity of 350–400 million years, many of the published accounts describe very fragmentary remains that convey correspondingly little information. A second special problem lies in the fact that our present knowledge of some of the plants has developed over the years through numerous scattered reports and some of these present constantly changing generic concepts. We want this account to be useful to as broad a range of readers as possible, and especially to teachers of botany, geology, and paleontology. We have therefore sorted out what seem to us to be the most significant and best known plants to describe. Many of the more fragmentary, and some of the more problematical, fossils have been omitted and we have condensed or left out many reports in the older literature in favor of more recent discoveries. Thus the specialists in Devonian botany may not find our text as encyclopedic as they might wish; to serve this end in some degree we have prepared a rather extensive bibliography.

SOME IMPORTANT SUMMARY ACCOUNTS IN DEVONIAN PLANT LITERATURE

We are citing here some of the more informative accounts which should at least be known to anyone with a serious interest in the Devonian floras. Our starting point, and a landmark of real distinction, is E. A. Newell Arber's little book *Devonian Floras* which appeared in 1921. Arber made numerous contributions to Paleozoic paleobotany and had an especially good firsthand knowledge of the geology of Devon.

Arber's *Devonian Floras* was edited by his wife, Agnes Arber, with some aid from three great men of the time—D. H. Scott, A. C. Seward, and W. T. Gordon—Newell Arber himself having died at the age of forty eight in 1918. Arber recognized two more or less distinct floras, the *Psilophyton* flora of the Early Devonian, and the *Archaeopteris* flora of the Late Devonian. It is a classic work in fossil botany and is worth thumbing through to understand the status of Devonian paleobotany at that time and see the tremendous advances that have been made over a period of sixty years.

In 1927, E. W. Berry wrote a rather brief account entitled "Devonian Floras" which appeared in the *American Journal of Science* (vol. 14, pp. 109–120). And in 1937 Ove Arbo Høeg brought out "The Devonian Floras and Their Bearing upon the Origin of Vascular Plants" in the *Botanical Review* (vol. 3, pp. 563–592.)

By far the most comprehensive summary of the plant life of the past has been brought together in the several volumes that have been published in the *Traité de Paléobotanique* by Masson et Cie, under the general editorship of Edouard Boureau. The treatment of Devonian plants is distributed through them as follows: Volume II (1967) *The Psilophyta,* by O. A. Høeg; *The Lycophyta* (in part) by W. G. Chaloner; Volume III (1964) *The Sphenophyta* by E. Boureau; and Volume IV (1970) includes those plants that in a broad sense were regarded as ferns and the newly established progymnosperms, by H. N. Andrews, C. A. Arnold, E. Boureau, J. Doubinger, and S. Leclercq.

This is a vast source of information, copiously and beautifully illustrated and includes a detailed documentation. It will be an indispensable reference source for many years to come. Since our approach here is somewhat different, the reader requiring detailed information on certain groups not included here may find it in the more encyclopedic *Traité.*

With the rapidly accumulating information that was being dug out of the Devonian rocks it was becoming evident by mid-20th century that the amount of data at hand exceeded the clarity of its understanding and arrangement! No one was more aware of this than Harlan P. Banks, and the revised classification that he presented in "The Early History of Land Plants" at a Yale University symposium in 1968 has brought considerable order out of chaos, at least for the present. Banks was concerned particularly with the more primitive plants of the Early and Middle Devonian, the term *psilophyte* having been applied to a wide assemblage of plants. He presented a new classification of this assemblage and confined the term *psilophyte* to one segment.

In 1979, Chaloner and Sheerin prepared a very useful discussion entitled "Devonian Macrofloras" in which they show, in chart form, the stratigraphical record of many Devonian plants as well as chronological charts that record the first appearance of numerous vegetative and reproductive features. This study contains a great deal of highly condensed data and is a valuable source of information for anyone concerned with the more detailed aspects of the evolution of plant life in Devonian times.

Several other summary accounts have appeared which deal with more specialized aspects of the whole story and they will be referred to in the appropriate chapters that follow.

SOME HISTORICAL REFLECTIONS

A few words about the historical development of Devonian plant studies may be of interest to note the sharp rise in interest immediately after Arber's *Devonian Floras* appeared. The Canadian naturalist and educator, Sir J. W. Dawson, had recorded in numerous publications between 1850–1890 much basic information on Devonian plants based on his extensive explorations in the Maritime Provinces. His studies generated little interest outside the circle of a few specialists, in part because the plants he described were so unlike anything that was living or that occur in younger sediments.

The discovery of the beautifully preserved Early Devonian petrified plants at Rhynie, Scotland in 1912 and their subsequent study by Robert Kidston and William H. Lang in 1917–1921, has been credited with the revival, or rather initiation, of a real and continuing interest in Devonian plants. This is to some extent true but it also was a result of the appearance in Britain and on the European Continent of a class of remarkably competent botanists many of whom had a strong interest in paleobotanical-evolutionary studies. Kidston (1852–1924) and Lang (1874–1960) collaborated very effectively and after the former's death, Lang engaged in a series of studies of plants found in the Old Red Sandstone of Scotland.

Lang was one of the foremost morphologists of his time. Kidston left his place of employment in a Glasgow Bank in his late twenties to pursue his career in paleobotany. He was to a considerable degree self-made as a botanist; he was intelligent, hard-working, and a tireless collector. Knowing his botanical limitations, he collaborated in some studies with men such as David Gwynne-Vaughan and Lang. Kidston was probably the foremost authority in his day on the stratigraphic distribution of Carboniferous plants in Great Britain.

William Lang had two students of special note with whom he collaborated and who carried on his fine standards: Isabel Cookson (1893–1973) who is especially well known for her work with the Late Silurian *Baragwanathia* flora of Australia and William N. Croft whose important studies of early Devonian plants from Wales were cut short by his death at the age of thirty-eight.

On the continent, Richard Kräusel (1890–1966) and Hermann Weyland (1888–1974) laid some of the important foundations on which our present knowledge of Devonian plants is based. Of special note are their three papers entitled *Beiträge zur Kenntnis der Devonflora* (1923–1929). These were based largely on German compression fossils which to a degree complemented the Rhynie studies; they developed new techniques including an improved one for dealing with fragmentary but highly important pyritic petrifactions. Kräusel published in many areas of paleobotany and was long associated with the University of Frankfort and the Senckenberg Museum. Weyland was employed by Bayer and Co. (I. G.

Farben) as a chemist for many years and carried on much of his paleobotanical work in his "spare time."

Farther north, Alfred G. Nathorst (1850–1921) initiated investigations of Devonian plants in the Arctic regions, notably Bear Island and Spitsbergen. Nathorst's paleobotanical studies were diverse and abundant and he was responsible for the establishment of the distinguished Paleobotanical Department of the Natural History Museum in Stockholm. Ove Arbo Høeg, for many years a professor at the University of Oslo, extended the studies in Spitsbergen, his volume on "The Downtonian and Devonian Flora of Spitsbergen" (1942) being one of the most informative contributions to Arctic paleobotany.

Belgium has produced two of our most productive Devonian paleobotanists. Suzanne Leclercq, for many years a professor at the University of Liège, started her career with coal ball studies and soon transferred her interests to the Devonian. Her studies of the Middle Devonian plants from the quarry at Goé have contributed a wealth of information and much remains to be done there. François Stockmans, who was associated with the Natural History Museum in Brussels, has gathered together there one of the finest collections of Devonian plants in the world. His monographic volumes reflect his strong interest in the stratigraphic use of fossil plants while Miss Leclercq's studies have been evolutionarily oriented.

In this country we will confine our brief historical comments to a few words on Chester A. Arnold (1901–1977) who was the modern pioneer in studies of Devonian floras in the eastern United States. He is to be remembered especially for his work on plants that are now assigned to the progymnosperm group.

We have drawn our information for this account from the sources noted above and from many others and to some degree from our own investigations. The chapters that follow are the result of the labors and thoughts of many people; we acknowledge their contributions with thanks and appreciation.

2 / A Timetable and Classification

TIMETABLE

We will frequently make reference to the time-wise (stratigraphical) occurrence, the geographical distribution, and classification of the fossil plants that we are describing. We are therefore giving in this chapter certain basic information that may serve as a useful reference source.

Our timetable (Table 2.1) is taken in part from the very useful chart that has appeared in several editions from Elsevier in Amsterdam. It will be observed that two divisions of the Devonian are shown, one with the basic Lower, Middle and Upper divisions in general use on this side of the Atlantic and the somewhat more detailed European one. However, the various authors, whether European or North American, are not necessarily confined to using one system or the other. And in virtually all cases where fossil plants have been properly reported the Formation date is given. For further information on such detailed stratigraphy the reader is referred to works such as:

Cooper, G. A., et al., 1942. Correlation of Devonian sedimentary formations of North America. *Bull. Geol. Soc. Amer.* 53: 1729–1793.

House, M. R., J. B. Richardson, W. G. Chaloner, J. R. L. Allen, C. H. Holland, and T. S. Westoll. 1977. A correlation of Devonian rocks of the British Isles. *Geol. Soc. Lond.,* Spec. Report No. 7, 110 pp.

House, M. R., Scrutton, C. T., and Bassett, M. G. (Eds). 1979. The Devonian System. *Special Papers in Palaeontology* 23: 1–353.

TABLE 2.1. Geologic Timetable for the Periods Covered in the Present Account.[a]

| System | Series/Stage | | Time Began (my) | Duration (my) | Plant Megafossil Generic Assemblage-Zones al. Banks, 1980 |
	N. American	European			
Mississippian	Lower	Visean			
		Tournaisian Tn 1b			
		? Tn 1b (part) ? Tn 1a	345		----- ? -----
	Upper	Famennian	353	8	VII. Rhacophyton
		Frasnian	359	6	VI. Archaeopteris
Devonian		Givetian	365	6	V. Svalbardia
	Middle	Eifelian	370	5	IV. Hyenia

			Age (Ma)		Floral zones
	Lower	Emsian	374	4	III. *Psilophyton*
		Siegenian	390	16	II. *Zosterophyllum*
		Gedinnian	395	5	
Silurian	Upper	Pridolian	405	10	I. *Cooksonia*
		Ludlow	415	10	
	Lower	Wenlock	425	10	
		Llandovery	435	10	

[a](Derived from the following sources: The Geological Timetable, 3rd Ed., Elsevier, Amsterdam, 1975, compiled by F. W. B. Van Eysinga; Banks, H. P., 1980, Floral Assemblages in the Siluro-Devonian, *in* D. L. Dilcher and T. N. Taylor, Eds., *Biostratigraphy of Fossil Plants*, Dowden, Hutchinson and Ross; T. N. George *et al.*, 1976, A correlation of Dinantian rocks in the British Isles, *Geol. Soc. London Spec. Report. No. 7*; and M. House *et al.*, 1977, A correlation of Devonian rocks of the British Isles, *Geol. Soc. London Spec. Report No. 8.*

Martinnson, A. (Editor) 1977. The Silurian-Devonian Boundary. *Int. Union of Geol. Sci. Ser. A.* No. 5, 347 pp., E. Schweitzerbart'sche Verlagsbuchhandlung, Stuttgart.

Oswald, D. H. (Editor). 1968. *International Symposium on the Devonian system.* Alberta Soc. Petrol Geol., 2 vols, 1377 pp., Calgary.

It is important to emphasize that there are many areas in the world where research on Devonian stratigraphy is very active and there is much that remains to be learned about the correlation of Devonian rocks over disjunct areas. This may mean a problem of correlation through a distance of only a few miles, such as Devonian horizons in northern Maine and nearby New Brunswick or Quebec (where the present authors have worked for some years), or over much greater distances as, for example, between Wales on the east side of the Atlantic and New York State on this side, both being areas that we will refer to in the following chapters.

Although much geological and botanical information is now available, the matter of dating any particular horizon where fossil plant material is found varies greatly. If the Formation or deposit is bracketed by others whose age is established the dating may be quite precise but if the Formation is isolated the dating becomes more difficult.

Sufficient information has now accumulated so that when a flora of several well preserved species is encountered its age can be determined with considerable precision. But if the fossil plant content consists of very fragmentary remains, or even a small number of well preserved ones (representing few species), the degree of precision will be correspondingly limited. We still have inadequate information on the longevity (time range) of most Devonian plants. In recent years the application of spore studies to problems of age determination and correlation has contributed to greater accuracy.

It is worth noting that geologists have instituted rules, following the Code of Stratigraphic Nomenclature, governing how to refer to rocks, fossil assemblages, and time. For our purposes, the terms Lower, Middle and Upper Devonian are time-stratigraphic units, essentially referring to *rocks* which serve as a record for a specific interval of geologic time. The terms Early, Middle and Late Devonian refer solely to *time* based on the rock record, i.e., a specimen may be of Early Devonian age.

To know the precise position of a fossil-bearing horizon, relative to those below and above it, is all-important if one is trying to understand the racial sequence or development of the various plant groups. But it is also of some interest to have some understanding of the age in years, and people who have some interest in the life of the past will often ask "How old is it?" in reference to a particular fossil plant or animal. The age in years is determined by several radiometric methods which are described in most textbooks on paleontology or stratigraphy. Thus the figures in the column at the right of Table 2.1 will give at least an approximate age for the Devonian plants described here.

In spite of the difficulties mentioned above, some attempts have been made to characterize major units of Devonian time based on plant megafossils. As noted earlier, Arber (1921) recognized an Early Devonian assemblage characterized by *Psilophyton* and a Late Devonian one characterized by *Archaeopteris*. Kräusel (1937) added a Middle Devonian assemblage characterized by *Hyenia* and S. Leclercq (1940) suggested substituting *Protopteridium* (now *Rellimia*) for *Hyenia* because it seemed to be more widespread.

Banks (1980b) presents the most current Devonian megafossil assemblage zonation, including lists of the plant structures typical for each level, such as type of vascular strand, branching pattern or leaf (if any). This should be of use in establishing a tentative age for a sample where fossils are too poorly preserved to be identified to genus. The megafossil assemblage zones are also compared to current dispersed spore assemblage zones of McGregor (1977) and Richardson (1974).

CLASSIFICATION

There are several distinctive, if not unique, problems involved in trying to classify Devonian plants and it is important to understand these when dealing with the literature on the vegetation of this period.

We are concerned with plants that for the most part have no close relatives in the living flora. The plants of Late Silurian and Early Devonian age are small and simple in form as compared with most living vascular plants. Their aerial parts are not composed of distinct stems and leaves and only rarely do we have adequate information concerning their underground parts; where this is available (in the case of the more primitive plants), it suggests there was little structural difference between the subterranean and aerial axes. *Psilotum*, the living Whisk fern, is one of the very few living plants that offers at least a superficial resemblance to those of the Early Devonian and the consensus of opinion is that *Psilotum* is not related to any of the plants described here.

Beginning with the description in 1917 of the early Devonian flora of Rhynie, Scotland, an accelerating interest and flow of publications has continued to the present time. Although many of the plants are of simple morphology as compared with modern vascular plants, they do present considerable variation in branching patterns, form and arrangement of the spore-bearing organs, and presence (and variety) or absence of emergences. Most of the paleobotanists who published their findings during the half century after 1917 were more intent on reporting new and fascinating plants than on any critical analyses of their interrelationships. Thus a great many of these fossils were simply referred to as "psilophytes," a name that unfortunately became firmly established and much too inclusive.

A quick look at the historical origin of the name *psilophyte* is of interest although it perhaps adds to the confusion. In his *Devonian Floras* of 1921, Arber wrote:

> It is now clear that in Devonian times, two terrestrial floras, quite distinct as regards affinity, existed, one in the earlier part, and one in the later portion of the Devonian period. The former will here be termed the *Psilophyton* flora; it consisted largely, as we hope to show, of Thallophyta belonging, for the most part, to a group now quite extinct which we propose to term the Procormophyta or Propteridophyta. (1921, p. 1)

In their treatise of 1917, in which *Rhynia* was first described, Kidston and Lang proposed a new taxon, the Psilophytales, distinguished as follows:

> This Class is characterised by the sporangia being borne at the ends of certain branches of the stem without any relation to leaves or leaf-like organs. For this Class we propose the name Psilophytales. This name is derived from that of the earlier described though less perfectly known genus *Psilophyton*, and further suggests the resemblance between the plants of this class and the existing Psilotales. (p. 779).

The authors included two genera in their new class, *Rhynia* and *Psilophyton*. It should be noted that the origin of the common name "psilophyte" stems from both a fossil and a living plant. Since then the presumed relationship between early Devonian vascular plants and extant Psilotales has been generally discounted as one of apparent simplicity of form and structure.

In his survey of 1968, *The Early History of Land Plants*, Banks confined the common name *psilophyte* to the plants that he included in his subdivision Rhyniophyta. In view of the diversity of types that this name has been used to encompass, this was a move in the right direction. It would seem better still if the name could be discarded completely, one reason being that the plants included in the genus *Psilophyton* constitute the focal point of another subdivison. Of greater importance is the *general morphological diversity* mentioned above of the plants now included in the three subdivisions established by Banks. Probably the most significant aspect of this diversity centers in the variety of sporangial structure including: the shape and size of the sporangia, their arrangement and the way in which they are borne on the plant, and their mode of dehiscence where this is known. When preserved, the vascular anatomy of these plants is of diagnostic value also.

Any classification of Devonian plants in use at present must be understood to be tentative and this will probably hold true as long as "new" plants continue to be reported. But some kind of classification is necessary. We think it may prove most useful to present here a very brief outline of the major groups, with a few notations, and then give as detailed a classification as is appropriate in the respective chapters where they are described.

Division	Tracheophyta
Subdivision	Trimerophytina
	Rhyniophytina
	Zosterophyllophytina
	Lycophytina
	Sphenophytina
	Pterophytina
Class	Cladoxylopsida
	Coenopteridopsida
	Filicopsida
	Progymnospermopsida
	Pteridospermopsida

The first three subdivisions, the trimerophytes, rhyniophytes, and zosterophyllophytes, compose what is now a rather large assemblage of early Devonian plants that, by any standards, are primitive and simple in their morphology. This assemblage includes the plants that were collectively referred to as "psilophytes" prior to 1968.

With the continued discovery of new plants since 1968 the sharpness of the identity of each of these three has tended to decrease. This is in no way a censure on the classification but rather reveals an increase in our knowledge of the interrelationships and evolutionary trends in the assemblage as a whole.

We are somewhat concerned with the "taxonomic significance" of these three subdivisions in comparison with the three that follow, especially the Pterophytina. It is of course an old problem as to how one should employ taxa (species, genus, family, etc.) in groups of plants of widely diverse morphology. The plants included in these first three subdivisions constitute the beginnings of the whole land vascular flora and, as might be expected from their relative simplicity, the number of structural characters is small. But whether or not their relative value as subdivisions is valid we regard it as a useful classification for the present.

It is our assumption at this time that it is from the trimerophytes, rhyniophytes, and zosterophyllophytes, that the later groups of pteridophytes and early seed plants evolved. The Lycophytina (lycopods) and Sphenophytina (horsetails), represented respectively today by genera such as *Lycopodium* and *Equisetum*, present distinctive and vexing problems. It is futile, and probably violates a rational concept of plant evolution, to try to identify the "first lycopod" or the "first articulate." However we are very much concerned with the origins of these two groups. In the case of the lycopods we have several early Devonian plants that give some clues to their beginnings. The origin of the articulates, which become so abundant and diverse in the Carboniferous, is very obscure. They are clearly identifiable in the Upper Devonian, but fossils referred to the group from earlier horizons are questionable.

The Cladoxylopsida, Coenopteridopsida, and Filicopsida include plants that in a very general and unfortunately vague fashion seem to represent the origin of the ferns as the latter appear in later geologic periods. The term "fern" has been applied to a vast array of plants, living and fossil, and we defer further discussion for a later chapter.

The Progymnospermopsida was proposed in 1960 for plants that display a combination of pteridophytic and gymnosperm characters. This is one of the major contributions that paleobotany has made to elucidating plant evolution, but we are in the early stages of understanding just what plants should be called a progymnosperm.

One of the most significant developments in Devonian botany has been the recognition of seed-plants in this period. In view of the considerable amount of knowledge that we now have on the pteridosperms (seed-ferns) in the Carboniferous, we are extending our discussion of this group into the Lower Carboniferous.

In addition to those fossils that can be assigned with some degree of confidence to the groups listed, there are many reports of plants, some very well preserved, that cannot be satisfactorily classified. However some of them give us a significantly broader understanding of the diversity of Devonian vegetation, as well as glimpses of plant groups that probably will be better known in the future. We intend to be highly selective in dealing with these problematical fossils and conservative in commenting on their possible affinities.

3 / Invasion of the Land:
Pre-Devonian Plants

Two very critical and problematic phases in the development of plant life on the earth that have interested many workers in recent years are the origin of unicellular organisms and the subsequent development of a land vegetation of vascular plants. The first of these is quite outside our thesis but leads to the second, and it may simply be noted here that it remains something of a puzzle as to why about two billion years or more elapsed between the two events. Similarly, information on events leading from the presumed algal ancestors to land plants remains unknown although numerous theories have been postulated.

We are concerned with the evidence bearing on the earliest vascular plants. Numerous articles have appeared in recent years which speculate and offer evidence, or presumed evidence, concerning the time at which this took place. It is doubtful that it would be profitable for us to review all of the literature even if space permitted. We will cite a few representative studies which will help to refer the interested reader to the trend of affairs. It is our opinion that the first significant megafossil land plant remains (which *may* be vascular) presently known come from the lower Upper Silurian (Ludlow-Edwards et al., 1979). With this as a basis, it is of interest to note what several others have concluded:

Suzanne Leclercq (1954) wrote an article entitled "Are the Psilophytales a starting or a resulting point?", stating "Despite prevailing scepticism regarding the existence of precursors of the vascular plants in the Cambrian, I am inclined to admit that possibility." (p. 313).

Gray and Boucot (1977) wrote: "The evidence we have presented can be construed nevertheless as strongly favoring the presence of a terrestrial flora well prior to the first appearance of vascular plant megafossils." (p. 171).

W. G. Chaloner (1967) wrote: "The palynological evidence is consistent with a Silurian origin of land plants followed by a relatively slow diversification, rather than an early Paleozoic or Precambrian origin suggested by some authors." (p. 83).

H. P. Banks (1972) wrote: "*Cooksonia* in the Downtonian (uppermost Silurian) of Wales is currently the oldest proven vascular plant macrofossil." (p. 365).

PROBLEMATIC MICROFOSSIL REMAINS

Although there is very little controversy over the lack of significant macrofossils below the Wenlock (mid–late Silurian), there is a great deal of disagreement concerning the meaning of the microfossil remains, some as ancient as the Cambrian. Banks (1975b) has prepared a concise summary of the major types of evidence, which may be referred to for more detailed information. Briefly there are three categories of microfossils that have been discussed as evidence of the existence of pre-Pridolian, land, vascular plants: sheets of cells or the cuticular remains of them, tubular structures that to some degree resemble tracheids, and spores. We offer a few comments on these taken in part from Banks' summary and in part from other more recent papers.

Sheets of cells

Several authors have described fragments appearing to represent sheets of cells or cuticle with impressions of cells (Fig. 3.1C). These are commonly found in Llandoverian or Wenlock strata and in a presumed Upper Ordovician horizon of Libya. They conceivably could have come from axes of land vascular plants. However, they are known only as isolated fragments and lack stomata. Alternative explanations exist, such as their representing some type of algal, bryophyte, or animal remains.

Tubes

Some of these are especially intriguing, occurring isolated (as in Pratt, et al., 1978) or from the compressions which Lang (1937) described under the name *Nematothallus* (Fig. 3.1A, B). The latter consists of intermixed large (up to 25 μm) and small (2.5–5.0 μm) tubes, covered by a cuticular layer similar to those discussed above, and with spores present.

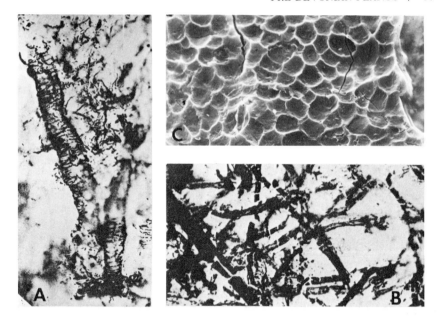

Figure 3.1. A,B. Fragments of tubes, banded and unbanded, described by Lang as parts of *Nematothallus*. A. × 250. B. × 245. C. SEM of part of a sheet of cuticle-like material. × 480. (A,B. from Lang, 1937, Phil. Trans. Roy. Soc. London 227B. C. Courtesy of Dianne Edwards.)

The large tubes have annular thickenings (Fig. 3.2D, E) and to some degree resemble tracheids of vascular plants such as *Asteroxylon.* So far as we are aware, they lack end walls and in some cases they divide. The small tubes are smooth-walled (Fig. 3.2B, C). Isolated tubes resemble the two types described by Lang. Some have been reported as associated with chitinozoans or graptolites which may or may not be significant. There is no positive proof that they were derived from land vascular plants such as those discussed in the present treatise.

Spores

There are numerous accounts describing spores from pre-Devonian strata which have a resistant wall and triradiate mark (Fig. 3.2F) or are in tetrads. Spore tetrads (Fig. 3.3) date from the Upper Ordovician (Gray et al., 1982), while isolated spores occur from the Llandovery on. It is possible that some of these early types of spores may have been produced by bryophytes, algae, or highly problematical plants such as *Parka.* Gray et al. (1982) also suggest possible affinities with vascular plants. The lack of any association with significant macrofossils lends strong doubt to any of these possibilities at the present time.

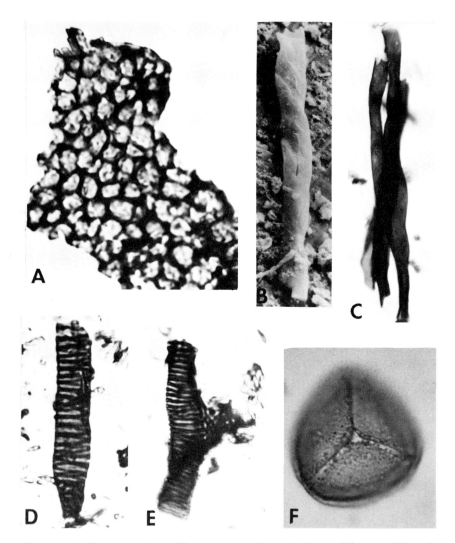

Figure 3.2. Remains of possible land plants from the Lower Silurian of Virginia. A. Cuticle-like remains. × 530. B,C. Smooth-walled tubes. B. × 542 C. × 511. D,E. Banded tubes, one branched. both × 665. F. A trilete spore from the same preparation. approx. × 1000. (A–E. From Pratt et al. 1978, Rev. Palaeobot. Palynol. F. Courtesy of L. Pratt)

The recent reports of these types of remains from continental-derived sediments by Pratt, et al. (1978) and Strother and Traverse (1979) support earlier arguments (Lang, 1937) for a possible terrestrial habitat for the organisms which they represent.

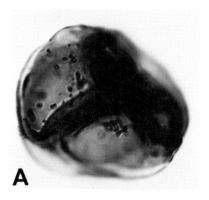 

A   B

Figure 3.3. Spores from the Ordovician of Libya. A. A tetrad. × 1500. B. A single spore interpreted as being trilete. × 1500. (Both from Gray et al. 1982, Geology.)

The evidence presently at hand is not convincing as to the existence of land vascular plants prior to the Ludlovian, although the fragments discussed above may represent land dwelling nonvascular plants. However, it is only fair to add that the door should not be closed to such a possibility. We have experienced many surprises in paleobotany and the time period with which we are concerned has had its share. Many "new" and strange plants have turned up in the past two decades. Perhaps of special importance is the paucity of known transitional forms of plant life between the algae and vascular plants.

The earliest occurrence of a probable vascular plant is from the late Ludlow of Wales (Edwards, et al., 1979), where they record fertile specimens of *Cooksonia* sp. and slender bifurcating axes. Similar but less well-preserved ones are recorded from the late Wenlock of Ireland (Edwards, et al., 1983b). The latter exhibit dark central strands but no actual vascular tissue has been obtained. The authors propose that the plants lived in a lowland environment, since they were preserved in near-shore marine sediments.

COOKSONIA: THE EARLIEST LAND VASCULAR PLANT

Current evidence suggests that plants placed in the genus *Cooksonia* presently offer us the most authentic information that we have concerning the nature of the oldest, simplest and most primitive land vascular plants. Banks (1972) refers to *Cooksonia* as "as simple a vascular plant as one can imagine " (p. 365); this seems to us to be an accurate statement and several other genera included in the rhyniophyte group come close to it in simplicity (see Chapter 4). It is therefore vital to take a close look at what is known about the plants that have been described in this genus.

Cooksonia was founded by W. H. Lang (1937) on numerous small and fragmentary specimens collected from several localities of the Downton Series in Wales. Holland and Richardson (1977) and House, et al. (1977) equate the Downton Series with the Pridolian or Upper Silurian; the Silurian–Devonian boundary apparently lies near the top of the Downtonian but the exact boundary position is still debatable. In regard to the plant remains found there, Lang's application of appropriate study techniques, his descriptions, and his extensive knowledge of living plant morphology resulted in an example of excellence that any paleobotanist might well emulate.

Two species of *Cooksonia* were erected in his 1937 account, the type species *C. pertoni* and a second one, *C. hemisphaerica*. The genus is named for Isabel Cookson who collected some of the more critical specimens on a visit to a quarry at Perton Lane with Lang in 1934. *C. pertoni* is based on smooth dichotomizing stem compression fragments of about 2.5 cm long and up to 1.5 cm wide, some of which bore terminal sporangia that are noticeably broader than long (Fig. 3.4A, B), measuring about 2×1 mm. The sporangia show no evidence of a specialized dehiscence region. Lang demonstrated the presence in these of smooth spores with a triradiate mark measuring about 32 μm in diameter (Fig. 3.4F). Some of the axes showed a central strand of elongate cells but the wall thickenings were not well preserved. The second species, *C. hemisphaerica*, came from the Targrove Quarry (now known to be Gedinnian in age) and is based on comparable fragments but with hemispherical terminal sporangia measuring 2 mm in diameter. These did not yield spores. Lang was able to demonstrate the presence of a slender vascular strand consisting of annular tracheids in associated vegetative axes he believed represented *C. hemisphaerica*. None have yet been found in fertile specimens; despite this, *Cooksonia* is regarded as being a vascular plant (see Edwards, et al., 1979, for a complete discussion).

Thus based on numerous specimens obtained from several localities, *Cooksonia* is revealed as a plant that was certainly small, probably only a few centimeters tall, and possessing a cutinized epidermis, triradiate spores, and most probably vascular tissue of annular tracheids.

Cooksonia has been reported from approximately equivalent age strata at several other localities. In 1962 Obrhel gave a brief description of some fossils he referred to as *Cooksonia cf. C. hemisphaerica* and *Cooksonia* sp. from the Late Silurian (Pridolian) near Beroun, Czechoslovakia. They are relatively large specimens, one measuring 6.5 cm long and showing equal dichotomies and terminal sporangia. Although preservation did not reveal spores and vascular tissue, there is no doubt

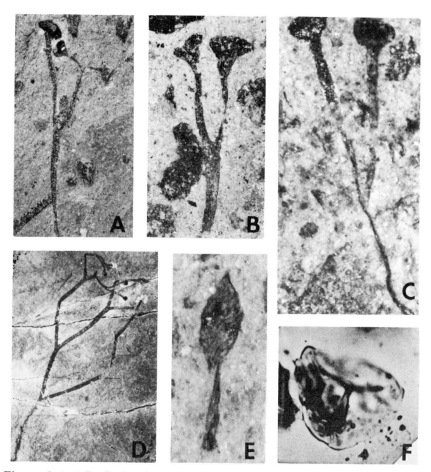

Figure 3.4. A,B. *Cooksonia pertoni* from the Upper Silurian of Wales. Fertile specimens showing dichotomizing axes and terminal sporangia. A. × 3. B. × 6. C. *Cooksonia cambrensis* from the Upper Silurian of Dyfed, Wales. × 21. D. *Cooksonia cf. hemisphaerica* from the Pridolian (uppermost Silurian) of Bohemia. × 1. E. *Tortilicaulis transwalliensis*, a plant of uncertain affinities. Note ovoid sporangium. From the Upper Silurian of Wales. × 15.5. F. A spore removed from a sporangium of *Cooksonia* by Lang. × 1000. (A,B,F from Lang, 1937. Phil. Trans. Roy. Soc. London 227B. C,E. from Edwards, 1979, Palaeontology. D. from Obrhel, 1962, Geologie.)

that they are correctly referred to as *Cooksonia* and they add to our understanding of the gross aspects of the plants.

In 1973, Banks described *Cooksonia* sp. from the Late Silurian Bertie Formation of Herkimer County, New York. *Cooksonia* here is known from sparsely dichtomizing axes 1.8–2.2 mm in diameter which bear terminal sporangia about 2.5 mm wide and 1.5 mm long. Banks notes that "the cluster of axes with some dichtomies indicates that the axes may represent a small tufted habit of growth" (p. 233).

Itschenko (1975) reports *Cooksonia pertoni* and *C. hemisphaerica* from the Late Silurian of Podolia, USSR. There seems to be little doubt that the fossils illustrated in the photos are *Cooksonia* but the rather poor preservation reveals no cellular details.

Dianne Edwards (1979) described and illustrated a considerable collection of *Cooksonia* specimens from the Late Silurian of South Wales. This is based, as are all reports of *Cooksonia*, on very fragmentary material, none of which shows vascular tissue. Spores were obtained from quite a number of sporangia. This report is of special interest in that enough material was available to treat statistically and the specimens exhibit considerable variation in the morphology of sporangia (shape and attachment of stalk). Similar variation was described by Edwards, et al. (1983) from the late Wenlock of Ireland. One cannot help but wonder what a "species" is in such simple plants where the number of significant characters are so few. The absence of vascular tissue raises the additional question of whether *Cooksonia* is a good genus. Is there a possibility that some of the "species" were vascular and others not? It seems very likely that here we may be approaching the borderline between vascular and nonvascular plants.

In this and other accounts, Edwards emphasizes the need for precise stratigraphic information and detailed sedimentary analysis in order to better understand the possible environment in which these early plants may have lived (also discussed by Lang, 1937).

Cooksonia is also known from Early Devonian sediments where some species exhibit greater complexity than the late Silurian ones.

Cooksonia caledonica (Fig. 3.5) was described by Edwards (1970a) from specimens obtained from the Dittonian stage (earliest Devonian or Gedinnian) at Aberlemno near Angus, Scotland. It is known from plant fragments 6–7 cm long that dichotomize with a slight tendency toward being pseudomonopodial. The terminal sporangia range from oval to reniform to globose and are approximately 2 mm in diameter. The preservation, although far from perfect, suggests that they dehisced along the distal margin, much the same as in *Renalia* (p. 71). Although identifiable tracheids were not observed, film-pull transfers of the stem reveal slender elongate cells. In this case, and in many others, fossils have been interpreted as remnants of primitive vascular plants that produced trilete spores even though neither vascular tissue nor spores are preserved. Such statements are based on comparisons with presumably related fossil plants in which comparable structures are preserved.

Iurina (1964) reported a new species, *Cooksonia crassiparietalis*, which comes from the Lower Devonian of central Kazachstan, USSR. It apparently differs from other species in the larger size of the sporangia, which are round–elliptical and about 5.5 mm wide and 3.5 mm long, and in their obviously specialized dehiscence region. They contain finely orna-mented spores 50–65 μm in diameter. Iurina's photographs show a very slender black line running through the center of the axes and in the case of fertile branches, this extends into the base of the sporangium. It may represent a vascular strand of a few tracheids.

Again, the sporangia of *C. crassiparietalis* strongly resemble those of *Renalia*.

Ananiev and Stepanov (1969) have described specimens of *Cooksonia* that are attributed to *C. hemisphaerica* and *C. pertoni* from the Early Devonian of Western Siberia. Their specimens show a generally more complex morphology (Fig. 3.6); for example, the specimens and restoration figure of *C. hemisphaerica* indicate a plant in excess of 8 cm tall with a pseudomonopodial axis bearing side branches. These divide at least three times and some ultimate branchlets terminate in sporangia. It seems problematical as to whether these fossils represent significantly more complete specimens than have been found at other localities or whether they represent a distinct genus or species which may have evolved from the smaller and less complex plants generally referred to as *Cooksonia*.

OTHER POSSIBLY VASCULAR PLANTS

Cooksonia thus far is the only undoubted vascular plant found in Late Silurian times. That others may have existed is suggested by some fragmentary plant remains illustrated by Lang (1937) and Edwards (1979) from the Downtonian of South Wales and by Itschenko (1969, 1975) from the Ludlovian and Downtonian of Podolia.

The other plant remains described by Lang (1937) include slender small bifurcating axes referred to *Hostinella* (see p. 59), a vegetative axis with H-branching tentatively identified as cf. *Zosterophyllum* (from the Ditton Stage) and plants lacking vascular tissue and of problematical affinities named *Prototaxites, Pachytheca,* and *Nematothallus.* Lang discussed the setting in which these plants might have grown, postulating on the basis of lithological data that they represent land-dwelling plants probably living on or near a tidal mudflat.

Edwards (1979) added some other plant remains to the Late Silurian flora of South Wales, recording a new genus *Tortilicaulis*, which might represent a bryophyte or a vascular plant. It consists of slender axes 0.1–

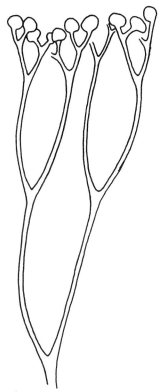

Figure 3.5. A restoration of *Cooksonia caledonica.* × 1.5. (From D. Edwards, 1970a, Palaeontology.)

0.4 mm wide and up to 10 mm long, each axis terminated by an elongate–fusiform or oval body interpreted to be a sporangium (Fig. 3.4E). No evidence of a dehiscence region or spores could be demonstrated nor could the presence of vascular tissue be shown.

A plant occurring in the Upper Silurian of South Wales and the lowest Devonian of the Welsh Borderland is called *Steganotheca*; with it occur sterile axes, some with large spine-like emergences on them and some with rather unique forms of branching (H or K types). These remains, although presently not referable to any known plant genus, suggest that Late Silurian land floras were more diverse than was previously thought. This is also supported by current evidence of diverse but simple trilete spores found dispersed in Middle and Late Silurian sediments.

A flora ranging from Upper Ludlow to Downtonian age was described from the Skala Beds of Podolia (USSR) by Itschenko (1969, 1975) which

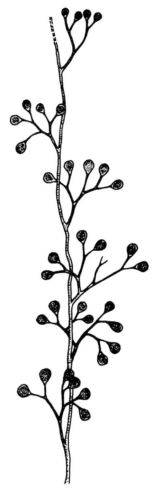

Figure 3.6. A restoration of *Cooksonia hemisphaerica* as envisioned by Ananiev and Stepanov, based on specimens from Siberia. × 1. (Redrawn from Ananiev and Stepanov, 1969. Tomsk State Univ. Trudy.)

included several plant types in addition to some referable to *Cooksonia*. These are generally poorly preserved and fragmentary, but she records the following: a plant with terminally borne, rather elongated sporangia called *Eorhynia*, which is somewhat reminiscent of *Rhynia* or *Salopella*; vegetative axes referred to *Zosterophyllum*; axes covered with fine hairs called *Lycopodolica* and some extremely fragmentary remains referred to *Prototaxites, Primochara,* and *Prehepaticites.*

SILURIAN-DEVONIAN NONVASCULAR PLANTS

Some of the nonvascular plants not discussed in detail above but which commonly occur in association with *Cooksonia* are of considerable interest in their own right. These include *Prototaxites, Pachytheca,* and *Parka.* They may represent intermediate types between aquatic (algal?) ancestors and plants with vascular tissue, and, as to habitat, it has been speculated that they may have been aquatic, terrestrial, or intermediate.

Prototaxites Dawson, 1859

The genus *Prototaxites* was established by Dawson in 1859 for large silicified trunks found at several Lower Devonian localities in Gaspé, Quebec, Canada. Some of the trunks were one meter wide and up to two meters long (Fig. 3.7). Thin sections reveal the trunks to consist of intermixed large and small tubes, 13–35 μm and 5.3 μm or less in diameter respectively. Based on Dawson's description, and a later more complete one by D. Penhallow (1889), the tubes appear aggregated in some areas, tubes of both sizes may branch one to several times, and the small tubes may be septate. Dawson initially interpreted *Prototaxites* as a primaeval gymnosperm, having mistaken the clustering of tubes in some areas to be growth rings and some light colored regions lacking cellular detail as medullary rays. He also was influenced by the external bark-like pattern and apparent spiral thickenings in some large tubes. He named the Gaspé specimens *Prototaxites logani* and allied the plant with the taxads.

The English botanist Carruthers, after examining some of Dawson's preparations, disagreed with his interpretation; instead, he allied *Proto-taxites* with the algae, seeing some similarity with some large brown algae. He renamed the plant *Nematophycus* and provided some accurate illustrations of the large and small tubes so characteristic of the plant.

After many years of controversy, Dawson finally adopted Carruther's view, and in his *Geological History of Plants* (1892), renamed *Prototaxites Nematophyton* and allied it with the algae. *Nematophyton* was used by subsequent workers; however, despite its inappropriate connotation, the name *Prototaxites* is valid due to its prior date of publication.

Up to thirteen species of *Prototaxites* are now known, ranging from the Late Silurian to Late Devonian, most being Silurian and Early Devonian in age. Occurrences of the genus are recorded for England, Scotland, Wales, Germany, Libya, and North America (Ohio, eastern Canada). Most of these are listed in Kräusel and Weyland (1934) and it appears that some species distinctions are based on minor differences

Figure 3.7. *Prototaxites* Dawson. A. A view of a large specimen lying along the bedding surface of sediments exposed in a roadcut near Cross Point, Quebec. B–D. *Prototaxites southworthii* Arnold from the Upper Devonian of Kettle Point, Ontario. B. Longitudinal section showing large and small tubes. Large tubes can be quite long. × 22. C. Transverse section showing large and small tubes. × 85. D. Detail of C. × 339. E. TEM showing septal pore structure in a small tube. × 3500. (A–C. NCUPC collections. D. From Arnold, 1952b, Palaeontographica. E. From Schmid, 1976, Science.)

that might ultimately be shown to be preservational. In addition to petrifactions, some compressions of varying size also have been found (Lang, 1937; Kidston and Lang, 1921).

Well-preserved specimens of *Prototaxites* were described by C. A. Arnold (1952b) from the Upper Devonian of Ontario as *P. southworthii*. This species is represented by a large axis consisting of thick-walled tubes more or less longitudinally oriented and 19–50 μm in diameter which are surrounded by thin-walled tubes 4.75–9.5 μm in diameter. The latter are sometimes branched and have been called hyphae. Some of the small tubes exhibit vesicle-like swellings in which septa occur. The large tubes may be surrounded by a sheath composed of small tubes one to several layers thick. Numerous scattered, light-colored areas were noted, corresponding to similar areas referred to by Penhallow and others as medullary spots. They seem to be a region where the small tubes or hyphae are partially disintegrated (Arnold, 1952b).

In 1976, Schmid examined desilicified fragments of *P. southworthii* using TEM, noting that the thick wall of the large tubes contains anastomosing filamentous components. The small tubes possess a septum with a complex pore; the septum is two–layered with a space in between and an elliptical aperture located in each of the two layers. These septal pores are reminiscent of those of some modern algae and fungi but differ in details from all of them.

Several workers have suggested that *Prototaxites* represents a land-dwelling organism, others that it is marine or freshwater-dwelling. Sedimentological and chemical data support mainly the former. Some reconstructions have shown it with presumed reproductive organs and leafy appendages (Corsin, 1945, Jonker 1979) but these ideas have been based on poorly preserved material not in organic connection and are very doubtful. It also has been depicted as a vertical plant with spreading roots, or as a horizontal, massive, trunk-like structure. Its position in the sediments might support the latter, but further investigation is needed to clarify its mode of growth and reproduction.

The affinities of *Prototaxites* remain problematical. It still is compared with large brown algae, such as some Laminariales, or red algae (Jonker, 1979) but convincing evidence is lacking. Its organization is unique among the plant kingdom and it might best be regarded as a thallophyte, perhaps representing a type of plant quite different from any known today.

Pachytheca Hooker, 1861

When first discovered, *Pachytheca* was thought to represent the remains of some animal; later it was regarded as a seed and only near the

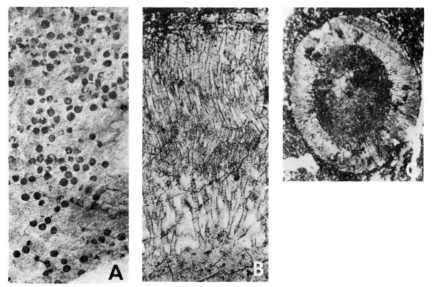

Figure 3.8. *Pachytheca* Hooker. A. Surface of rock showing numerous individual pachythecas. × 1. B. Section through cortical region and outer portion of medulla of *P. fasciculata*, showing tubes. × 90. C. Section of specimen showing both medulla and cortex of *P. fasciculata.* × 11. (A. NCUPC Collections. B,C. From Kidston and Lang, 1924, Trans. Roy. Soc. Edinburgh 53:603–14.)

end of the 1800s was it accepted as representing an alga. An individual consists of a single spherical body 1.5–5 mm in diameter and is composed of an inner zone called the medulla and an outer zone or cortex (Fig. 3.8). First named by Sir Joseph Hooker in 1861, the later descriptions by him (1889) and Barber (1889, 1890) provided the first extensive details of its structure. Accounts by Kidston and Lang (1924) and Lang (1937) demonstrated some variation at the species level.

The medulla of *Pachytheca* is composed of multicellular filaments extending in all directions. These filaments change direction at the periphery of the medulla and extend into the cortex where they are radially aligned with spaces in between. The filaments are encased in a sheath (the tubes) and may branch in the outer cortex. In most species, only one filament is present in each sheath, while in *P. fasciculata*, up to 10 filaments may occur within each sheath. The filaments range from 2–20 µm wide, being different widths in different species. The outermost region of the cortex is delimited by a clear area in well preserved specimens.

Some spheres exhibit a small papilla; Kidston and Lang (1924) illustrate one in section view which shows the papilla to represent only the medulla. It is believed to represent a site of attachment.

Since its internal structure became known, *Pachytheca* has been regarded as an alga, most similar to some blue-greens and greens. Niklas (1976) compares its mode of growth to the green alga order Chaetophorales, and especially to *Coleochaete*, which often occurs on lake bottoms or as an epiphyte on aquatic plants.

Pachytheca has been found in the Late Silurian and Early Devonian of Great Britain, Germany, Belgium, Spitsbergen, Australia, and North America. The locality cited by Dawson in southeastern Canada is Dalhousie, New Brunswick, not Gaspé as mentioned by Kidston and Lang (1924). The authors have collected *Pachytheca* from Emsian age sediments located between Campbellton and Dalhousie, New Brunswick.

Parka Fleming, 1831

Another unique plant which was originally attributed to various groups of plants or animals is *Parka decipiens*, best known from the Early Devonian of England, Scotland, and Wales, but occurring in the Late Silurian as well. It might also be present in Germany but its occurrence there is doubtful (see Kräusel and Weyland, (1930).

Parka was first named by Fleming in 1831 and regarded as a fruit or group of fruitlets. Later it was suggested to represent a mass of mollusc or eurypterid eggs and later still, Dawson and Penhallow suggested it represented the fertile regions of a fern allied to *Marsilea* and *Salvinia* (water ferns).

Don and Hickling (1917) clarified the nature of *Parka* as a dorsiventral thallus with clusters of spores present in it and suggested that it was more closely related to the algae. Their account represents another example of careful and critical observation and interpretation of a problematic organism. It also depicts the resolution of overlapping research interests; after Don and Hickling discovered each had independently studied *Parka*, they decided to publish their results jointly. Sadly, A. W. R. Don was killed in action in Greece in 1916 while the joint paper was still in press.

Parka is a circular to ovoid thallus 0.5–5 cm in size with an apparent upper and lower surface. The thallus is cellular, composed of an upper and lower epidermis and intervening parenchyma (Fig. 3.9). Circular to oval bodies 1–3 mm in diameter occur on or within the thallus which represent masses of spores. The latter are apparently cuticularized, 28–34 μm in diameter but lack any indication of being borne in a tetrad. Don and Hickling (1917) interpreted the spore masses as being embedded in the mid-region of the thallus, but Niklas (1976) interprets them as occurring in sporangia on the surface of the thallus. He also suggests that

Figure 3.9. *Parka decipiens* Fleming. Ventral surface view showing typical sporangial pattern of thallus organization. × 2.2. (From Niklas, 1976. Trans. Roy. Soc. Edinburgh 69:483–99.)

Parka was attached by a ventral holdfast, that its thallus form resulted from growth analogous to dichotomizing filaments and that it was at least two cell layers thick.

Parka is quite a distinctive plant; Lang (1945) regarded it as "anomalous," along with *Pachytheca* and some other genera. Don and Hickling (1917) cautiously suggest it is a "low spore-bearing plant," possibly a "thallophyte with Algal affinities" (p. 661). Niklas supports the idea that it is an alga and again draws comparison with some greens.

CONCLUDING REMARKS

Returning to more general considerations, the picture emerging at present concerning land plants in Silurian and Early Devonian times is one including the existence of very small, simple plants, some vascular (*Cooksonia*) and some of problematical, possibly bryophytic or algal affinities. An unusually large form existed in *Prototaxites*. These plants apparently inhabited lowland, near-shore environments.

Two other aspects regarding the advent of land vascular plants require brief consideration at this point, namely contributions of micro-fossil studies to our understanding of early land plant evolution, and some controversial reports of fairly complex presumed pre-Devonian plants.

As mentioned at the start of this chapter, the dispersed spore record, which can provide supplemental information on past plant life, generally supports a similar sequence of events as the macrofossil record, namely that vascular land plants arose in the Late Silurian and increased in diversity and complexity throughout the Devonian. Studies by several palynologists in the past few years have shown that a greater diversity of trilete spores occur in Late Silurian sediments than was known earlier (Richardson and Lister, 1969; Richardson and Ioannides, 1973; McGregor and Camfield, 1976). These resemble younger forms in their general morphology although differing in details.

Strother and Traverse (1979), Gray and Boucot (1971), and Gray, et al. (1982) illustrate spores ranging from Late Ordovician to Silurian that they suggest might have been produced by land plants. These need not all have been vascular however. Lastly, studies of *in situ* spores of Late Silurian and Devonian plants (Gensel, 1980; Allen, 1980, 1981) show that relatively few morphological types are known from macrofossils, leaving a large number of dispersed spores of unknown affinities—again they need not all have been produced by vascular plants.

Reports of Silurian plants of greater complexity than those previously discussed have been made by several workers. These include the flora containing *Baragwanathia, Yarravia*, and other taxa from Victoria, Australia; some of the sediments containing these plants (or ones very similar to them) are interpreted as Late Silurian by Garratt (1978) on the basis of associated graptolites. Klitszch, et al. (1973), Boureau, et al. (1978) and others have reported a variety of megafossils from sediments regarded as Late Silurian in Libya whose morphology and complexity is more comparable to forms known from the Middle Devonian elsewhere. In both cases, questions exist regarding the exact age of the sediments and identification of the fossils present, and the outcome is not yet known. If correctly dated and interpreted, these would greatly alter our conception of Late Silurian and Early Devonian vascular plant evolution.

4 / Early Land Vascular Plants and the Origins of Diversity

We will describe in this chapter those vascular plants that probably evolved from the complex represented by the fragmentary but significant fossils that were described in Chapter III. The records that we have for vascular plants in the early part of the Lower Devonian (Gedinnian) are very few in numbers of species. The present evidence suggests that vascular plants of the *Cooksonia* type, small and very simple in gross morphology, struggled to obtain a foothold on the land for some 15–20 millions of years. It was probably the most difficult phase in the evolution of the land flora as we know it today, involving as it did the development of an adequate soil. But once achieved, the diversity and complexity of types and increase in size began to accelerate.

By the end of the Early Devonian, at least 35–40 million years later than the oldest vascular plant known to date, we find a great variety of vascular plant genera, many of which can be assigned to the trimerophytes, rhyniophytes or zosterophyllophytes. As discussed in Chapter II, these groups were established by Banks (1968a) to more accurately reflect the great diversity in form and structure and evolutionary potential of the plant types until then referred to as "psilophytes."

Students of the Devonian floras since Arber's time have created a considerable number of new taxa mostly in a sincere if somewhat desperate effort to establish order among the strange new plants that keep turning up—and there are certainly more to come. The choice, as usual, is whether to "lump" or to "split." But the choice here is unique in that we are concerned with so many plants that have no correlative living ones. Our policy here, as elsewhere, is to introduce any fossil that is well enough preserved to display significant features but to be very cautious about its classification; where this cannot be determined with reasonable assurance it will simply be regarded as "problematical." Hopefully this will reveal the fascinating variety in Devonian plants without forcing them into a classification from which they will have to be extricated in the near future.

A Note on General Morphology and Terminology

In this chapter we are dealing with plants that either show little differentiation into stem, leaf, and root, or only the beginnings of such morphological diversity. The aerial shoot system requires special consideration since it cannot be adequately described in terms of modern plant terminology; we therefore insert the following basic glossary:

Types of axial branching

dichotomous—in which the apical meristem divides to form two branches which may be equal (isotomous) or unequal (anisotomous).

trichotomous—in which the apical meristem apparently divides to form three branches, the central one usually being the strongest; it is not always possible to distinguish this from two closely spaced dichotomies.

monopodial—in which there is a dominant central axis with lateral branches arising from buds more or less distal to the shoot apex, the buds usually being in axils of leaves (in higher plants); this stage apparently had not been attained by any of the plants described in this chapter.

pseudomonopodial—in which the apical meristem seems to divide to form two branches, one of which is dominant to a greater or lesser degree resulting in an upright main axis and distinct side branches.

Next, a few comments concerning the distinction between stem and leaf seem to be in order. The development of a pseudomonopodial habit was quickly followed in some plants by the development of branch systems that were profusely divided and three-dimensional. In some plant groups that we will encounter later in the Devonian, we find branch systems that are flattened (essentially two-dimensional). A distinct lamina came into existence apparently by the fusion of several terminal branch units or telomes. These apparent stages in the evolution of a mega-phyllous leaf will be elaborated on later in the text.

The earliest and simplest of our plants were sparsely dichotomous and the entire branch system was naked. In other plants we find the aerial axial system adorned with a considerable variety of emergences. These are structures that are best described as cortical outgrowths and they vary, both on individual plants, as well as in different species and genera, from a few cells to structures several mm long. In many cases they are sharp-tipped and generally referred to as spines. In a few instances they appear to have had a glandular tip. For the most part these emergences seem to have had no regular arrangement. Emergences were probably the precursors of some microphyllous leaves (although some arguments for a telomic ancestry are made) which may be distinguished by their regular arrangement and the presence of a vascular strand.

Roots rarely have been found in attachment in early Devonian plants. It is not certain that they were lacking, as in some modern pteridophytes such as *Psilotum* and *Tmesipteris*, simply were not preserved, or perhaps have not as yet been recognized as such. Some instances of plants with roots will be cited in later sections.

The classification outline below is based on that given by Banks in 1975c. The rather abundant evidence that we now have suggests that the plants included in the Trimerophytina and Rhyniophytina are more closely related to each other than they are to those included in the Zosterophyllophytina. Chiefly to indicate this distinction in dealing with a considerable number of genera we have divided the chapter into two sections accordingly.

Trimerophytina

Plants that usually have a central (pseudomonopodial) axis with branches that dichotomize or divide into three divisions; the axes are smooth or with emergences varying in abundance and form in different species; fertile branches much divided and terminating in dense clusters of fusiform sporangia that dehisce longitudinally; xylem a protostelic centrarch strand of scalariform tracheids.

Trimerophytales
  Trimerophytaceae
    Psilophyton Dawson, 1859
    Trimerophyton Hopping, 1956
    Pertica Kasper and Andrews, 1972
    Dawsonites Halle, 1916
    Hostinella Stur, 1882 *pro parte*
    Psilodendrion Høeg, 1942
    Psilophytites Høeg, 1952

Rhyniophytina

Plants with a dichotomizing shoot system lacking emergences; some branchlets terminated by a single sporangium; xylem a protostelic, centrarch strand.

Rhyniales

Rhyniaceae

Rhynia Kidston and Lang, 1917

Horneophyton Barghoorn and Darrah, 1938

Cooksonia Lang, 1937

Steganotheca Edwards, 1970

Salopella Edwards and Richardson, 1974

Dutoitea Høeg, 1930

Eogaspesiea Daber, 1960

Questionable Rhyniophytina

Taeniocrada White, 1903

Nothia Lyon, 1964

Renalia Gensel, 1976

Yarravia Lang and Cookson, 1935

Hedeia Cookson, 1935

Zosterophyllophytina

Aerial shoot system pseudomonopodial or dichotomous, naked or bearing emergences; sporangia borne laterally, reniform or globose and dehisce along their distal, convex margin; xylem strand protostelic, elliptical or terete in transverse section, exarch.

Zosterophyllales

Zosterophyllaceae

Sawdonia Hueber, 1971

Zosterophyllum Penhallow, 1892

Gosslingia Heard, 1927

Crenaticaulis Edwards ex Banks and Davis, 1969

Rebuchia Hueber, 1970

Bathurstia Hueber, 1972

Serrulacaulis Hueber and Banks, 1979

Oricilla Gensel, 1982

Koniora Zdebska, 1982

Hicklingia Kidston and Lang, 1923

PART I. THE TRIMEROPHYTES AND THE RHYNIOPHYTES

The Trimerophytes

The type for this group is *Trimerophyton robustius* (Dawson) Hopping, 1956. Although presenting some distinctive morphological features it is known from a specimen showing but a small part of the plant. But in general, considerable information has accumulated on the plants assigned to this group and they seem to compose a natural unit in which some very interesting trends are evident. We begin with the genus *Psilophyton* which, for several reasons, affords a good starting point.

Psilophyton (Dawson) Hueber and Banks, 1967

As a genus, *Psilophyton* is widely dispersed geographically and probably existed throughout the Devonian, although our most significant information comes from Lower to Middle Devonian horizons. The different species range considerably in size and complexity, the more advanced ones seeming to present a focal point from which subsequent lines of evolution originate. In brief, the knowledge that we have concerning the several species that have been assigned to *Psilophyton* is among the most impressive and significant in paleobotany.

Psilophyton dawsonii Banks, Leclercq & Hueber, 1975.

Psilophyton dawsonii Banks, Leclercq and Hueber, 1975, is known from petrified specimens found in calcareous cobbles that occur in sandstone cliffs along the south shore of Gaspé Bay, Quebec, and in northern Ontario, Canada, the age being late Early to early Middle Devonian.

The fossils consist of well-preserved petrifactions which were studied by preparing serial peels, as well as examining portions of the plant that were macerated out of the matrix. The restoration (Fig. 4.1) reveals a representative portion of the plant; the glabrous axes form a three-dimensional pattern and attained a diameter of about 2–4 mm. The sterile branches end in forked, blunt-tipped branchlets and the fertile ones terminate in a rather dense cluster of paired sporangia (Fig. 4.2C, E) which are 3 to 5 mm long and 1.0 to 1.5 mm in diameter. The cluster includes about thirty-two pairs and each sporangium contains large numbers of smooth-walled spores.

The vascular system (Fig. 4.2A) consists of a terete protostele, only primary xylem being present. The centrally located protoxylem consists of spiral and scalariform elements; the metaxylem, which has been described in greater detail by Hartman and Banks (1980), consists of scalariform tracheids and some that display circular bordered pits (Fig. 4.2B, D). Especially distinctive are the interconnections between the scalariform bars which range from simple slender strands to variously reticulate structures. The better preserved portion of the extra-stelar tissues includes an inner cortex of thin-walled, loosely arranged parenchyma and an outer cortex of thick-walled collenchyma interrupted by sub-stomatal chambers.

Figure 4.1. Reconstruction of part of the plant of *Psilophyton dawsonii*, from Banks, Leclercq, and Hueber, 1975. × 0.75. (Reproduced with permission of the Paleontological Research Institution, Ithaca, New York.)

We have a considerable wealth of information on other species of *Psilophyton* and, as a genus, it probably stands now as the most informative one we have to present a general picture of early land vascular plants. The features present in *P. dawsoni* that seem to best typify the genus are: the dichotomous to pseudomonopodial form of the plant as a whole, the comparatively massive centrarch protostele; the dense clusters of sporangia borne in terminal pairs and the high number of spores per sporangium. The other species that we will describe are characterized by: their overall size, the larger ones tending to be strongly pseudo-monopodial; the diversity in the form of their emergences where present; differences in branching pattern; and the size of the sporangia.

Psilophyton crenulatum Doran, 1980

The specimens assigned to this species present one of the most remarkable cases of fine morphological preservation in the Devonian

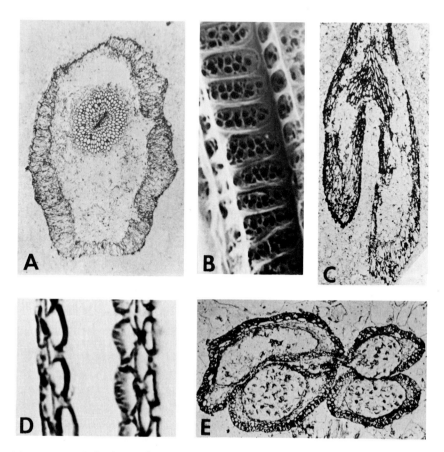

Figure 4.2. *Psilophyton dawsonii.* A. Transverse section of stem showing xylem strand and outer cortex. × 15. B. SEM of tracheids as they appear from the outside showing reticulum of strands that connect the scalariform bars. × 1200. C. Oblique longitudinal section of a pair of sporangia, × 29. D. Longitudinal section showing overarching pit borders and intercellular membrane. × 1306. E. Section through a group of two pairs of young sporangia showing spores. × 26. (A,B,D from Hartman and Banks, 1980, Amer. Journ. Bot. C,E. from Banks et al. 1975, reproduced with permission of the Paleontological Research Institution, Ithaca, New York.)

floras. They were found in a secondarily silicified Lower Devonian tuff in a highway exposure in Atholville, New Brunswick, Canada. The matrix is quite different from the coastal shales of northern New Brunswick and the Gaspé peninsula where several other well-preserved plants have been found that will be described below. The naturally weathered surfaces of the rock did not appear very promising but Doran subjected them to a bulk maceration treatment using hydrofluoric acid with the result that numerous specimens, some constituting a large portion of the plant (as shown in the accompanying illustrations) were released intact.

The plant as a whole displays considerable variation in its branching patterns. The main axes (Figs. 4.3, 4.4A) which are 2 to 3 mm in diameter may be pseudomonopodial or branch dichotomously or trifurcate. They bear emergences that are spine-like and up to 6 mm long; a unique feature of the species is the variability in the morphology of the spines (Fig. 4.3), which may be undivided, bifurcate, or trifurcate. The abundance of these emergences varies greatly on different branches. Some of the axes also bear numerous minute (about 0.1 mm high) crenulations, from which the specific name is derived.

As in other species of *Psilophyton* the paired sporangia (Fig. 4.4C) are borne in dense terminal clusters; individually they are fusiform, 3 to 5 mm long, dehisce longitudinally and contain large numbers of spores that range from 40 to 120 μm in diameter (Fig. 4.4B).

A Note on the Preservation of Devonian Plants

A considerable portion of our knowledge of Devonian plants comes from *compression* fossils, sometimes referred to as *coalified compressions* or *mummified* remains. This is the usual mode of preservation in sedimentary rocks where the plant materials were covered by sediments that later became consolidated to form shales and sandstones. If covered quickly, thus reducing the degree of decay, much of the original plant material may be preserved and in many cases fossils of this kind can be removed from the rock by dissolving away the matrix and then applying oxidizing (clearing) techniques. Thus, such cellular structures as cutinized epidermis, sporangia with spores and sometimes carbonized tracheids are obtained. Also, portions of axes may be permineralized, wherein individual cells are surrounded and/or impregnated by minerals (such as iron sulfide, calcium carbonate, silica). These yield excellent structural detail.

It is especially important to note that we are dealing with three-dimensional organisms and many Devonian plants are preserved as such in the rock. When a specimen is split open the exposed surfaces may reveal only a small part of the plant, the rest of which may be followed through the rock by degaging—micro-excavating with a small hammer and needles, or with very hard sediments an electric engraver is useful.

One advantage in working with such well-preserved compression fossils, as compared with petrified ones, is the greater ease with which the gross form of the plant may be made clear. The relative toughness or fragility of compression fossils varies greatly. In some cases only a black amorphous material remains when one attempts to remove them, while *Psilophyton crenulatum* presents the other extreme where it is possible to

Figure 4.3. A fertile specimen of *Psilophyton crenulatum* Doran showing the massive sporangial aggregates, dichotomizing main axis, spines (some of which are bifurcate) and marginal crenulations. × 2.8. (From Doran, 1980, Canadian Journ. Bot.)

Figure 4.4. *Psilophyton crenulatum.* A. A specimen suggesting the plant was profuse in its branching pattern and that some branches may have been wholly sterile (vegetative) while others were abundantly fertile. × 1.0 B. Part of a dehisced sporangium showing the abundant spores still contained within. × 45. C. Terminal portion of a fertile branch showing the paired and twisted sporangia, all of which apparently had dehisced. × 6.6. (From Doran, 1980. Canadian Journ. Bot.)

remove almost complete plants and see them very much as they were in life.

Psilophyton princeps Dawson, and a Brief History of the Generic Concept

In what might be the most logical presentation of our present knowledge of the genus *Psilophyton*, the discussion that now follows should be given before the description of *P. dawsoni* and *P. crenulatum*. But in view of the very complex nature of the development of the present generic concept we felt that it would be most helpful to the uninitiated to describe two rather well known species first. Thus the following is a summary of the history of the generic concept—one of the most tangled and controversial in paleobotany.

In 1859 J. W. Dawson established the genus *Psilophyton* and he described two species: *P. princeps* and *P. robustius*; we will be concerned only with the former here; *P. robustius* is included now under the name *Trimerophyton robustius*.

The diagnosis that Dawson (1859) gave for *Psilophyton* reads as follows:

> "Lycopodiaceous plants, branching dichotomously, and covered with interrupted ridges or closely appressed minute leaves; the stems springing from a rhizoma having circular areoles, sending forth cylindrical rootlets. Internal structure: an axis of scalariform vessels, surrounded by a cylinder of parenchymatous cells, and by an outer cortical cylinder of elongated woody cells (prosenchyma). Fructification probably in lateral masses, protected by leafy bracts." (p. 478)

His description was accompanied by a sketchy restoration and in 1870 he presented a somewhat different one which is shown here (Fig. 4.5). Although his initial description mentioned the presence of "rudimentary leaves" on the aerial axes, both restorations show the plant as naked (glabrous). The chief modification in the later figure is the presence of fructifications which he referred to as " . . . oval sacs or spore-cases, borne gracefully on slender bending stalks." This restoration has been reproduced by numerous later writers including one of the present authors (Andrews, 1947).

It is well to remember that Dawson was one of the early pioneers in North American paleobotany. Travel in Quebec and New Brunswick was very different in the mid 19th century than it is today, in Quebec being chiefly by boat along the coast. He deserves much credit for his explorations and transporting his collections under difficult conditions. The result, however, which has a direct bearing on the problems that have developed, is that we have no exact records of Dawson's collecting sites

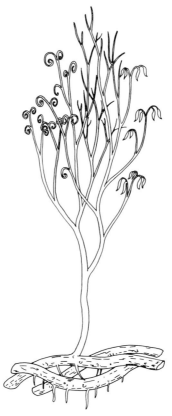

Figure 4.5. Dawson's restoration of *Psilophyton princeps* in 1870. (From Dawson, 1892.)

and in view of the very rapid erosion that takes place along the coastal area every year it is probable that most of the outcrops from which he did collect have long since disappeared. And many of the specimens that he collected have been misplaced or lost.

In his 1871 account, Dawson also described a varietal form *Psilophyton princeps* var. *ornatum*, the axes of which are densely spiny. An outcrop on the north shore of Gaspé has produced considerable quantities of this plant in recent years and it is now known under the name *Sawdonia ornata*; we will return to it on a later page.

Little attention was given to this aspect of Dawson's work until the latter years of the past century. In 1895 Solms-Laubach pointed out that Dawson's specimens bearing fructifications were not attached to spiny stems, and T. G. Halle followed up this observation (based on a study of specimens from several areas) in 1916a with his binomial *Dawsonites arcuatus* for fragmentary, naked axes bearing the terminal paired sporangia.

Suspicion has also been directed to the correlation of the upright shoot system with the creeping "rhizomes" in Dawson's restoration. The latter may be referable to *Taeniocrada dubia* as is believed by Banks and others (see Banks, *et al.*, 1975). In 1924 W. N. Edwards made a significant contribution to the understanding of the varietal form, *ornatum*, which he called *P. princeps*, with cuticular studies. W. H. Lang, in 1931 and 1932, with his usual precision and effective techniques, added to the knowledge of the cuticular structure of both smooth and spiny axes. And in 1942 W. N. Croft and Lang gave us a major breakthrough in demonstrating that a specimen similar to the varietal form, *ornatum*, bore lateral sporangia. This was confirmed by Hueber in 1964 based on specimens from the Abitibi River. It was then clear that the complex of specimens that had been dealt with over the years (more than a century) included at least two very different kinds of plants, one that bore sporangia terminally in pairs and the pairs were grouped in dense clusters, and another kind in which the sporangia are borne laterally.

It is perhaps understandable that Dawson lumped these plants together. He placed considerable significance on studying fossils in the outcrop, where larger and intact ones could be observed (mentioned in several of his papers). We now know that these plants occur rather close to one another in at least one horizon along the north shore of Gaspé Bay. A few years ago, one of us found a large block about 4 ft. square and 3 ft. deep lying on the beach just below the horizon where abundant axes of *Sawdonia ornata* occur. The latter is just out of reach so the fallen block promised easier collecting than normal. As we worked on the block, we found the upper side covered with numerous specimens of *S. ornata* while a layer midway through the block yielded axes referable to *Psilophyton* and an even lower layer yielded remnants of plants called *Psilophyton robustius* by Dawson and now referred to as *Trimerophyton* or *Pertica* (see p. 49). It would be quite easy to develop a composite view of plants as a result of such a juxtaposition at a time when very little was known about the morphology and variation in early vascular plants. As a final note, the plant types actually occur in the opposite sequences *in situ*—the block must have turned over as it fell.

In 1967 Hueber and Banks selected a neotype for *Psilophyton princeps* from the remaining collections in the Redpath Museum in Montreal which had sparsely spiny axes bearing terminal sporangia; and in 1971 Hueber assigned a new binomial, *Sawdonia ornata*, for the densely spiny specimens (Dawson's *Psilophyton princeps* var. *ornatum*) in which the sporangia are borne laterally. It is also important to add that in

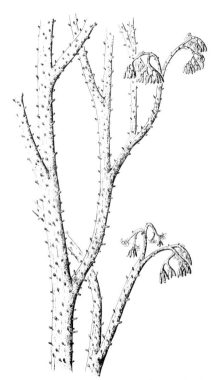

Figure 4.6. A restoration of *Psilophyton princeps*. (From Andrews, 1974. Annals Missouri Bot. Garden.)

a more detailed description of *P. princeps* by Hueber (1967), the plant's emergences are described as being ". . . blunt, 2.0 to 2.5 mm long, rigid, and randomly arranged, tipped by cup-like disc."

A restoration showing a portion of the plant of *Psilophyton princeps* is given here in Fig. 4.6. This shows the distinctive form of the emergences which seem to be best described as peg-like. A further note on these may be appropriate. Hueber's 1967 account is based on specimens from the Gaspé and New Brunswick areas; specimens have also been found (Kasper et al. 1974) in Maine in which the emergences are more abundant as well as somewhat shorter and broader than those in the Canadian plants (Fig. 4.7). Whether this merits variety distinction is questionable but it adds in a small way to our knowledge of the evolution of these structures.

Thus the general organization and affinities of these important plants is now much better clarified, after a century of confusion, but unfortunately agreement on the nomenclature falls short of complete success. Fossils that are quite clearly referable to Dawson's "var. *ornatum*" have

Figure 4.7. *Psilophyton princeps.* Part of an axis of a specimen from the Trout Valley flora, Maine, showing the peg-like emergences. × 3. (From Kasper, Andrews, and Forbes, 1974, Amer. Journ. Bot.)

been found in Siberia and an account of them, accompanied by a good restoration, has been given by Ananiev and Stepanov, in 1968. Most subsequent authors dealing with material of "var. *ornatum*" have referred it to *Sawdonia* however.

Psilophyton dapsile Kasper, Andrews, and Forbes (1974).

This is the smallest and simplest species found thus far, indeed it seems to approach the minimum in these features. It was found as rather thickly matted remains forming a thin band in the Trout Valley Formation (late Lower or early Middle Devonian) of Maine. The plant attained a height of about 30 cm; the axes are dichotomously branched, not more than 2 mm in diameter and bear no emergences; the fertile branchlets terminate in the usual dense clusters of paired sporangia that are only 2 mm long (Fig. 4.8).

Psiliphyton charientos Gensel (1979).

Psilophyton charientos is known from partially petrified specimens found in blocks along the north New Brunswick coast. The main axes attained a diameter of 33 mm and the plant seems to have been predominantly pseudomonopodial but not strongly so; the main axes,

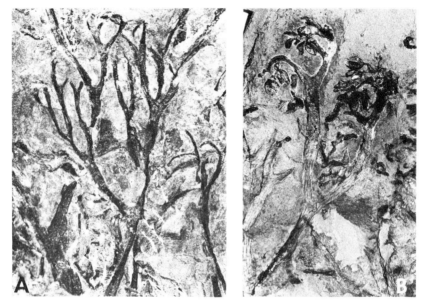

Figure 4.8. *Psilophyton dapsile.* A. Sterile specimen showing dichotomous branching, × 2.5. B. Terminal branches bearing numerous paired sporangia, × 2.6. (From Kasper et al. 1974. Amer. Journ. Bot.)

which attain a diameter of 3 mm, show some weak vestiges of equal dichotomy (Fig. 4.9). The axes bear abundant and very delicate spine-like emergences that are about 2 mm long and taper to a very sharp point. The xylem strand is well preserved (Fig. 4.10B), about 0.5 mm in diameter with a central protoxylem, and the scalariform tracheids show circular to irregularly shaped pit-like openings in the wall between the bars. The sporangia are 3.0 to 4.5 mm long, fusiform, contain large numbers of spores, and dehisce longitudinally (Fig. 4.10A, C).

Psilophyton forbesii (Andrews et al.) Gensel, 1979.

Psilophyton forbesii was first described (Andrews et al. 1968) from specimens found in the Trout Valley Formation of Maine and additional information was forthcoming from better preserved specimens found on the Gaspé coast (Gensel, 1979). It is a plant with a rather stout (pseudomonopodial) main axis that attained a diameter of 9 mm and, like other species in which some anatomy is preserved, it contains a centrarch protostele. The axes are naked, there being no evidence of emergences of any kind, (Fig. 4.11A, B), and it is probably one of the taller species in the genus. It certainly attained a height of 60 cm and very possibly more.

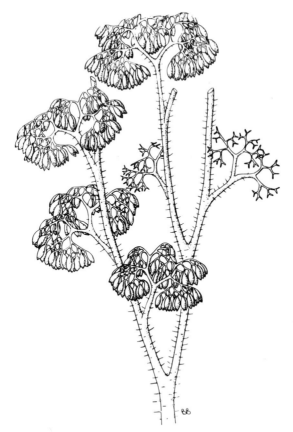

Figure 4.9. *Psilophyton charientos.* Restoration of a representative portion of a plant, about × 1.1. (From Gensel, 1979, Palaeontographica.)

The fertile lateral branches dichotomize up to six times and terminate in pairs of ellipsoidal sporangia 3.5 to 5.0 mm long which contained large numbers of spores 53 to 96 μm in diameter (Fig. 4.11C, D).

Spores preserved in sporangia in different species of *Psilophyton* are nearly identical, being about 50–80 μm in size, trilete, and with a partially enclosing granular layer that often is partially detached (Gensel and White, 1983).

Trimerophyton robustius (Dawson) Hopping, 1956.

The plant from which the group name Trimerophytina is derived was described by C. A. Hopping in 1956. It is based on only one small

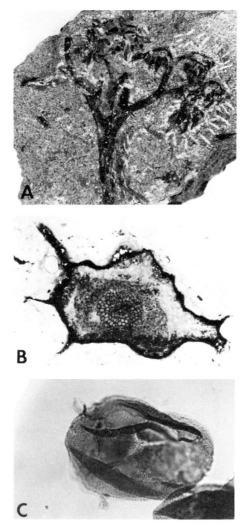

Figure 4.10. *Psilophyton charientos.* A. Specimen bearing terminal branch system. × 1.9. B. Transverse section of axis showing protostele, × 26. C. Spore, × 421. (From Gensel, 1979, Palaeontographica.)

specimen found in the collections of the Hunterian Museum of Glasgow University and bore the label *Psilophyton robustius* Dawson, Gaspé; Hopping adds "It was probably identified and named by Sir J. W. Dawson himself." The fossil itself consists of an axis fragment about 12 cm long and 9 mm in diameter, its most distinctive feature being the trifurcation of the branches. As a point of reference it seems appropriate to quote the generic diagnosis here:

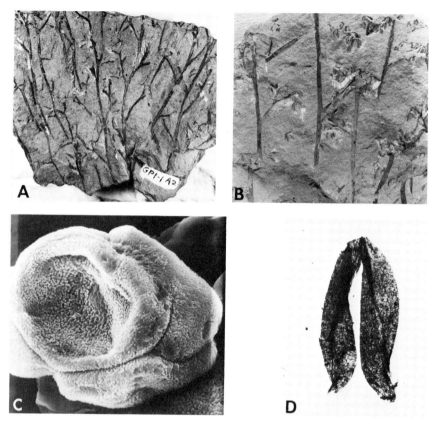

Figure 4.11. *Psilophyton forbesii.* A. Specimen showing parts of several axes with lateral branches. × .24. B. Fertile specimen. × 0.46. C. SEM of tetrad of spores with fine ornament. × 976. D. A pair of sporangia. × 10. (A,D. From Gensel, 1979. Palaeontographica. B,C. From Gensel, 1982. 3rd NAPC Proceedings, vol. 1, p. 199, figs. 1,2.)

"*Trimerophyton* (nov. gen.). The main axis bears lateral first order branches which are spirally arranged and are in three orthostichies. They leave the axis and trifurcate close to the main axis into second order branches of apparently equal thickness. Each one of these second order branches divides again unequally. Two third-order lateral branches arise from each median second-order branch, which continues and dichotomizes into two terminal third-order branches. These third-order branches dichotomize twice into fertile stalks or pedicels bearing terminal clusters of sporangia. [The sporangia are 4 to 5 mm long and contained smooth-walled spores averaging 52 μm in diameter.] The main axis and ramifying branches are smooth and without leaves."

Both of the present authors had an opportunity to examine the specimen in recent years. It has been considerably dissected, apparently by Hopping, who seems to have extracted a maximum amount of information from it. One of his text figures, which we have reproduced in Fig. 4.12, shows the distinctive features. In view of the importance of the specimen as a nomenclatorial starting point for the Subdivision it is unfortunate that we do not have more information about the plant as a whole. However, collections have been made recently in Maine and the Gaspé which reveal closely related plants described under the name *Pertica*.

Pertica Kasper and Andrews, 1972.
***Pertica quadrifaria* Kasper and Andrews, 1972.**

Pertica quadrifaria, the type species for the genus, was based on a large collection of specimens found in the banks of Trout Brook, Baxter Park, Maine. The plant attained a height of at least a meter, the main axis (stem) ranging up to 1.5 cm in diameter.

The stem bore side branches in a tetrastichous pattern which actually form a clockwise spiral. Some of these are fertile and others sterile, the two being intermixed, although some specimens suggest a concentration

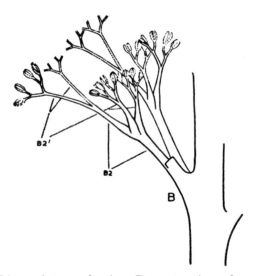

Figure 4.12. *Trimerophyton robustius*. Reconstruction of a lateral branch showing the initial trifurcation, second-order trifurcations and terminal dichotomies. (From Hopping, 1956, Proceed. Roy. Soc. Edinburgh, 66:1–28.)

of fertile ones toward the distal end of the plant (Fig. 4.13). In both cases the individual branch dichotomizes profusely; the successive divisions are at right angles to each other resulting in a very dense, three-dimensional pattern. Both the main axis and the branches bear minute papillae less than 0.4 mm long. However, these are so small as to give the plant a glabrous appearance when observed with the naked eye.

When a specimen is first split open only a small part of the whole branch complex is revealed; it is thus a good example of the way in which many of the Devonian plants are preserved and is illustrative of the kind of information that can be obtained by carefully following the branches down into the matrix using both parts of the specimen.

The fertile branches terminate in very dense, spherical clusters of sporangia (Fig. 4.13B, C) which are ovoid, measuring about 3 by 1 mm. The number per cluster ranges from a minimum of 32 to an estimated 256. Although not well preserved, spores were obtained from some of the sporangia; they are quite uniform in size, measuring about 64 μm in diameter, which suggests rather strongly that the plant was homosporous.

Pertica varia Granoff, Gensel, Andrews, 1976.

This second species provides considerable information and is also based on abundant and well preserved specimens. They were obtained from the late Lower Devonian Battery Point Formation on the north shore of Gaspé Bay. Some of the main axes of these specimens were up to a meter long being only fragments of the whole plant. It is estimated that the plant may have attained a height of 3 meters.

The main axis bore branches in a decussate arrangement with the successive subopposite pairs rotating through 90°. The sterile branches generally tend to be pseudomonopodial (Fig. 4.14A), but equal dichotomies as well as trichotomies are present. They range up to 17 cm in overall length with a diameter of 8 mm at the proximal end.

The fertile branches are only one half to one third the length of the sterile ones (Fig. 4.14B). They are more uniformly dichotomous than the sterile branches and terminate in paired sporangia forming clusters of a total of eight to sixteen pairs (Fig. 4.14C). The individual sporangia are ovoid-fusiform and 3 to 5 mm long; they dehisced longitudinally and their walls bear some minute, sharp-pointed emergences that are about 200 μm long. The spores, of which there must have been several thousand per sporangium, are 56 to 90 μm in diameter and their walls bear discrete coni or grana up to 1 μm high (Fig. 4.14D, E).

Emergences are present on the axes, some being small and papillate and others conical to hair-like and they range from 0.04 to 0.4 mm long.

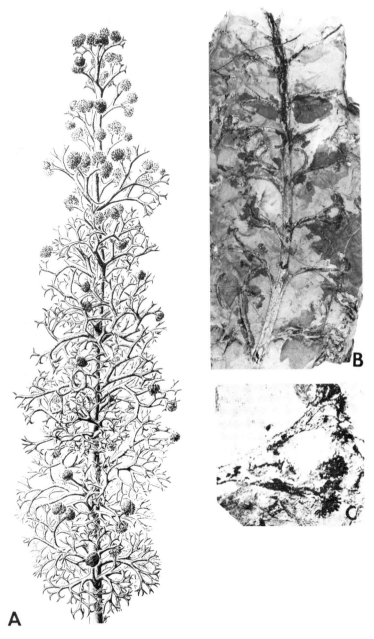

Figure 4.13. *Pertica quadrifaria.* A. Reconstruction based on many fertile and sterile specimens. About × 0.20. B. Holotype specimen showing main axis and some of the fertile branches, about × 0.22. C. Distal portion of a fertile branch showing sporangial clusters, about × 0.9. (From Kasper and Andrews, 1972, Amer. Journ. Bot.)

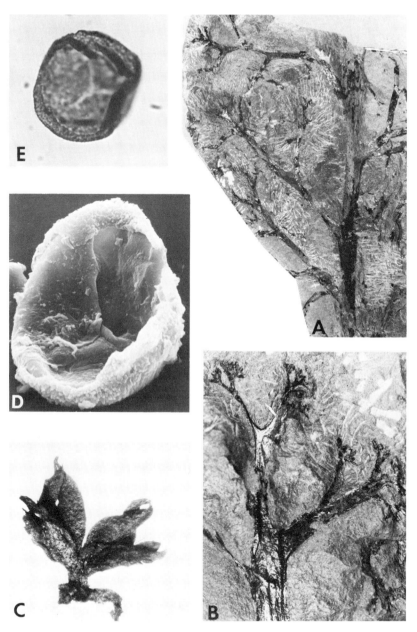

Figure 4.14. *Pertica varia.* A. Specimen showing main axis and some lateral branches, × 0.5. B. Fertile lateral branch with sporangial clusters, × 0.9. C. Cluster of eight sporangia, isolated by maceration, × 5.2. D,E. SEM and light micrograph of spores isolated from sporangia. D. × 998. E. × 500. (A–D. From Granoff et al. 1976, Palaeontographica. E. from Gensel, 1980, Rev. Palaeobot. Palynol.)

At this point we wish to comment on several items of general or summary interest including the following: problems of nomenclature, evolution in the Trimerophytina, and a few notes on the localities from which the fossils described thus far have come.

A Comment on Nomenclature

Devonian plants generally present special problems of naming and morphological interpretation. This is due in part to the fact that they cannot be compared with, or described in the same way that most living plants can. It is also due to the fact that the amount of information we have varies greatly from species to species.

Trimerophyton as a generic name, which is also used to typify the subdivision, presents a problem that is not unique. The branches in the single specimen available to Hopping are characterized by their initial two trichotomies and spiralled arrangement. However, the specimen is clearly a fragment of the whole plant. The species that is described as *Pertica varia*, and is based on a large collection of specimens, displays some occasional trichotomous branching. The question arises as to whether *Trimerophyton robustius* could be a portion of a plant of *Pertica varia* in which this type of branching was consistent for a few nodes. The present authors have collected extensively along the Gaspé and New Brunswick coasts in recent years without finding fossils in which this feature is uniform but we see no reason to doubt the possibility. We will encounter other examples of this general problem.

Evolution in the Trimerophytina

The plants we have considered thus far show varying degrees of complexity from the small dichotomously branching *Psilophyton dapsile* to the larger species of *Pertica*. We hesitate to try to align the known species in a definite pattern but several trends seem evident:

The development of a strong pseudomonopodial axis from a dichotomously branching system; this made possible plants of much greater height.

Branch systems which are regularly arranged on the main axis; here one may envisage the dense three-dimensional branches of *Pertica quadrifaria* as a starting point of a megaphyllous leaf.

The great variety of emergences may represent an early stage in the evolution of at least some of the regularly arranged and vascularized microphyllous leaves that we will see in other, presumably more advanced, plants. Recent speculation that some microphylls may have evolved from reduction of more extensively divided branch systems has been put forth as well (Bonamo and Grierson, 1981; Zdebska, 1982).

The distinctive sporangial clusters may point toward the much denser sporangial aggregates found in the coenopterid ferns of the Carboniferous.

The Localities

In an account of this kind we feel that it will be of interest to occasionally insert a few notations on the general nature of the more important localities. Many of the plants described thus far have been obtained from northern Maine and the coastal areas of southeastern Canada (Fig. 4.15). In some places the sedimentary rocks are quite rich in fossil plants and there is every reason to suppose that they will continue to yield new plants for some years to come.

Most of our productive digging in northern Maine has been along the banks of Trout Brook and some of its tributaries. The brook cuts through the Trout Valley Formation for a distance of about two miles; the age of

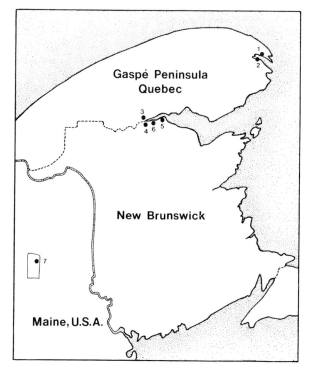

Figure 4.15. Map showing Lower Devonian plant localities. 1,2. North and south shores, respectively, of Gaspé Bay; 3–6, localities along Restigouche River at the head of Chaleur Bay; 7, Trout Brook, Baxter Park.

the formation, based on all available information, is either very late Early Devonian or early Middle Devonian. The plants occur along the banks in outcrops, some of which are ephemeral, having been exposed by a heavy spring runoff and disappearing in the summer. At other places the rock is very hard, yet can be split with stout chisels and hammers, and fine "float" specimens are even found in the stream bed. For the most part the number of species at any one place is small, often only one. We have encountered very little fossil material here with cell structure preserved but a great deal of information has been obtained on the gross morphology of the plants. The entire area is heavily forested and one can but wonder about the plant deposits that presently are hidden from view.

In southeastern Canada our explorations, as well as those of others, have been carried on chiefly in two areas: 1) on both sides of the Restigouche River, the northern banks being in Quebec and the southern ones in New Brunswick. The southern side has proven to be the most productive thus far, especially between Dalhousie and Campbellton. The exposures here are regarded as belonging to the Campbellton Formation of Middle Devonian age on older geological maps. However, analysis of dispersed spores from these sediments by D. C. McGregor of the Geological Survey of Canada suggests they are late Early Devonian in age, ranging from early to late Emsian (McGregor, pers. comm.). 2) along the north and south shores of Gaspé Bay, our own experience having been chiefly on the north shore. This is a spectacular coast with several thousands of feet of sediments exposed immediately back of the beach (Fig. 4.16). They fall within the York River Formation to the east and the overlying Battery Point Formation to the west. In a recent detailed study of the spores of these rocks, McGregor (1977) has identified the Emsian-Eifelian boundary as about midway through the Battery Point Formation. Most well-preserved plant fossils are located below the boundary.

In both of these Canadian localities the coastal rocks are exposed to violent storms and erosion is rapid. The plant deposits tend to be found as small lenses that are exposed one summer and gone the next. Most of our digging has been done in exposures immediately back of the beach but in the Dalhouse-Campbellton area some offshore productive rocks are available only at the lowest tides.

Localities yielding these and other early Devonian plants are known from several other parts of the world (see D. Edwards 1973, 1980a for more comprehensive accounts) but few exactly contemporaneous deposits have yielded quite as extensive a flora as in northeastern North America. Plant-bearing sediments ranging from Gedinnian to latest Emsian (and younger) age occur in Germany, Belgium, Great Britain (Senni Beds of South Wales, Rhynie and other Old Red Sandstone localities in Scotland), and Norway. Early Devonian plants also have been described from several

Figure 4.16. A view of the north shore of Gaspé Bay showing some of the Lower Devonian sediments.

areas in Russia, as well as in South America, Africa, Australia, and China. More investigations are needed to better understand some of the plants being found in these areas.

The prospects for adding to our fund of knowledge in the future are good. Studies of the dispersed spore flora indicate a higher speciation than the macrofossils have yielded thus far and we have fragmentary macrofossils suggesting species and genera of which we know very little at present.

Some Problematical Fossils Probably Related to Trimerophytes

The generic names recorded below have been used frequently enough in Devonian plant literature to require some explanation. It seems only fair to say that they represent the borderline aspects of significant information; they are fossils that tell us just enough to make it difficult to ignore them and we include them as a source of reference.

Hostinella Barrande ex Stur 1882 and *Dawsonites* Halle, 1916.

If not of great significance as biological entities, these names are of interest in revealing some of the problems and vagaries of paleontology.

The generic name *Hostinella* was taken by Stur from a manuscript of Barrande and has been used for a variety of naked axes with dichotomous or pseudomonopodial branching with, occasionally, bud-like protuberances in the upper angle of the dichotomy. Some of Stur's specimens have since been referred to other genera including *Pseudosporochnus* and *Protopteridium*. The name itself, spelled *Hostinella* and *Hostimella*, presents a minor othographic problem. It was taken from the town of Hostin which is near Beroun, Czechoslovakia. According to Obrhel (1961) it was spelled Hostin until the latter part of the 19th century, when it was changed to Hostim. Since this is not an orthographic mistake the original spelling *Hostinella* should be considered valid.

In searching through the fossil plant collections in the Swedish Natural History Museum, Banks (1968b) found petrified fragments of stems referable to *Hostinella*. The axes, which range from 2 to 3 mm in diameter, reveal a small cylindrical protostele with a centrarch protoxylem. As to its natural affinities, it is very possible that these are pieces of more distal axes of *Psilophyton goldschmidtii* as described by Halle.

In 1925, W. H. Lang described several apparently fertile species of *Hostinella* from the Middle Old Red Sandstone of Cromarty, Scotland. His *Hostinella pinnata* is based on small fragments of incurved axes which bore sporangia 3 mm long and 0.5 mm wide; these fragments are now referred to the genus *Rellimia* (or *Protopteridium*) which is discussed in Chapter 8. Another one, *H. racemosa*, is of interest in showing the variety of fossils that have been placed in this generic entity. It consists of an axis 2 mm wide and 14 cm long; this bears short lateral branches each about 5 mm long and terminated by an oval body 5 to 6 mm long and 2 mm broad. These were described as sporangium-like bodies although maceration yielded no spores. In the second volume of the *Traité de Paléobotanique* Høeg saw fit to transfer this to *Dawsonites*, but it does not conform to the concept of that genus either.

Dianne Edwards (1980b) has described petrified axes under the name *Hostinella heardii* from the early Devonian of South Wales. The xylem consists of a terete strand of tracheids of about 0.5 mm in diameter and with close-set annular thickenings. It is pointed out that tracheids of an almost identical pattern are found in some members of the Zosterophyllophytina. There also is a possibility that this fossil may represent *Krithodeophyton croftii* which is found at the same locality.

As to *Dawsonites*, this generic entity was established by T. G. Halle in the course of working out the controversial history of *Psilophyton princeps*, for very fragmentary, smooth axes bearing terminal fusiform sporangia. His specimens came from the Lower Devonian rocks at Röragen, Norway, and it seems very likely that they are terminal portions of a *Psilophyton* plant.

In 1972 Chaloner described some fragmentary but interesting fossils from early Middle Devonian age rocks of Fair Isle which lies about 120 km north of the Scottish mainland. Given the name *Dawsonites roskiliensis*, the species is based on axis fragments up to 7 cm long and 1.0 to 1.5 mm broad, with sporangia borne terminally or on side branches in pairs or in clusters. The sporangia, 4 mm long and 1 mm broad, probably dehisced longitudinally, and present a resemblance to *Hedeia*.

Høeg (1967) included in this genus plants with a variety of sporangial types and in differing positions; it has been suggested by several workers that these should be excluded from *Dawsonites* and the genus restricted to clusters of terminal fusiform sporangia of a *Psilophyton* type.

These examples of problematical fossils that are presumed to be trimerophytes are but a sampling from the literature. For the most part they are so fragmentary as to lend little confidence in assigning them to a natural genus. It is a matter of judgment as to whether such remains should be brought into print and if an admonition is allowed we suggest that one should be very conservative in doing so.

The Rhyniophytes

Rhynia Kidston and Lang, 1917.

Probably the most publicized Devonian plant locality known today is the one found near the village of Rhynie, Scotland. It is remarkable in several ways; our knowledge of it continues to increase and a brief note on its discovery seems appropriate here. Rhynie lies west of Aberdeen in northeast Scotland and the fossil deposit is located about a half mile from the village. The first detailed description of the area that we are aware of is that of William Mackie (1914) who was a regional Medical Officer of Health for Elgin and a geologist on the side. According to this account he was studying and collecting in the area as early as 1910 and he made several visits there in 1912 and 1913 with W. R. Watt. It seems most likely that it was not until 1912 that he observed fossil plants in the cherty rock. Albert Long writes (in a personal letter to W. S. Lacey), " . . . I always understood that he was sitting near the drystone dyke ['dyke' = wall or bank in Scotland—WSL] eating his lunch when he spotted a piece of mottled chert containing *Rhynia*."

Mackie figures a transverse section of a *Rhynia* (?) axis as well as the stele of an *Asteroxylon* specimen in his 1914 paper. He also indicates in a footnote that at that time, "Dr. Kidston is engaged on an investigation of the fossil plants."

Mackie did not find the plants *in situ* and he notes "Up to the date of my last visit to the Rhynie area, which was on the 13th October 1912,

Figure 4.17. The renowned fossil plant locality, the Rhynie Chert Beds, occur in the left center of the photo, which also shows part of the village of Rhynie. The hill in the upper left is the "Tap O'Noth", crowned by its Iron Age vitrified fort. (Photo courtesy of Dr. A. G. Lyon.)

these cherts had not been found anywhere *in situ* . . ." Credit for finding the plants *in situ* seems to go to Mr. D. Tait, a fossil collector for the Geological Survey who in 1913 was working under the direction of Dr. Flett of that organization. The peaty chert was found when he dug trenches, " . . . in the lower field between Easaiche Bridge and Windyfield farm-house—the field in which corn was grown last year." (p. 223)

As there was some disagreement concerning the age of the beds, the British Association established a committee under the chairmanship of Dr. J. Horne who was the Government Geological Survey Officer in Scotland at the time of Mackie's discovery. Considerable credit is due Dr. Horne and in Part II of their series of studies, in which Kidston and Lang (1920) describe *Hornea lignieri,* they note: "We regard it as a distinct, though allied genus, and have named it after Dr. Horne, to whose energy and interest the successful discovery of the Rhynie peat-bed *in situ* is largely due." (p. 611)

The generic name later proved to be invalid, having been used previously for a modern flowering plant and *Horneophyton* was substituted by Barghoorn and Darrah in 1938. The specific epithet is for Prof.

Octave Lignier (1855–1916), a French botanist noted for his studies of Mesozoic plants and astute speculations on plant morphology.

Thus a considerable number of people entered into the initial phases of discovery and investigation of the Rhynie Chert plants.

With regard to the deposit and the plants it contained, Kidston and Lang (1917) have given a good summary description, as follows:

"The whole history of the formation of the Rhynie Chert Zone, at least of that portion from which our specimens were taken, can be clearly read. One can in imagination see a land surface, subject at intervals to inundation, covered with a dense growth of *Rhynia Gwynne-Vaughani.* By the decay of the underground parts of *Rhynia* and the falling down of withered stems (for this plant had no leaves) a bed of peat was gradually formed varying from an inch to a foot in thickness. The peat was then flooded and a layer of sand deposited on its surface. Again the *Rhynia* covered the surface, and this process of the formation of beds of peat, with the deposition of thin layers of sand, went on till a total thickness of 8 feet had accumulated.

After the formation of 8 feet of alternating peat and sand local physical conditions must have altered, for water with silica in solution, possibly discharged from fumaroles and geysers, poured over the peat and sealed it up. Thus the whole was converted into a band of chert, the structure of the plants being preserved in many cases in great perfection." (p. 764)

Although originally considered to be of Middle Devonian age, the Rhynie deposit is now regarded as being Early Devonian (Chaloner, 1970; Westoll, in House et al. 1977).

In the four parts of their series that deal with the vascular plants, Kidston and Lang (1917–1921) recorded three genera, *Rhynia, Horneophyton* (originally *Hornea*) and *Asteroxylon.* These have attracted a great deal of attention and their restoration figures have been reproduced many times. It may, therefore, seem unnecessary to treat them in detail here but this is not the case. Little more was done with the Rhynie chert for several decades after 1921, perhaps because it was assumed that little remained to be known, but in more recent years an interest has again focused on the locality. These studies have resulted in significant changes in, or additions to, our knowledge of all three genera. The revisions that have resulted to date are perhaps more than anything a reflection on the difficulty of following the plant parts, and reassembling them into their life-form, through the very hard silica matrix. The preservation is extraordinarily good in parts of the deposit but the material is not always suitable for satisfactory use of the peel method. The difficulties of working out an accurate restoration, as compared to one derived from good compression fossils as described above, are enormous.

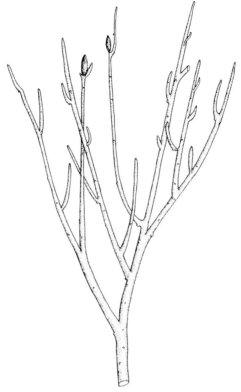

Figure 4.18. Restoration of *Rhynia gwynne-vaughanii*. (From D. Edwards, 1980, Rev. Palaeobot. Palynol.)

Kidston and Lang's initial contribution (1917) on the Rhynie plants dealt with *Rhynia gwynne-vaughanii*. In their second paper (1920) additional information was given and a second species, *R. major*, was described.

On the basis of these two reports, *Rhynia gwynne-vaughanii* was presented as a plant attaining a height of about 20 cm consisting of dichotomizing rhizomes which produces upright shoots (Fig. 4.18). The parts identified as the rhizome bore non-septate rhizoids. The dichotomizing upright axes are leafless, that is, naked with the exception of the presence of small structures called "hemispherical projections" some of which formed the seat of development of lateral branches (described as "adventitious branches"). The latter are distinctive and problematical in that their central vascular strand is not connected to that of the main axis.

The axes attained a diameter of 3 mm, bore some stomates and contained in the center a small cylindrical stele of annular tracheids. In some cases these are smaller in the center, probably representing the protoxylem. Some of the axes terminate in sporangia that are cylindrical and measure about 3 mm long and 1.5–2.5 mm in diameter (Fig. 4.19); they have a multicellular wall and contain large numbers of spores with a triradiate mark. They are on average about 40 μm in diameter. No obvious dehiscence mechanism has been determined. An abscission layer forms at the sporangium base and whole sporangia are shed. This results in few being found attached. Kevan et al. (1975) describe the occurrence of an ancient type of arachnid inside a *Rhynia* sporangium and discuss the possibility that spores were dispersed by these organisms.

The second species, *Rhynia major*, as described in the 1920 study, appears as a generally larger plant but of similar habit. The axes are described as attaining a diameter of 6 mm and have a correspondingly larger protostele (Fig. 4.20) than that of *R. gwynne-vaughanii*. The cylindrical sporangia measure 12 × 4 mm and contain spores that are 65 μm in diameter. The differences between the two are not easily recognizable, especially when attempting to separate the various isolated parts in a matrix such as the Rhynie chert.

As an interesting addendum to this brief discussion of *Rhynia*, we will

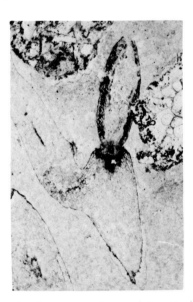

Figure 4.19. Section showing sporangium and its attachment (and abscission) point of *Rhynia gwynne-vaughanii.* × 8.5. (From D. S. Edwards, 1980, Rev. Palaeobot. Palynol.)

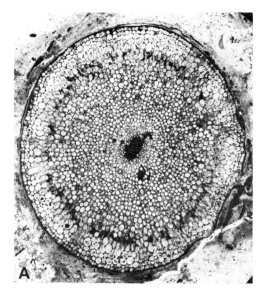
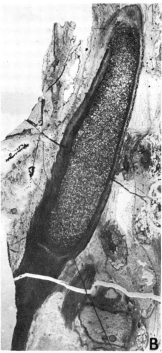

Figure 4.20. *Rhynia major.* A. Transverse section of an aerial axis, × 20. B. A slightly tangential longitudinal section through a sporangium containing a large number of spores. × 5.5. (From Kidston and Lang, 1917, Trans. Roy. Soc. Edinburgh 51:761–84.)

add a note on several studies that have been made recently which suggest that *R. gwynne-vaughanii*, or at least some of the specimens attributed to it, were gametophytes, probably of *R. major*. Pant (1962) discusses the matter in some detail and Lemoigne (1968) gives photos of structures that are partially immersed in the axes and suggestive of being archegonia. Opinions differ as to the identity of these structures, which hopefully will be clarified by further study.

We have chosen to include here a restoration of *R. gwynne-vaughanii*, as prepared by David Edwards (1980) based on his study of the original slides as well as serial peels and ground thin sections from new collections (Fig. 4.18). He found that the plant is more complex in its branching than originally interpreted. He identified no sexual organs (i.e., presumed archegonia or antheridia) and rejected the interpretation that the plant represents the gametophyte of *R. major*. He tends to accept the fossils as representing two distinct species of *Rhynia* but does add: ". . . although

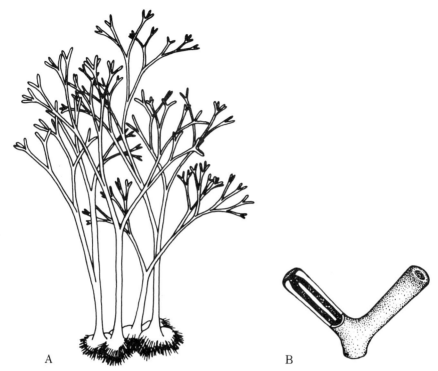

Figure 4.21. *Horneophyton lignieri.* A. A suggested reconstruction of the whole plant, about 0.4 ×. B. Reconstruction of a once-branched sporangium, the left branch cut away to show the columellate structure. × 4.0. (Redrawn from Eggert, 1974. Amer. Journ. Bot.)

the existence of aerial sporangium-bearing axes of *Rhynia gwynne-vaughanii* sensu Kidston and Lang has been confirmed, this itself does not rule out the possibility that some of the prostrate axes (rhizomes of Kidston and Lang) may have been gametophytic. . . ." (p. 185)

Other Rhyniophytes
Because of its terminally borne sporangia, slender protostele and dichotomizing axes, *Cooksonia* (discussed in Chapter 3) is placed in this subdivision.

The plants described below also are included, tentatively, in the subdivision. In one of her recent studies, Dianne Edwards (1976) sums up the status of the rhyniophytes rather well: "The group is of enormous botanical importance in that it contains the earliest vascular plants known. There has, however, been a tendency to assign plants to this taxon rather uncritically, with the result that it contains relatively few plants which are well described and many which are fragmentary and problematical" (p. 173). Some of the included plants are of great interest insofar as they are known but much remains to be determined about their interrelationships.

Horneophyton (Kidston and Lang, 1920) Barghoorn and Darrah, 1938.
Horneophyton ligneri (Kidston and Lang) Barghoorn and Darrah.

Horneophyton was a small plant probably attaining a height of 20–25 centimeters (Fig. 4.21A). The rhizome is distinctly lobed (Fig. 4.22A, B); the lobes themselves lack a continuous vascular strand and from them arose upright naked axes which dichotomized and contained a small central vascular strand probably similar to that of *Rhynia*. Some of the axes terminated in small, approximately 2–5 mm long sporangia, with flat, concave or convex apices and no apparent dehiscence mechanism (Fig. 4.21B). They are distinguished by having a central parenchymatous columella around which the triradiate spores were produced (Fig. 4.22D, E).

More recently several paleobotanists have studied the plant with particular respect to the nature of the spore-producing structures. These have stemmed in part from Kidston and Lang's report that in some cases the sporangium displayed a "lobing or branching of the columella." (Fig. 4.22C). Eggert (1974) has described the fertile branch endings as being composed of two or four distinct lobes with a continuous sporogenous zone, which implies a branched sporangium. The spores were apparently released by a small apical slit at the tip of each branch. Eggert, and El-Saadawy and Lacey (1979b) discuss the significance of the apparently branched sporangium relative to the origin of synangia; we will return to this on a later page. Bhutta (1972) described a possible dehiscence mechanism, involving drying out of sporangial wall cells forcing splits in the originally convex apical region.

Spores found in sporangia of *Horneophyton* are the best preserved *in situ* spores in the Rhynie chert. They occur in tetrahedral and decussate tetrads and are trilete, curvaturate, about 50 μm in diameter, and covered with small apiculi on the distal hemisphere. They compare best with some species of the dispersed spore genus *Apiculiretusispora* (Streel) Streel.

Nothia Lyon, 1964.
Nothia aphylla Lyon, ex Høeg, 1967.

This seems to be an appropriate place to introduce another Rhynie chert plant and comment on its history in the literature. In their contribution of 1920 Kidston and Lang described *Asteroxylon mackiei* (see page 117 of this work) as a densely leafy plant and they found associated

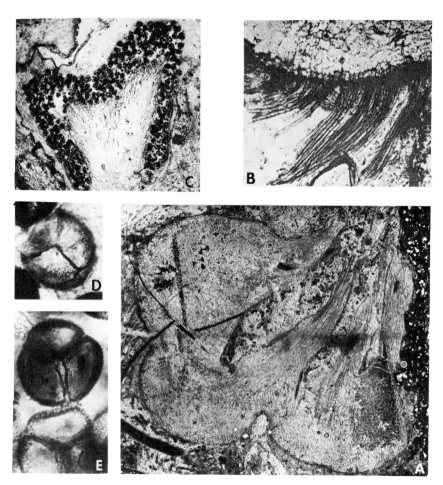

Figure 4.22. *Horneophyton lignieri.* A. A vertical section through a 'proto-cormous rhizome' with three lobes and with an aerial stem arising from each one. × 9.6. B. The lower part of the rhizome showing rhizoids. × 33. C. Longitudinal section through a divided or 'branched' sporangium. × 28. D,E. Spores, photographed from Kidston Slide #2438. D × 335, E. Approx. × 340. (A–C. From Kidston and Lang, 1920. Trans. Roy. Soc. Edinburgh 52:603–627. D. from Gensel, 1980, Rev. Paleobot. Palynol. E. Photographed from Kidston Slide 2454, Hunterian Museum).

with it certain naked axes bearing sporangia. In their well-known restoration these fertile axes were shown as possibly having been borne as terminal branches. It has since been established that the leafy plants bore sporangia of a very different nature on short axes scattered among the

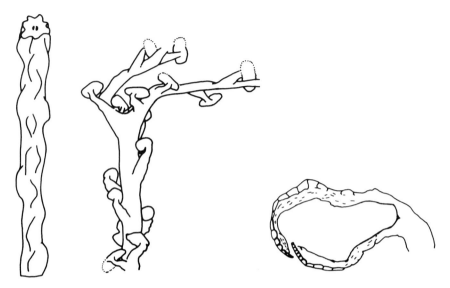

Figure 4.23. *Nothia aphylla* A. A restoration of a stem showing central strand, undulations of outer regions. × 7.6. B. A restoration of a branched fertile specimen with a total of 16 stalked sporangia. × 2.6. C. Median longitudinal section of a sporangium showing the annulus-like epidermal cells. × ca 2.3. (Redrawn from El-Saadawy and Lacey, 1979a, Rev. Paleobot. Palynol.)

microphylls. This discovery was made by Lyon in 1964, and at that time he proposed the name *Nothia aphylla* for the naked fertile axes (which clearly had nothing to do with *Asteroxylon*). The first formal description was given by Høeg in 1967 in the *Traité de Paléobotanique.* A more detailed account has been given by El-Saadawy and Lacey (1979a) and the information presented here is taken chiefly from that source.

The accompanying illustrations reveal the general morphology and some of the variability in the way the sporangia are arranged (Fig. 4.23). The available specimens apparently represent terminal parts of a plant. These are slender, naked axes which rarely exceed 2.5 mm in diameter, have a protostele that is elliptic in transverse section and probably a central protoxylem. Again, typical tracheidal thickenings have not been observed. The axes bear sporangia in approximate whorls of three, forming a rough spiral, but there is considerable variation in the arrangement in different specimens. The sporangia are essentially reniform, about 3 mm wide by 2 mm long by 1.4 mm thick and are borne on a stalk ranging from 1 to 4 mm long. Dehiscence into two slightly unequal halves was by an apical transverse slit (Fig. 4.24), and specialized cells apparently representing an annulus have been observed. The spores average 65 µm in diameter and in unopened specimens as many as 5000

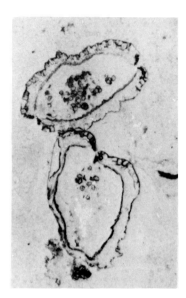

Figure 4.24. *Nothia aphylla.* Two sporangia in section view, with spores. × 20. (NCUPC collections.)

spores have been counted. The name *Nothia* is derived from the conical hill called 'Tap O'Noth' which appears in the background of the photo of the Rhynie area (see Fig. 4.17). It is notable for the presence of one of the finer examples of an Iron Age vitrified fort crowning the summit.

Renalia Gensel, 1976.
Renalia hueberi Gensel, 1976.

This is an intriguing, simple and primitive little plant (Fig. 4.25) and provides an excellent example of the kind of information that can be obtained from a good compression fossil. It occurs in great abundance in a dark gray shale outcropping in back of a small beach along the north shore of Gaspé Bay, Quebec. The sediments at this point fall within the Battery Point Formation in the late Early Devonian.

The stems appear on the exposed surfaces in great abundance and are aligned in one direction. Since no other plants were found associated with it there is a strong suggestion that it grew as a pure stand nearby or was actually covered in its place of growth. Large quantities of study material could be isolated by placing pieces of the shale in hydrofluoric acid.

The plant was erect with a slender, pseudomonopodial main axis 1.0 to 1.5 mm in diameter; stem fragments up to 11 cm long were obtained and it seems evident that the plant attained a height of 20 to 30 cm. The

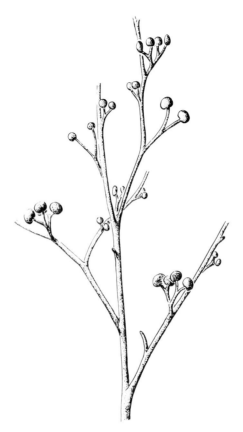

Figure 4.25. Restoration of *Renalia hueberi*, × 7. (From Gensel, 1976, Rev. Palaeobot. Palynol.)

lateral branches dichotomize up to four times and many of them are terminated by a reniform sporangium with a width of about 3.5 mm and a length of 2.5 mm (Fig. 4.26). They produce large quantities of smooth-walled spores 46–70 µm in diameter and dehisce along their distal margin. Partial maceration of the fossils not only reveals the structure and content of the sporangia but also the presence of a slender vascular strand in the axes (Fig. 4.26E) composed of a few dozen tracheids with spiral to scalariform thickenings.

    Renalia hueberi is regarded as having evolved from a *Cooksonia*-type ancestor. Although a small plant itself, it is considerably larger than any of the Cooksonias and differs from them in the upright axis. *Renalia*, and to some extent *Nothia*, combine features of the rhyniophytes and zostero-phylls; the terminal position of the sporangia conforms to the former

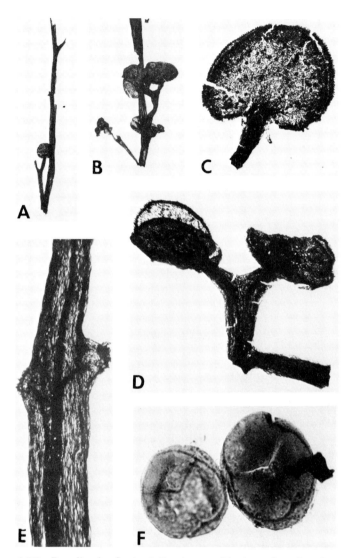

Figure 4.26. *Renalia hueberi.* A,B. Axes with branches bearing terminal sporangia. A. × 2.8, B. × 3.3. C. A single sporangium. × 17. D. Part of a fertile branch showing dehisced sporangia. × 12. E. Part of a cleared axis showing the central vascular strand, × 29. F. Two spores, × 478. (From Gensel, 1976, Rev. Palaeobot. Palynol.).

group and their shape and mode of dehiscence are like that in the better known members of the zosterophylls.

Hsüa robusta Li, 1982

An interesting new plant was described by Cheng-sen Li (1982) from the Emsian age Xujiachong Formation of the Qujing district in Yunnan, China. *Hsüa robusta* is considered to be related to *Cooksonia* but is clearly more robust and more complex in its branching pattern. Main axes 6–10 mm wide with a central vascular strand bifurcate at intervals and also bear much-divided lateral branches which are either fertile or sterile. These are about three times divided, up to 11 cm long, with circinately coiled apices. Fertile lateral branches terminate in one or more round to reniform sporangia. Tubercles occur on the axes near the point of departure of lateral branches, as do structures termed "rhizophore-like" and "root-like." All exhibit a dark central strand interpreted as representing vascular tissue and the root-like structures are three-times divided.

Features of the epidermis were described, including the presence of stomata. Some anatomy is preserved, showing the plant to have a centrarch protostele; tracheids are described as having scalariform to reticulate thickenings, but some could possibly be close-spaced annular. The sporangia, ranging in width from 1–8.2 mm and in height from 1–4.2 mm, dehisce along a thickened distal margin similar to *Renalia* and the zosterophyllophytes. Smooth trilete spores 18–36 μm in diameter were obtained from sporangia.

This plant is of great interest in appearing to be much more complex than *Renalia,* yet clearly allied to it and to *Cooksonia*. The striking similarity of its branching pattern to some zosterophylls and trimerophytes demonstrates the necessity of careful examination of early vascular plants for both vegetative and reproductive structures before assessing their affinities.

Steganotheca Edwards, 1970a
Steganotheca striata Edwards, 1970a.

This was apparently a very small plant and preserved only as carbonized remains with little significant structure but with some interesting features. The fossils were found in the abandoned Capel Horeb Quarry between Trecastle and Llandovery, South Wales, and are of Lower Devonian age.

The plant is known from very slender axes which range in width from 0.5 to 0.7 mm, dichotomize at a wide angle, and attain a maximum length of 5 cm. In view of the very slender nature of the axes it is doubtful if the plant was much taller than that in life. As to the anatomy of the axes Edwards notes: "A faint central line is sometimes seen when the rock is

Figure 4.27. *Eogaspesiea gracilis.* Restoration showing the slender and sparsely branched axes. × 1.2 (From Daber, 1960, Geologie.)

held at a suitable angle to the incident light." (p. 452) It seems likely that it represents a slender vascular strand but this is uncertain.

The branches are terminated by individual sporangia that are cup-shaped, measuring 2.5 mm long and 1.5 mm broad; the coaly compression is described as being much thicker at the distal end suggesting a "lens-like structure." Spores were not obtained from these terminal bodies but their close comparison with the spore-bearing organs of other rhyniophytes leaves little doubt as to their identity as such.

Edwards and Rogerson (1979) have reported *Steganotheca* from a late Silurian (Whitcliffian) horizon in Wales. These older specimens are more fragmentary and differ in having larger sporangia. There is a general resemblance of this plant to *Cooksonia* but the difference in sporangial morphology, essentially the cylindrical rather than broader-than-long shape as in *Cooksonia* or obovoid as in *Rhynia*, seems to justify generic distinction.

Eogaspesiea Daber, 1960
Eogaspesiea gracilis Daber, 1960.

The specimens of this plant were collected by Rudolf Daber at the time of a field trip in conjunction with the 1959 International Botanical Congress. They were found in an exposure a half mile east of Cap-aux-Os

Figure 4.28. *Eogaspesiea gracilis.* Part of a specimen, enlarged about × 1.2. (From Daber, 1960, Geologie.)

on the north shore of Gaspé Bay, being of late Early Devonian age.

The plant consists of very slender axes 8 to 9 cm long and not over 0.5 mm in diameter (Figs. 4.27, 4.28); they occasionally dichotomize at an acute angle and some of the branches are terminated by a sporangium 2.5 mm long and 1.0 mm in diameter. A few smooth-walled spores were obtained by maceration from one of the sporangia. The slender upright axes tend to be tufted at the base and were found associated with axial fragments somewhat larger in diameter which are thought to be the rhizomes although the two were not found in organic connection. Cells which appear to be annular tracheids have been obtained by maceration of the "rhizomes" but no such tissue was found in the upright axes.

Fossils such as *Steganotheca* and *Eogaspesiea* bring us to the "lower borderline" of plants that may be properly included in the Rhyniophytina. They are probably vascular plants but we do not have decisive evidence. They were certainly small and simple in their organization and seem to be representative of the very early invaders of a land habitat, and fragmentary as they are, they offer us some understanding of the beginnings of land vegetation.

PART II. THE ZOSTEROPHYLLOPHYTES

For the most part the plants included in this subdivision may be distinguished from the rhyniophytes and trimerophytes by the lateral attachment of the sporangia, directly or by a short stalk, to the axis. Vascular anatomy, where preserved, is an exarch protostele.

Zosterophyllum Penhallow, 1892.

Zosterophyllum nominally typifies the group. It is known from descriptions of at least eight species, several of which are founded on very fragmentary fossils. One species, *Z. myretonianum,* is now very well known; our present knowledge of it is based on data derived from the efforts of several investigators.

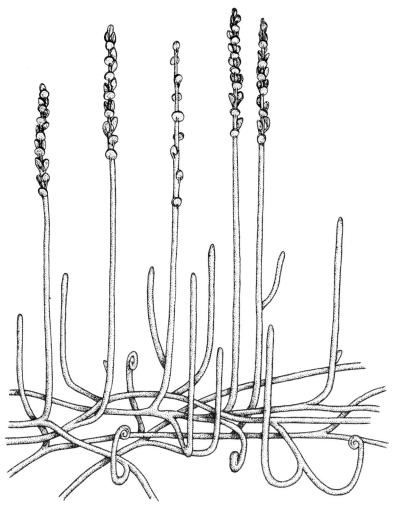

Figure 4.29. Suggested restoration of *Zosterophyllum myretonianum.*

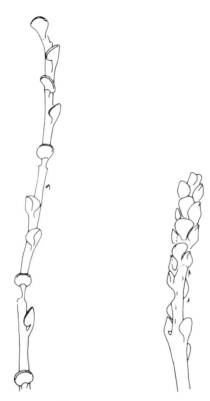

Figure 4.30. *Zosterophyllum myretonianum*. Line drawings of fructifications showing sporangial distribution. Approx. × 1. (From Edwards, 1975, Trans. Roy. Soc. Edinburgh 69:251–65.)

Zosterophyllum myretonianum Penhallow, 1892.

The plant was first described by David Penhallow from specimens that were sent to J. W. Dawson and Penhallow from Scotland by a Mr. Reid of Blairgowrie. Penhallow thought the plant to be an aquatic with a general resemblance to the modern grass *Zostera* whence the origin of the generic name. The specific name is taken from Myreton, a now derelict quarry near Dundee, the age of the rocks being basal Lower Devonian.

W. H. Lang became interested in the plant and in 1927 he added much to our knowledge of it and gave a historical sketch. He noted that specimens described as early as 1831 by J. Fleming (although of course not under Penhallow's binomial) were referable to it, and in 1841 Hugh Miller in his book, *The Old Red Sandstone*, described some specimens. More recently, as additional specimens have been found, it has been described in numerous publications which are well summarized in those of Walton (1964) and Dianne Edwards (1975). From these various studies the following concept emerges.

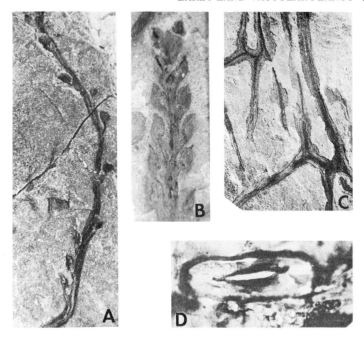

Figure 4.31. *Zosterophyllum myretonianum.* A. Specimen showing fairly widely spaced, spirally arranged sporangia. × 1.4. B. Specimen with densely spaced sporangia. × 1.5. C. Axes with H-branching, probably part of basal rhizomatous region. × 2. D. Detail of axis cuticle, showing a stomate. × 500. (A,B from Edwards, 1975. Trans. Roy. Soc. Edinburgh 69:251–65. C, from Lang 1927, Trans. Roy. Soc. Edinburgh 55:443–455. D. from Lele and Walton, 1961, Trans. Roy. Soc. Edinburgh 64:469–75.)

Zosterophyllum myretonianum was a plant of rhizomatous habit as shown in our restoration drawing (Fig. 4.29), with upright shoots departing from an apparently dense horizontal branching system; when compressed, these formed the so-called H-branching. Photographs in Lang's 1927 account show these features quite well and have been reproduced here (Fig. 4.31C). The upright shoots branched rather sparsely and some of them terminate in spikes of laterally borne sporangia.

In 1975 Dianne Edwards described a rather large collection consisting, for the most part, of fertile spike compressions from the Scottish Aberlemno Quarry. Her information was obtained by carefully degaging the specimens and an examination of a few of her drawings reveals the variation in size, distribution, and arrangement of the sporangia. Although they often appear to be arranged spirally, some seem to be aligned in two rows. In addition, some sporangia are located quite close together and others are more distantly spaced (Figs. 4.30, 4.31A & B). Her study of these numerous fertile specimens and a comparison with those of other species leads her to conclude that the size of the sporangia and their arrangement on the axis are precarious characters in delimiting species. She adds: " . . . sporangium and stalk characteristics [shape of sporangia], including the angle of insertion on the fructification axis are probably the most useful characters for species delimitation." (p. 264).

Cuticular features, including the presence of stomates, and details of branching were reported in an account by Lele and Walton in 1961. Epidermal cells are elongate to nearly isodiametric, the latter usually being located near the departure of branches. The stomata, the oldest known from a vascular plant, are elliptical to rectangular in shape. They originally were interpreted as a single large cell surrounding an elongate slit with subsidiary cells absent (Fig. 4.31D). Edwards et al. (1983a) suggest however that two guard cells were present, and cuticular flanges outlining them simply were not preserved. The stomata are not evenly distributed but are found chiefly on the erect branches.

Small protuberances interpreted as undeveloped lateral branches, occur on apparently mature axes. Similar bumps have also been found on axes of another species, *Z. divaricatum*, by Gensel (1982a), where they are similarly interpreted.

It is our opinion that although the identification of species in the genus *Zosterophyllum* is open to considerable question, the generic concept is quite clear. Additional significant information has been published, under other species names, which we include chiefly as adding to the generic concept.

In 1969 Edwards brought out two studies which clarified several aspects of internal structure. One of these (1969a) is concerned with partially petrified specimens of *Z. llanoveranum* from the Lower Devonian Brecon Beacons Quarry in Wales (Fig. 4.32). This binomial had been erected by Croft and Lang (1942) in their treatise on the flora of the Senni Beds of South Wales for smooth axes with sporangia borne apparently in two rows located to one side of the axis. They described several fragments of the terminal fertile parts of the plant and one of their figures which shows the arrangement of the sporangia quite well is reproduced here (Fig. 4.33). Edwards demonstrated that the vascular strand of the axes appears in transverse section as terete to slighty elliptic and is composed of scalariform tracheids with exarch protoxylem. The petrified sporangia contained spores and the cellular structure of the wall is well enough preserved to allow a restoration drawing showing the structure of the dehiscence region (see Fig. 4.34.).

The specimens described in Edwards' second article (1969b) are assigned tentatively to *Z. fertile* and also come from the Brecon Beacons Quarry. The vascular strand is quite small, being composed of about

Figure 4.32. A view of the Brecon Beacons Quarry, an abandoned quarry of Lower Old Red Sandstone age (Senni Beds) between Brecon and Merthyr, South Wales. (Photo courtesy of Dianne Edwards.)

thirty tracheids; it is described as terete to elliptic but the latter shape seems to be chiefly a result of elongation near the point of departure of a branch strand. One of her illustrations is a section through a petrified sporangium which is impressive in showing the large number of spores, probably several thousand.

Zosterophyllum divaricatum Gensel, 1982a.

Zosterophyllum divaricatum was described by Gensel (1982a) from the late Early Devonian of New Brunswick, Canada; this species, and some similar specimens from Germany (*Z. spectabile* Schweitzer, 1979), have fertile axes with spirally arranged sporangia which are all oriented to one side of the axis by a twisting of their stalks (Figs. 4.35, 4.36). On several specimens, major axis bifurcations occur within fertile regions so sporangia are located both above and below them.

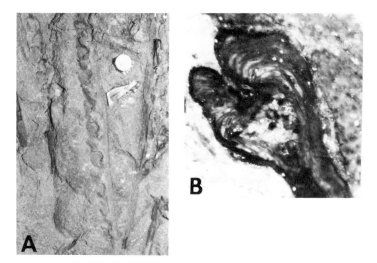

Figure 4.33. *Zosterophyllum llanoveranum.* A. Specimen showing sporangia oriented to one side of the axis. Approx. × 1. B. Section through sporangium showing dehiscence region. × 85. (A. From specimen V62516a illustrated in part by Croft and Lang, 1942. Phil. Trans. Roy. Soc. London 231B. B. from Edwards, 1969a, Amer. Journ. Bot.)

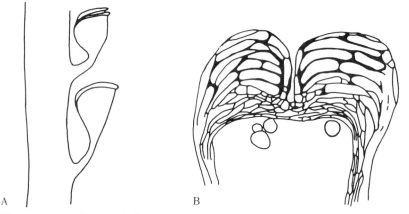

Figure 4.34. *Zosterophyllum llanoveranum.* A. Drawing of an axis with two sporangia showing distal dehiscence. × 4.4. B. Drawing showing margins of valves of sporangium. × 79. (Both redrawn from Edwards, 1969a, Amer. Journ. Bot.)

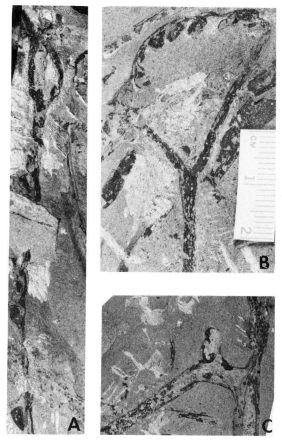

Figure 4.35. *Zosterophyllum divaricatum.* A. A specimen bearing sporangia with divided axis in fertile region. × 1.3. B. Fertile axis showing bifurcation below fertile region. × 1.4. C. A broad axis tentatively referred to this plant with a K-branching pattern; the lower half of the second dichotomy resulted in a long axis and the upper half is undeveloped and in a crozier stage. × 1.6. (From Gensel, 1982a, Amer. Journ. Bot.)

Well-preserved compressions of sporangia of *Z. divaricatum* demonstrate their shape and attachment (Fig. 4.37A, C). They are basically reniform, but often assume a variety of shapes resulting from compression during preservation. The sporangia are attached by a fairly long stalk (up to 4 mm) which departs from the major axis at an acute angle. The sporangia contain smooth trilete spores 50–90 μm in diameter, which are curvaturate (Fig. 4.37D). Axes exhibit circinately coiled apices (Fig.

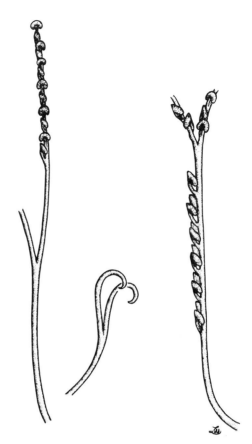

Figure 4.36. Restoration of *Zosterophyllum divaricatum.* × 1.2 (From Gensel, 1982a, Amer. Journ. Bot.)

4.37 E) and some also bear laterally borne bumps which may be aborted apices. Cuticles and tracheids are similar in morphology to those of *Z. myretonianum* although preserved stomata were not observed. Somewhat broader, entirely vegetative axes were found in association with the fertile ones which exhibit interesting H- and K- patterns; in one of them the upper axis resulting from two dichotomies of the main axis is still in crozier form (Fig. 4.35 C). These larger axes may represent the rhizomatous regions of the plant. It was noted in this study how little is actually known about the basal, vegetative regions of many *Zosterophyllum* species and thus the extent to which branching patterns may vary in the genus.

Some excellently preserved specimens of *Zosterophyllum* (Fig. 4.38) have been described from the Devonian of China by Li and Cai (1977). Although those authors assigned them to over a dozen species, fewer may

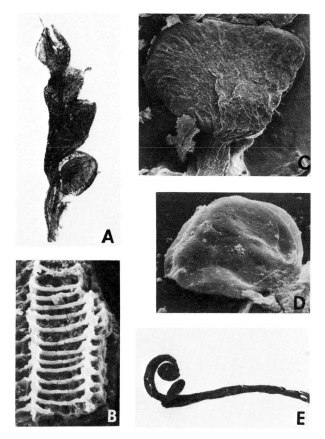

Figure 4.37. *Zosterophyllum divaricatum.* A. A fertile fragment showing relatively long sporangial stalk and a tendency to twist around the axis; lowermost sporangium shows marginal dehiscence. × 4.5. B. SEM of tracheid showing regularly spaced annular thickenings. × 489. C. SEM of a sporangium. × 9. D. SEM of a spore obtained from sporangium. × 533. E. Distal region of a bifurcating axis with circinately coiled tips. × 2.6. (All from Gensel, 1982a, Amer. Journ. Bot.)

actually be represented. They also indicate the wide range of diversity that is becoming apparent in the genus.

Gosslingia Heard, 1927.
Gosslingia breconensis Heard, 1927.

This was reported briefly by Heard in 1925 under the name *Psilophyton*; he brought out a detailed account under the new generic name in 1927. Additional information was contributed by Croft and Lang in 1942, and by Dianne Edwards in 1970b based on fossils from the Brecon Beacons Quarry, South Wales.

Figure 4.38. *Zosterophyllum sinense.* A well-preserved specimen with numerous aerial axes each bearing a terminal cluster of sporangia, from Cangwu, Kwangsi. About × 0.5. (From Li and Cai, 1977, Academia Sinica.)

The plants probably attained a height of a half meter and consisted of axes with a maximum diameter of 4 mm. The branching is dichotomous throughout but with a strong tendency to dominance of the "main" axes resulting in a slightly imperfect pseudomonopodial habit (Figs. 4.39, 4.40). The tips where preserved are circinate. The axes are smooth to occasionally papillate. However, a short distance below each branching point there is a structure referred to as an "axillary tubercle." This or apparently comparable structures have been reported in other early Devonian plants. In *Gosslingia*, it is essentially a very short branch (up to 1.3 mm long) that is vascularized. The position of these structures is indicated by a small circle below each dichotomy in the restoration figure.

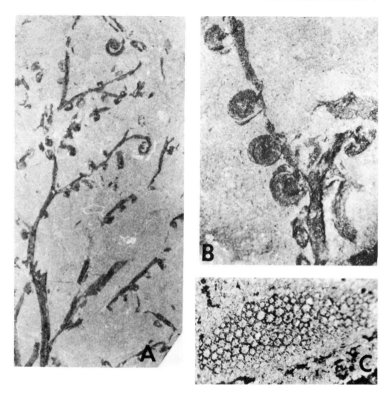

Figure 4.39. *Gosslingia breconensis.* A. Fertile axes showing circinate tips of branch endings. × 1.1. B. Detail of a fertile axis showing the almost globose sporangia and their mode of attachment. × 4.7. C. Transverse section of the elliptical xylem strand. × 48. (From Dianne Edwards, 1970b, Phil. Trans. Roy. Soc. London 258B.)

It has been variously interpreted (in other plants) and Edwards, after reviewing some of the previous reports, offers this concept: "It is perhaps more likely that the tubercle represented the base of a branch, which was either abscissed before preservation or was lost during preservation." (p. 242).

As to stem anatomy, stomates have been observed in the epidermis; within this there is an outer cortex of thick-walled cells, two to four cells thick. The inner cortical tissues are not well preserved. The vascular strand is elliptic in transverse section (Fig. 4.39C) which is regarded as the natural shape and not the result of crushing. The wall structure of the tracheids is reticulate to scalariform; some of the peripheral tracheids are smaller than the inner ones, probably representing the protoxylem. In any event it seems certain that the protoxylem is not central.

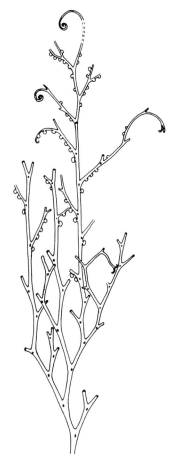

Figure 4.40. Restoration of *Gosslingia breconensis*, note especially its branching pattern. × 0.5 (Redrawn from Edwards, 1970, Phil. Trans Roy. Soc. London 258B.)

Heard's study of the stem anatomy is rather remarkable for the time, when little had been done with pyritic petrifactions. Concerning *Gosslingia* preparations, a few words from his account are of historical interest:

> "Twelve months' experimental work resulted in the obtaining of two unsatisfactory macerated fragments; namely, a bundle of tracheids, and a portion of cortical tissue with a few epidermal cells attached. ... Attempts to etch the polished surface were met with instant success, when concentrated nitric acid was employed as a reagent. On

the application of this acid to the polished surface of the pyrites, the internal structure of the plant was evolved in a manner similar to the development of an image on a photographic negative. The average time occupied in development was about 10 seconds." (p. 197–198).

The sporangia were borne along the main axis and lateral branches; they are reported to be on both sides of the main axis but on the outer side only of the lateral branches. Sporangia are both alternate and opposite, and most are borne on stalks which are oriented at right angles to the axis (Fig. 4.39B). The sporangia are oval to reniform, measuring approximately 2–2.5 mm in diameter and are mostly attached by stalks 0.5 mm long although a few are significantly longer. They open, as Edwards notes, along the distal margins like a cockle shell, and contain numerous spores 36–50 μm in diameter.

Specimens attributed to *Gosslingia* have been recorded also from the USSR (Petrosian, 1968), Germany (Schweitzer, 1979) and western North America (Tanner, 1982).

Oricilla Gensel, 1982.
Oricilla bilinearis Gensel, 1982.

This genus was described from specimens obtained from the upper Lower Devonian of New Brunswick. It consists of smooth axes 4–5 mm wide which are infrequently branched. Sporangia occur in two rows along the main axis and lateral branches, often in large numbers (Figs. 4.41, 4.42). Their stalks are less than 1 mm long and wide, and depart from the axis at right angles. The sporangia are 2–4 mm wide and slightly less in height (Fig. 4.41C). They are strongly reniform, opening along a specialized dehiscence zone as in other zosterophyllophytes. The cuticle of *Oricilla* is distinctive in that round epidermal cells surrounded by a ring of cells of generally similar shape are interspersed among the more elongate ones. Both *Serrulacaulis* and *Sawdonia* exhibit a similar type of cuticular pattern; however the function of these distinctive groups of cells is unknown.

Oricilla is distinguished from other zosterophyllophytes by the combination of smooth axes with two rows of very short stalked sporangia, the stalk and sporangium complex being oriented at right angles to the axis. In other genera, except *Gosslingia*, sporangia and stalks are oriented at acute angles so they nearly parallel the axes.

Hicklingia Kidston and Lang, 1923.
Hicklingia edwardii Kidston and Lang, 1923.

Our knowledge of this plant is of special interest and relates to the study of fossil plants in general. First, it presents an uncommon case in which the original description was based on a single extraordinary

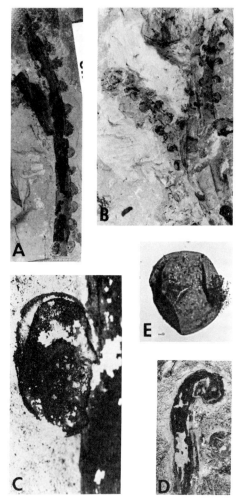

Figure 4.41. *Oricilla bilinearis.* A,B. Fertile axes showing the arrangement of sporangia in two rows. A. × 0.9. B. × 0.85. C. Detail of sporangium attachment by a short stalk at right angles to the axis; note distal dehiscence. × 10. D. A circinately coiled axis fragment. × 1.3. E. A spore. × 316. (From Gensel, 1982c, Rev. Palaeobot. Palynol.)

specimen which seems to show the entire plant, or nearly so, as though it were an herbarium specimen (Fig. 4.43). It was found by a Mr. G. Edward in the Middle Devonian (Middle Old Red Sandstone) of Waas, Caithness, Scotland, and in a short account of 1888, he referred to it with the rather charming description of: "a lovely bush-like plant with hundreds of ribbon or grass-like leaves that have pod, or pocket, or flower-like terminations."

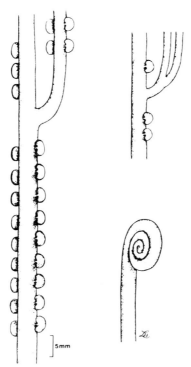

Figure 4.42. Restoration of *Oricilla bilinearis*. (From Gensel, 1982c, Rev. Palaeobot. Palynol.)

The generic name is for Professor Hickling who made a preliminary examination of the plant before turning it over to Kidston and Lang.

As described by Kidston and Lang, the plant became known as being very densely tufted, attaining a height of about 17 cm, the upright dichotomizing axes attaining a diameter of 2 mm. A "thread-like marking" was noted along the middle of some of the axes which probably represents a slender vascular strand but no cellular structure was actually observed. In their generic diagnosis, Kidston and Lang state: "sporangia large, oval, terminating the branches in the upper regions of the plant." This was apparently responsible for restorations prepared by later writers which show terminal sporangia. In "The Early History of Land Plants", Banks (1968) gives such a restoration and he included the plant in his Rhyniophytina. However in another part of their paper, Kidston and Lang stated: "The sporangia are most numerous on the distal regions of the branches in the upper part of the specimen, but are also present, apparently on lateral branches, lower down in the tuft." (p. 405).

Figure 4.43. *Hicklingia edwardii.* A photograph of the original specimen representing presumably an entire or nearly entire plant. × 0.6. (From Kidston and Lang, 1923, Trans. Roy. Soc. Edinburgh 53:405–7.)

Numerous paleobotanists have seen the specimen during the ensuing decades since 1923 and certainly several have had the urge to investigate it more thoroughly but have been prevented from doing so by the reluctance on the part of the curators to possibly damage so beautiful a museum piece. This policy is to some degree commendable but in this case it has contributed to a rather serious misunderstanding of the true nature of the plant.

In an account of 1976, Dianne Edwards has given us a revised interpretation of the plant based on a very careful examination of the original specimen as well as a second one that had been identified in the Kelvingrove Museum, Glasgow. Some of her photos reproduced here show very clearly the lateral attachment of the sporangia (Figs. 4.44,

Figure 4.44. *Hicklingia edwardii.* A terminal branch showing both lateral and terminal sporangia. × 2.3 (From Edwards, 1976, New Phytologist.)

4.45). These seem to be distinctive in the upright way in which the stalks depart from the main axis and the corresponding erect fashion in which the sporangia were borne. It is thus clear that although the sporangia were borne terminally there were also numerous ones borne laterally, probably in a spiral, around the distal parts of many of the upright axes. Unfortunately, spores have not actually been found in the presumed sporangia.

It is hoped that specimens showing cellular structure will ultimately be found, but with the evidence at hand the plant seems most appropriately placed in the Zosterophyllophytina and it would also seem that the relationship of *Hicklingia* to *Zosterophyllum* itself is a close one.

Sawdonia (Dawson) Hueber, 1971.
Sawdonia ornata (Dawson) Hueber, 1971.

It may be helpful at this point to refer back to our discussion of the early history of *Psilophyton* to recall that the plant that Sir J. W. Dawson described in 1871 under the name *Psilophyton princeps* var. *ornatum* was renamed *Sawdonia ornata* by Hueber in 1971. Fossils referable to *S. ornata* have been found in considerable abundance in recent years in a coastal outcrop along the north shore of Gaspé Bay, of late Early Devonian age. A representative specimen is shown in Fig. 4.47. We have indicated previously the difficulty in identifying the exact origins of the specimens Dawson worked with. We cannot be sure that the outcrop noted above was actually exposed a century ago, but his specimens

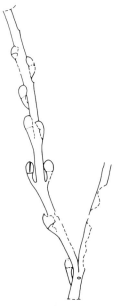

Figure 4.45. A line drawing of *Hicklingia edwardii* showing sporangial attachment. × 1.4 (Redrawn from Edwards, 1976, New Phytologist.)

compare in terms of lithology of the rocks and appearance of the plants to those obtained today; at the very least, the fossils present in it are clearly conspecific with those that Dawson described.

It is something of a paradox that, although numerous people have collected specimens of *S. ornata* from this locality in recent years, the plant has not, as of this writing, been formally and fully described on the basis of such collections, although data derived from this and other localities form the basis for the generic diagnosis.

In 1968, two Soviet paleobotanists, A. R. Ananiev and S. A. Stepanov, described fossils from the lower Devonian of western Siberia which are specifically identical with the Canadian plant (although they chose to retain the binomial *Psilophyton princeps* for their specimens). Their report is informative and includes a description and several illustrations of the laterally borne, spore-bearing organs (Fig. 4.46). The axes measure 4–5 mm wide and are at least 25 cm long. Some distal segments were found with circinately coiled tips. Emergences were apparently stiff, tapered, and about 4–4.5 mm long, varying in density from one part of the axis to the next. The sporangia occur along the sides of the axes on short stalks and are 2.5–3.0 mm in diameter. Spores were obtained in small numbers from some sporangia. Most are round, smooth, trilete, and 70–77 μm in diameter, while some others, which were

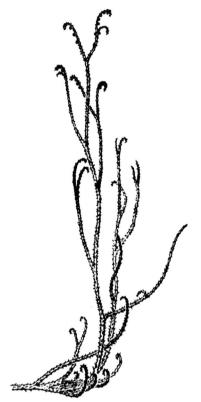

Figure 4.46. *Sawdonia ornata* (Dawson) Hueber, or *Psilophyton princeps* of Ananiev and Stepanov. See text for discussion concerning appropriate binomial. (From a restoration drawing by Ananiev and Stepanov, 1968. Tomsk State Univ. Trudy.)

interpreted as immature, were 20–25 μm in diameter, smooth and very thin-walled, and folded. The spores were not illustrated, but were compared to *Leiotriletes pullatus* Naumova and *L. minutissimus* Naum. respectively.

Based on the diagnostic data of Hueber (1971) and observations of the authors, *S. ornata* from Gaspé is essentially identical. The axes range from 1–4 mm in diameter and bear tapered spine-like emergences 0.5–4 mm long (Fig. 4.47). The emergences often appear slightly expanded at their tips and somewhat glandular. The axes dichotomize frequently, equally or unequally, and exhibit a strong tendency to a pseudo-monopodial habit. Sporangia are attached along the axis by very short stalks (less than 1 mm long) and are 3–3.5 mm in diameter. They contain smooth, trilete spores 54–65 μm in diameter, which McGregor (1973)

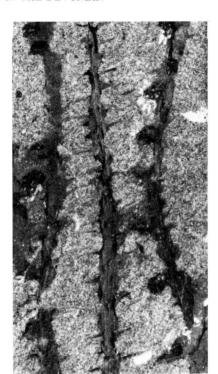

Figure 4.47. *Sawdonia ornata.* A specimen from the north shore of Gaspé Bay showing the laterally attached sporangia and spiny emergences. × 2.3. (NCUPC Collections.)

compared to the dispersed spore genus *Calamospora atava.* Anatomy consists of an elliptic, exarch protostele (Fig. 4.48D).

The cuticle of *Sawdonia ornata* is distinctive (Figs. 4.48A–C); elongate to isodiametric epidermal cells occur, the latter possibly located near the sites of branches. They exhibit papillae or dark, thickened areas representing papillae. Interspersed among these are groups of cells termed 'hair bases' by W. N. Edwards, which consist of a central circular cell surrounded by apparently elongate radiating cells. Hairs have not been found attached and it is equally possible that the central cell may have had a secretory, storage, or aerating function.

Stomata occur, sometimes quite densely and elsewhere rarely. In the Polish specimens described by Zdebska (1972), she notes: "The stomata are arranged longitudinally along the axis, they are sparse, and their distribution is irregular. A stoma (probably) consists of two guard cells and three to five subsidiary cells, the anticlinal walls of which form projections which overhang the guard cells and form a shallow stomatal

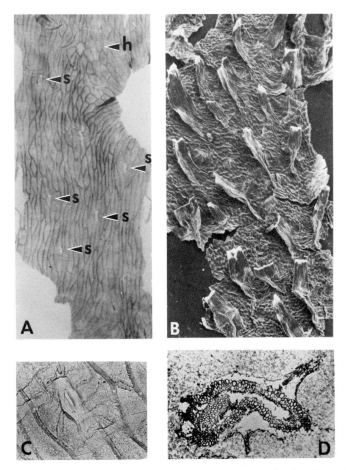

Figure 4.48. *Sawdonia ornata.* A. Fragment of axis cuticle showing epidermal cells, a hair base (h), and stomata (s). × 40. B. SEM of fragment of axis cuticle showing papillate epidermal cells and several broken emergences. × 22. C. Detail of stomate. × 106. D. Transverse section of stem showing the elliptic protostele. × 17. (A–C. NCUPC Collections. D. from Hueber and Banks, 1967, Taxon.)

pit." (p. 91). Edwards et al. (1983) suggest the stomata were slightly sunken and consists of two guard cells.

Sawdonia ornata was quite wide-ranging, both stratigraphically and geographically, being reported on the basis of vegetative characters from the lower Upper Devonian of New York State by Hueber and Grierson (1961) and from Lower Devonian strata of Poland by Zdebska (1972), southern England by Chaloner et al. (1978) and the USSR as noted earlier.

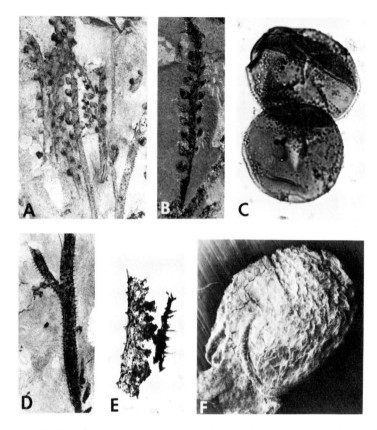

Figure 4.49. *Sawdonia acanthotheca.* A. Several fertile axes showing globose laterally borne sporangia. × 0.4. B. Single fertile axis with nearly oppositely arranged sporangia. × 0.75. C. Spores. × 446. D. Single axis showing emergences. × 0.5. E. Cleared axis cuticle showing variation in emergence morphology. × 4.2. F. SEM of sporangium showing small emergences on outer wall. × 15. (A,C–F. from Gensel et al., 1975; Bot. Gazette. B, from Gensel, 1982b, 3rd NAPC Proceed. vol. 1.)

Sawdonia acanthotheca Gensel, Andrews, and Forbes, 1975.

This species was described from well-preserved specimens (Fig. 4.49) found in late Lower Devonian rocks on the coast of northern New Brunswick, Canada. The occurrence of the available material is of some interest, consisting as it did of a single loose block (about two cubic feet of rock) and although its origin was clearly in the immediate vicinity, a careful search did not produce any more.

The axes attain a diameter of 8 mm and the longest specimens measured 18 cm; it seems probable that the plant attained a height of at least 1/2 meter. A distinctive feature lies in the variation in the

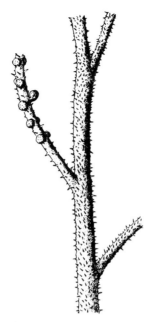

Figure 4.50. Restoration of *Sawdonia acanthotheca.* × 0.75. (From Gensel et al. 1975, Bot. Gazette.)

morphology of the emergences. They range in length from 0.13–1.8 mm and in form from hair-like to rather stout deltoid structures (Fig. 4.49E). Epidermal cells are elongate-polygonal, differing apparently in details from those of *S. ornata.* Although permineralized material was not available, macerations of axis fragments revealed a bundle of annular tracheids which at least demonstrate that it is a vascular plant.

The sporangia tend to be clustered on certain axes (Figs. 4.49 and 4.50); they are arranged singly or in pairs, are about 4 mm in diameter, and dehisce along a median line forming two nearly equal valves. The sporangia are covered with emergences (Fig. 4.49F) about 0.2 mm high which may be scale-like or blunt spines. The spores are about 56 μm in diameter; the wall is essentially smooth but partially covered with minute, round globules which may represent tapetal residue (Fig. 4.49C). As in so many of the early Devonian plants, the number of spores per sporangium was very high, being estimated conservatively at 50–100,000.

Serrulacaulis Hueber and Banks, 1979.
Serrulacaulis furcatus Hueber and Banks, 1979.

This is an important addition to our knowledge of the zosterophyllophytes, both in the distinctive features of its morphology and in its

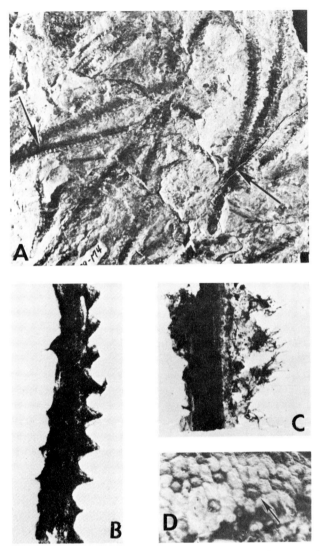

Figure 4.51. *Serrulacaulis furcatus.* A. Specimen showing bifurcating axes. × 0.5. B. Fragment of axis showing stout deltoid emergences. × 4.5. C. Fragment of axis from rhizome area showing rhizoids departing from emergences. × 4.6. D. Detail of sporangium wall showing distinctive cell arrangement. × 111. (From Hueber and Banks, 1979, Rev. Palaeobot. Palynol.)

origin in the early part (Frasnian) of the Late Devonian. The fossils were obtained from a horizon of that age in Schoharie County, New York.

The plant consists of sparsely dichotomizing axes that were probably oval-shaped in transverse section with a maximum diameter of 12 mm. Some of these were regarded as having been the rhizomes which are distinguished from the upright stems by the presence of abundant, non-septate rhizoids. Both the rhizomes and the erect stems bore two rows of emergences; these are arranged on opposite margins of the axes (Fig. 4.51), are 1.5–3.0 mm long, and are cone-shaped with an elliptic base which measures up to 2 × 3 mm. They are contiguous vertically giving the plant the appearance of having scalloped or saw-toothed margins. The overall height of the plant is estimated to have been not more than 60 cm.

Although permineralized material was not available, a coaly compression revealed a partially preserved xylem strand composed of helically thickened tracheids. Elongate epidermal cells are intermixed with isodiametric ones (Fig. 4.51D) which are thicker-walled than the rest. These are surrounded by radiating cells, and thus also resemble the 'hair bases' or 'rosette' cells of *Sawdonia ornata*.

The sporangia are borne alternately in two rows along one side of some of the aerial axes. They measure about 3 mm wide and 2.4 mm high and are attached by a short stalk about 1 mm long. Dehiscence is along the convex upper margin, resulting in two valves of equal size. Spores were obtained from some of the sporangia. They range from 42–66 μm in diameter and have a smooth to slightly granulate exine.

Crenaticaulis Banks and Davis, 1969.
Crenaticaulis verruculosus Banks and Davis, 1969.

This plant is known from compression specimens found in the now classic area along the north shore of Gaspé Bay, Quebec, a few hundred feet west of Seal Rock Landing, being of lower Emsian age. The axes are pseudomonopodial (Fig. 4.52B) but show a basic dichotomy in their development. The longest axes measure 22 cm and range up to 3 mm in diameter. They are papillate and further distinguished by having one or two rows of multicellular, tooth-like emergences (Fig. 4.52A) 0.1–0.9 mm long and 0.1–0.5 mm high. Each row was either single or double and with the latter arrangement, the teeth overlap irregularly. The epidermis of the axes was sufficiently well-preserved to reveal occasional stomates.

Subaxillary tubercles (which were mentioned as being found in *Gosslingia*) are present and some specimens have been found with a short branch attached, or rather the basal part of a branch (Fig. 4.52A). The

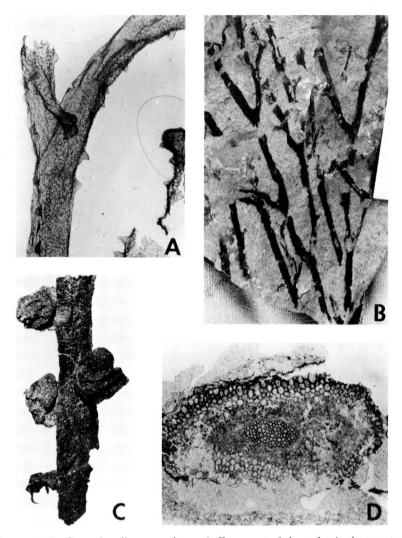

Figure 4.52. *Crenaticaulis verruculosus.* A. Fragment of cleared axis showing two rows of teeth, and part of a subaxillary branch. × 4. B. General view of several axes which are bifurcated. × 0.9. C. Fragment of fertile specimen showing attachment of sporangia. × 3. D. Transverse section of axis showing thick-walled outer cortical cells and elliptical protostele. × 23. (NCUPC Collections.)

authors suggest that the *Gosslingia* and *Crenaticaulis* axes may have borne rhizophore-like branches similar to those in living *Selaginella*.

Pyritized remains of the stem reveal a cortex of five to six rows of thick-walled cells and an elliptical strand of xylem (Fig. 4.52D). Most of

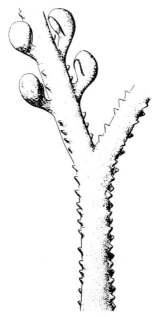

Figure 4.53. Restoration of *Crenaticaulis verruculosus*. (Redrawn from Gensel et al. 1975, Bot. Gazette.)

the tracheids are scalariform with some smaller peripheral ones, presumably of protoxylem, that appear to be annular or spiral. Thus the stele was evidently exarch.

The sporangia are arranged in loose groups, either opposite or subopposite along the axes, and none have been observed in a terminal position (Figs. 4.52C, 4.53). They are reniform to spherical and borne on an incurved pedicel about 1 mm long; dehiscence is from one side to the other, dividing the sporangia into two unequal segments. Spores were not observed.

Rebuchia (Dorf) Hueber, 1970.
Rebuchia ovata (Dorf) Hueber, 1970.

In 1933, Erling Dorf described some fertile spikes of a zosterophyllaceous type under the name *Bucheria ovata*, from the Beartooth Butte Formation, Lower Devonian, of Park County, Wyoming. The generic name proved to be invalid since it had been used previously for a genus of living angiosperms. The new name *Rebuchia* was proposed by Hueber in 1970, and he later (1972a) added to the knowledge of the plant based on more complete specimens collected from the original locality. Our summary is based chiefly on that 1972a account.

Figure 4.54. *Rebuchia ovata.* View of specimen with divided basal regions and terminal fertile areas. × 1.1 (From Hueber, 1972a, Rev. Palaeobot. Palynol.)

The plant apparently presented a densely branched, tufted aspect (Fig. 4.54) with several distinct upright axes, 10–15 cm tall, that dichotomized freely. The branches are 1.5–2.0 mm in diameter and smooth. Some of these branches terminate in a fertile spike composed of two rows of sporangia which are characteristically aligned in one direction. This results from the curvature of the sporangial stalks toward one side of the axis. The number of sporangia per spike ranges from about 15–20. The individual sporangia measure 2.8 mm high and 2.7 mm wide. Their dehiscence is described as basipetal resulting in the division of the sporangium into two essentially equal halves.

Spore masses isolated from one of the sporangia show the spores to be circular in outline, smooth-walled, and ranging from 68–75 μm in diameter. No triradiate mark was observed and it seems likely that the available spores were not completely mature at the time of fossilization.

Hueber also reported (1972b) another species of the genus, *R. capitanea*, from the Siegenian (middle Lower Devonian) of Bathurst Island in the Canadian Arctic. It is based on a part of a stem bearing a spike of 54 sporangia which are distinguished by their larger size, being

3.2–3.5 mm high and 3.6–3.9 mm wide. This rather fragmentary material bears some resemblance to *Oricilla bilinearis* in the way in which sporangia are attached and might better be assigned to that genus.

Koniora Zdebska, 1982.
Koniora andrychoviensis Zdebska, 1982.

This plant is known from compressions and pyritized pieces of axis obtained from two boreholes made in the Bielsko-Andrychow area of the Polish West Carpathians, the age being late Early Devonian (Emsian). Since the largest pieces are about 7 cm long, being limited by the size of the boring, some uncertainty about how they relate to one another must exist relative to the author's restoration. The largest axes are about 4 mm wide; they divide by nearly equal dichotomies, and terminate, both sterile and fertile ones, with several closely spaced divisions that are strongly curved to form hook-like endings (Figs. 4.55, 4.56).

There are two especially distinctive features of the axes; one being the presence of conspicuous spines (Fig. 4.55A–C) that tend to decrease in size from the more basal parts of the axes outward. However, there is a considerable mixture of spines of various sizes, the longest ones being 4 mm long and 0.3 mm wide at their base. The longer spines bear occasional minute teeth 12–16 μm long. All spines seem to lack any regular arrangement on the axes. The second distinctive feature of the axes is the presence of four longitudinal wings.

Some ten pyritized axis fragments were available for study. The best of them show a minute protostele that is slightly oval in transverse section (Fig. 4.55C); it apparently consists entirely of tracheids with some of the peripheral ones being protoxylem elements.

Although most of the distal branches were sterile some were found, after removal of the rock with hydrofluoric acid, to bear a sporangium at the final or penultimate dichotomy. The sporangia are attached slightly below the dichotomy either directly or by a very short stalk (Figs. 4.55C, 4.56). They are described as oval or reniform, typically about 2.5 mm wide, and composed of two equal valves. A few minute spines are borne on the surface. Spores were not observed.

One would like to have larger portions of a plant than those on which *Koniora* is based, but in view of the rather good quality of preservation and the numerous fragments we regard the restoration as a valid one. That the sporangia are correctly identified as such seems evident from their gross form as well as some resemblance to the way they are borne in *Gosslingia*. The closest comparison seems to be with the latter genus, but with *Koniora* being distinguished by its location of sporangia, uniform dichotomous branching, variation in size and distribution of the spines, strongly recurved and divided branch terminations, and axial wings.

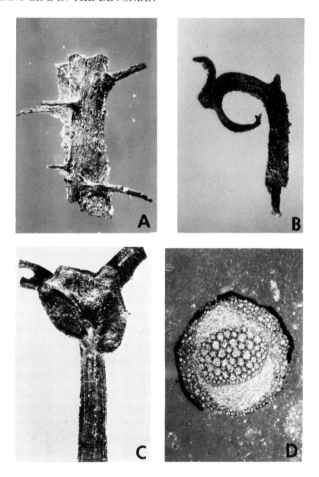

Figure 4.55. *Koniora andrychoviensis.* A. Fragment of axis showing long spines and wing, × 17. B. Apex of a branch showing short spines, × 16.5. C. A fragment of an axis with a sporangium at the point of dichotomy, × 15. D. Transverse section of axis showing elliptical exarch protostele. × 32. (All from Zdebska, 1982, Palaeontology.)

Zdebska regards *Koniora* as representing a link between the zosterophylls and the lycopods, especially in regard to the evolution of lycopod sporophylls.

THE QUESTION OF GAMETOPHYTES

The plants treated throughout this monograph, with few exceptions, are regarded as representing the diploid spore-producing or sporophytic

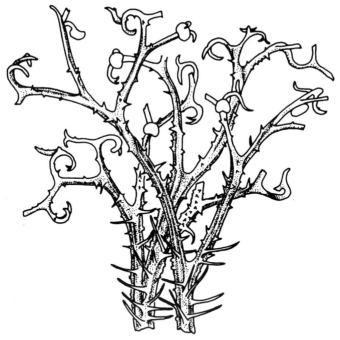

Figure 4.56. Restoration of *Koniora andrychoviensis*. approx. × 2.3 (From Zdebska, 1982, Palaeontology.)

phase of their life history and little evidence exists pertaining to their gamete-producing or gametophytic phase. However, accounts of possible gametophytes appear from time to time, especially of early Devonian age, and these are interpreted either as being the gametophytic phase of a known (sporophytic) taxon or they are regarded as being distinct from any known taxon or taxa and possibly representing plants more closely allied to bryophytes. Search for, and consideration of, gametophytes is important to our understanding of the entire nature and comparative life history strategies of early vascular plants in relation to their possible algal ancestors and other non-vascular plants. It also should contribute to a better understanding of the early history and evolutionary relationships of major bryophyte groups.

Anyone who has investigated the gametophyte phase of living pteridophytes is aware of the several problems attendant to recognizing or interpreting these structures from fossils: the gametophyte plant body is relatively undifferentiated with easily destroyed cells, it is variable in form, ranging from axis-like to prostrate and thalloid, it may or may not be vascularized, and gametangia (which would provide definitive evidence that a structure is gametophytic) may be partly or completely embedded in the thallus and thus not easily observed. Similarly, the plant bodies of liverworts or mosses are not always easily recognized as such from the fossil record. Of the limited and problematical reports of possible Devonian gametophytes, we will consider briefly selected examples which we regard as illustrative of the information that has been obtained thus far concerning these important, but elusive structures.

Problematical Gametophytes

It was noted earlier in this chapter that several studies have suggested that some or all of the axes attributed to *Rhynia gwynne-vaughanii* may be gametophytes (Merker, 1958, 1959; Pant, 1962; Lemoigne, 1968), but that conclusive evidence is still lacking as the structures (i.e. gametangia) supporting this interpretation could equally represent wound tissue or the sites of stomata and sub-stomatal chambers.

Lyonophyton W. and R. Remy, 1980.
Lyonophyton rhyniensis W. and R. Remy, 1980.

A somewhat more convincing case is made on the basis of specimens recently described from the Rhynie Chert as *Lyonophyton* (Fig. 4.57). These structures are interpreted as remains of upright gametophytes consisting of an axial portion which terminates in a bowl-shaped structure bearing rounded, short-stalked antheridia (sperm-producing gametangia) just inside the lobed margin. Other sections showed more centrally located structures with tube-like extensions which were interpreted as archegonia (egg-producing gametangia). Remy and Remy note that dark centrally located cells occur in the axial portion which may represent conducting cells such as are found in some mosses. Apparent packets of dark objects in the presumed antheridia are interpreted as developing sperm.

Remy and Remy note the similarity in size and structure of axes between *Lyonophyton* and *Rhynia* and *Horneophyton*. They also compare *Lyonophyton* closely with bryophytes in general, suggesting it is organized more like mosses than liverworts. In a later paper, W. Remy (1982) suggests that *Lyonophyton* might represent a member of a group of plants which immediately preceded in evolution both bryophytes and vascular plants. He simultaneously suggests that Devonian gametophytes were erect, radially organized entities not very different in form from their sporophytic counterparts. While most intriguing, these suggestions are very general and information is lacking on the specific nature of the sporophyte phase of *Lyonophyton*. A good understanding of relationships

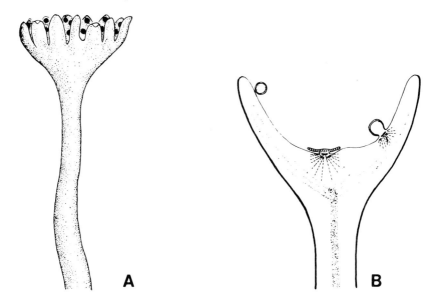

Figure 4.57. *Lyonophyton rhyniensis.* A. Suggested restoration drawing of distal part of the thallus. B. Section view showing position of presumed gametangia and conducting strand. (Redrawn from Remy and Remy, 1980).

between any possible gametophytes and genera included in the rhyniophytes, trimerophytes, zosterophyllophytes or any other group is still needed.

Sciadophyton Steinmann emend. Kräusel and Weyland, 1930.

Schweitzer (1980a, 1981), Remy et al. (1980), and Remy and Remy (1980) have advanced the suggestion that the lower Devonian plant *Sciadophyton* may be gametophytic (or predominantly so) rather than sporophytic.

Sciadophyton was the name applied by Steinmann (1929) to the compressions of curious rosette-like plants (Fig. 4.58) originally described from Gaspé by Sir J. W. Dawson (1871) as *Annularia laxa* and to similar plants from Germany. Kräusel and Weyland (1930) redescribed both the German and Canadian specimens, and established a second species, *S. steinmannii* for the German ones. Both species consist of numerous vascularized axes attached to a central disc-like structure and radiating from it in a star-like fashion. Axes vary in number per plant, length (a few mm to 10 cm) and width (1–4mm), and in the extent to which they bifurcate. The German specimens exhibit shield-shaped or funnel-like expansions at the tips of some axes which are lobed at the margins. Round

Figure 4.58. *Sciadophyton steinmannii.* A restoration drawing of plants showing the expanded area at the tips of some axes. (Redrawn from Kräusel and Weyland, 1930).

bodies 0.2–1.5mm in diameter occur on the upper surfaces of the expanded area, the smaller ones being located closer to the margin. Maceration has not yielded spores from these bodies; Kräusel and Weyland suggested they may represent vegetative propagules (brood bodies).

Schweitzer (1980a, 1981) and Remy et al. (1980) have suggested that the axes and distal expansion area represent gametophores and the round bodies represent gametangia. They compare *Sciadophyton* in terms of thallus organization and presence of vascular tissue to the axial, vascularized gemetophyte of the extant genus *Psilotum*. This is particularly problematical as in our opinion *Psilotum* is a highly reduced pteridophyte, sometimes allied to ferns and sometimes retained in its own class or division, which is not at all related to any early Devonian plants (see Gensel, 1977 for further discussion). Thus the interpretation of *Sciadophyton*, known only from compression fossils, as a gametophyte remains both intriguing and problematical.

Sporogonites-a Gametophyte/Sporophyte Tentatively Allied to the Bryophytes

Sporogonites exuberans Halle, 1916.

This plant was first described by Halle in 1916b from the Lower Devonian of Röragen, Norway; subsequently it has been found in the Lower Devonian of Belgium, Scotland and northern France. Andrews (1960) provided additional information based on the extensive collections of this plant from Belgium housed in the Royal Museum of Natural

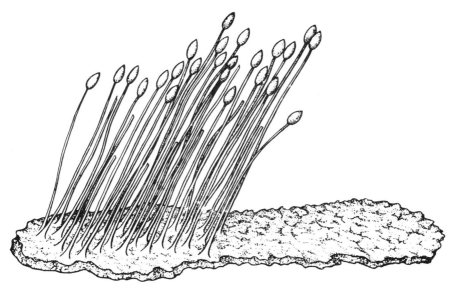

Figure 4.59. *Sporogonites exuberans*. Restoration showing basal thalloid region and numerous sporangium-bearing axes. (From Andrews, 1960).

History in Brussels, and his report will form the basis of our discussion here.

Sporogonites consists of slender, unbranched axes, each terminating in an oval-rounded sporangium 2–4 mm wide and 6–9 mm long (Fig. 4.59). The axes frequently are aligned parallel to one another; this feature led Andrews to examine the specimens more closely to determine if a common region of attachment existed. He found a thin carbonaceous thallus-like structure at the base of some axes. In some regions the sporangium-bearing axes departed directly from this thallus. No evidence of vascular tissue has been found in the thallus or sporangium-bearing axes.

The sporangia are grooved, especially in their distal regions but show no evidence of a specialized dehiscence region. A few of the sporangia studied by Halle were permineralized; sections of these showed the sporangia to consist of a wall several cell layers thick, inside of which occurs a basal sterile region and distal sporogenous region. Spores occur in tetrads and are 20–25 μm in diameter, being either reticulate or papillate. Halle (1916b, 1936) suggested the spores may have been shed through openings along the grooves.

The plant is of particular interest in having a moss-like sporangium terminating axes (similar to sporangiophores of mosses) which apparently were attached to a basal unspecialized thallus. On the basis of these characters, Andrews concluded: "In summary, it seems reasonable to assume that *Sporogonites* may be considered as a plant that had attained a bryophytic level of evolution, possibly related to liverworts or representing a distinct and independent group." (Andrews, 1960, p. 88).

It should be noted further that limited fossil evidence of leafy liverworts exists from the Devonian. Hueber (1961b) described a well-preserved liverwort from the Upper Devonian of New York State as *Hepaticites devonicus* which since has been renamed *Pallavicinites devonicus* because of its close similarity to the living liverwort *Pallavicinia* (Schuster, 1966). Other liverworts have been described by Walton (1925, 1928) from the Carboniferous; however, fossil evidence of liverworts and mosses in general is extremely scanty in the Devonian and Carboniferous and to date does not allow for a good assessment of evolutionary trends and relationships within or between those groups, or between bryophytes and vascular plants.

SUMMARY

The preceding chapter surveyed the three major groups of early vascular plants, the trimerophytes, rhyniophytes, and zosterophyllophytes, with emphasis on their basic characteristics and some probable intermediary forms. We hope that this survey demonstrates the great advances that have been made in unravelling the nature and evolutionary potential of those plants which for many years were lumped together as "psilophytes." Undoubtedly new finds will continue to modify delineation of these major groups and establish the existence of other, as yet unknown, ones. Subsequent chapters will describe other plant types that coexisted with these major ones; at the same time, our evidence thus far suggests these plants formed a major component of most Early Devonian floras. Rhyniophytes appear to have disappeared by the Middle Devonian, while some zosterophyllophytes extended into the Late Devonian (*Sawdonia, Serrulacaulis*). Trimerophytes probably were present in the Middle Devonian although no clearly delimited forms have been described. Scheckler (1976) described briefly a possible Late Devonian trimerophyte from Canada.

It is not uncommon to find in outcrops vast numbers of plants represented only by their vegetative remains. When preserved as compressions or impressions, it would be extremely difficult to assess their affinities—a glance at some of the early papers, e.g. those of Dawson, Halle, and Lang demonstrates the problems that arise when attempting to identify vegetative axes; they could belong to any one of these groups. Anatomy is helpful in assessing phylogenetic relationships but also not always definitive. It is extremely important to examine reproductive organs when trying to deal with these early vascular plant types.

Equally interesting is the question of why fertile remains are so sparse in some lenses? Is this chance, or might it indicate that these plants grew and reproduced vegetatively to a large extent?

As some of the more prominent plants of the Lower Devonian, these three groups also demonstrate levels of evolutionary advancement attained by the end of the Lower Devonian. Most are relatively unspecialized, consisting of predominantly bifurcating axes, some of which are strongly pseudomonopodial and show some differentiation of lateral branches. Vascular anatomy is mostly in the form of a terete to elliptical protostele with centrarch or exarch maturation. Bifurcation of axes, whether equal or unequal, results in daughter strands similar to, or only slightly different from, the parent strand. Sporangia in rhyniophytes seem to be indehiscent or with unspecialized dehiscence zones; those of trimerophytes are fusiform and split open longitudinally, and those of zosterophyllophytes dehisce along a specialized zone on their distal margin. No true leaves (except microphyllous ones in lycophytes) had appeared and thus far the presumed lycophyte *Drepanophycus* demonstrates structures which can be considered fairly conclusively to be roots and to resemble those of extant Lycopodiums. Dichotomizing structures occur in *Psilophyton dawsonii* (Banks et al. 1975) and in the presumed rhyniophyte *Hsüa robusta* Li, (1982) which are suggested to be root-like. Numerous Early Devonian plants are interpreted as rhizomatous (i.e. they have creeping or underground stems) or as having rhizoids or adventitious roots, and these may have served physiologically as roots. No Early Devonian plants seem to possess primary roots. They may be primitively rootless but this is not totally certain as most fossils reveal only a portion of a plant, and these seem to be mostly of aerial regions. Possible primary roots have been described for some progymnosperms and for a number of Carboniferous plants. Thus, the origin and early differentiation of these structures remains problematical and hopefully continued investigation of Devonian plants will reveal further information about them.

Lastly, we briefly consider some of the recent reports of possible gametophytes which may or may not be related to one or more of the plants placed in these groups or ones treated in subsequent chapters. Conclusive evidence of the gametophytic nature of fossil entities is difficult to obtain, as is a putative relationship to a given sporophytic taxon. This leaves the affinities, and to some extent the haploid nature, of some of these proposed gametophytes uncertain while others seem to represent plants more closely allied to mosses or liverworts.

5 / The Lycopods

The living lycopods include four or five genera of which *Lycopodium, Selaginella,* and *Isoetes* contain most of the species; these three genera are widespread, the first two being especially numerous in species of herbaceous terrestrial and epiphytic plants that are found in a wide range of habitats. The group as a whole has long attracted attention from botanists on several accounts; their history is a long one with quite clearly defined members of the group being known from the Middle to lower Upper Devonian. They apparently reached their peak of diversity and size in the Carboniferous where the vast forests were in many places dominated by tree-forms well over one hundred feet tall, and these were one of the major contributors of organic material that formed the extensive Upper Carboniferous coal beds. Although they declined in abundance and size toward the close of the Paleozoic their racial vigor has been by no means lost as evinced by the numerous living species.

Although the lycopods are classified in some texts as the Microphyllophyta, in reference to the numerous small and simple leaves so characteristic of many species of *Lycopodium* and *Selaginella,* the leaves are divided in some of the Devonian representatives and in some of the arborescent members of Carboniferous age they are slender grass-like organs up to a meter in length. Also the leaves of most lycopods have a single unbranched vein but in two living species of *Selaginella* a branched venation pattern has been reported (Wagner, et al. 1982). And in the late Devonian *Prolepidodendron breviinternodium* (described on a later page) there are two veins in each leaf.

The features that seem to characterize the lycopods as a whole are: the arrangement of the leaves in a dense, regular spiral pattern with resultant distinctive scar patterns in the deciduous members of the group; a sporangium borne on the upper surface of certain leaves (sporophylls) which may or may not be different from the vegetative leaves; and an axial xylary structure consisting of a solid, exarch or nearly exarch, cylindrical protostele in the smaller species, variously fluted or dissected with a central pith in the larger ones, and a cambium producing secondary wood in the great forest trees such as *Lepidodendron*. Some of the Carboniferous members achieved the seed habit, where the megasporangium became enveloped by the sporophyll; it is a unique type of seed and implies no close relationship to other Carboniferous gymnosperms such as the pteridosperms and cordaites.

In 1967 Chaloner and Boureau provided, in Volume II of the *Traité de Paléobotanique*, a detailed and well-illustrated summary of the entire fossil history of the lycopods as it was then known. Significant contributions have been added since that time, especially to the Devonian record with which we are primarily concerned here. We have drawn much information from their account and acknowledge it with thanks.

Assembling what seems to be the more significant records of the lycopods in the Devonian has been an especially interesting but somewhat frustrating task. In the best preserved specimens, very careful degagement, or the use of other techniques, may be necessary to reveal the gross morphology of both sterile and fertile leaves. And the greatly varying degrees of perfection in the preservation of axes of either herbaceous or deciduous species can reveal an endless range of morphology. This cannot be overemphasized and Grierson and Banks (1963) have summed up the problem very well as follows:

> "By far the greatest number of specimens of fossil lycopods of Devonian age are suitable only for inclusion in form genera. Many are good enough only to be labelled "lycopod fragment". Range extensions based on indiscriminate assignment of such specimens to well-defined genera weaken these taxa as well as detracting from their stratigraphic significance." (p. 221).

Thus numerous generic names that are based on decay stages of stems and which convey no significant information, or may actually be misleading, are omitted here.

Any classification of Devonian fossils that seem referable to the lycopod group must be tentative in the present state of our knowledge. We are reluctant to align them in formal categories such as families and orders but there does seem to be a sequence in development starting with plants from the basal part of the Devonian which present features that may be interpreted as primitive lycopods; by Middle Devonian times the characteristic features of the group become much more clearcut; and by the close of the Devonian the heterosporous stage of reproduction has been achieved and arborescent forms are becoming a significant part of the vegetation as a whole. It therefore seems to us most convenient and informative to arrange this chapter in three sections, as follows:

I. A discussion of what may be called *Pre-lycopods*. These are small, herbaceous plants composing a somewhat mixed assemblage in which the interrelationships are not entirely clear but which may represent the origins of the group.

II. Middle Devonian fossils that present the more clear-cut features of the group.

III. Chiefly Upper Devonia fossils that show a considerable increase in size of the plants and the earliest evidence of heterospory in the group.

I. PRELYCOPODS

As noted above, the plants described here do not present all of the morphological and anatomical features that identify the lycopod group. They seem closer to the lycopods than any other group and, in the present state of our knowledge, it is at least reasonable to regard them as representing the kind of plants from which the lycopods might have originated.

It is especially important to explain, or perhaps it is better to say record, an apparent anomaly in the order of our presentation. A well-known fossil described in Part II, *Baragwanathia longifolia*, that does seem to be a 'good' lycopod, comes from a horizon that is as old or very possibly older than any of the plants presented in this section. We can only say, for those who are not familiar with this problem, that the fossil record does not always present plants in the precise order in which we might expect them. It suggests very strongly that there are gaps in the record that future explorations will help to clarify.

Asteroxylon Kidston and Lang, 1920.
Asteroxylon mackiei Kidston and Lang, 1920.

This plant was first described by these authors in Part III of their treatise on the flora of Rhynie, Scotland. It is the largest and certainly one of the most interesting of the Rhynie plants. The vegetative parts (Fig. 5.1) consist of what was apparently a naked (lacking appendages) rhizome

Figure 5.1. Restoration of *Asteroxylon mackiei*. (From Chaloner and Macdonald, 1980. With permission Controller H.B.M. Stationery Office.)

system of axes up to 5 mm in diameter which gave rise to upright stems of pseudomonopodial habit, about 1 cm in diameter, which in turn bore dichotomizing side branches. This aerial shoot system was densely covered with microphylls about 5 mm long. Stomata that are described as "of ordinary form" are present, although not abundant, in the epidermis of the stem and leaves, both of which have a distinct cuticle.

The vascular tissue of the aerial axes is an actinostele (whence the origin of the generic name) which is quite deeply lobed in the larger ones (Fig. 5.2A). The protoxylem cells are immersed in the ends of the stelar arms and all of the tracheids have a secondary wall of very delicate spiral bands (Fig. 5.2C). Xylary foliar traces, which consist of a small centrarch strand of xylem enclosed by a zone of phloem depart from the stele but reach only to the base of the microphylls. The cortex is a broad zone of parenchymatous tissue, the most distinctive feature of which is a prominent middle zone of "trabecular arrangement of the cells with wide intercellular spaces between the rows." This seems indicative of a boggy or semi-aquatic habitat in which the plants grew.

In their 1920 account, Kidston and Lang described leafless, dichotomizing branches bearing terminal sporangia that were associated with the leafy shoots, but they were careful to point out that the two were not found in organic connection. Their restoration showing this association has been reproduced many times and seems to have been almost accepted

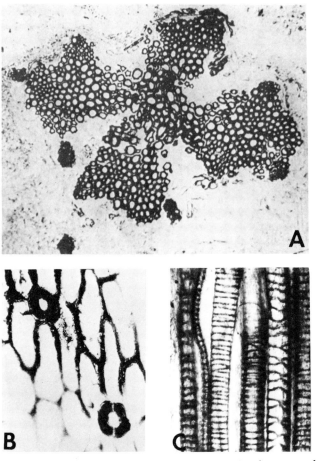

Figure 5.2. *Asteroxylon mackiei.* A. Cross section of xylem strand including several leaf traces. × 39. B. A view of epidermal cells and stomates. × 130. C. Longitudinal view of tracheids. × 226. From Kidston and Lang, 1920. Trans. Roy. Soc. Edinburgh 62:643–680.

as fact but it was not until 1964 that A. G. Lyon correctly described the sporangiate organs. His report is brief but several paleobotanists especially concerned with Devonian plants have examined his preparations and concur with his findings. Lyon found sporangia borne terminally on short stalks that were interspersed among the microphylls (Fig. 5.3A). In his words:

"The rather large sporangia (up to 7 mm in width when empty) are tangentially flattened and are borne laterally and independently on the

Figure 5.3. *Asteroxylon mackiei.* A. Longitudinal section through leafy fertile shoot showing two stalked sporangia. × 0.6. B. Restoration of same. (A. From Lyon, 1964. Nature. B. From Chaloner and Macdonald, 1980. with permission, Controller H.B.M. Stationery Office.)

> shoots between the leaves. . . . Although most of the sporangia are empty, a few have been found containing spores which are about 50 μm in diameter, and it is evident that these were shed through an extended marginal slit.
>
> Each sporangium is borne on a short vasculated pedicel which expands distally into a pad of tissue over which the spore-sac extends."
> (p. 1083).

In a very readable publication, *Plants Invade the Land*, issued from the Royal Scottish Museum (Edinburgh) in 1980, Chaloner and Macdonald include some fine illustrations of the Rhynie plants, including a restoration of a fertile terminal shoot of *Asteroxylon* which we have included here (Fig. 5.3B).

As noted above, one of the characteristic features of lycopods is the direct association of a sporangium with a sporophyll, the sporangium being borne on its upper surface. In *Asteroxylon* the stalked sporangia are borne along the stem but with no apparent regular relationship to the (sterile) leaves. Thus we have here a relationship that seems more closely comparable to the lycopod concept than the kinds of early land vascular plants described in Chapter IV, but it obviously does not fulfill the 'typical' lycopod morphology.

It may be noted that the naked sporangiate branches that Kidston and Lang reported as being associated with the leafy shoots of *Asteroxylon* have been named *Nothia aphylla* and are described elsewhere in this account.

Drepanophycus Goeppert, 1852.

The genus *Drepanophycus* is a particularly vexing one, representing one of the more commonly occurring, yet relatively poorly circumscribed, taxa of Devonian plants. Ranging from the Lower to the Upper Devonian (Siegenian-Frasnian), it is broadly defined as consisting of robust axes (up to 4 cm in diameter), some creeping and some upright, bearing stout emergences with broad bases and tapering extremities (Fig. 5.4). The emergences range from short and thorn-like to about 2 cm long and often are curved. They possess a vascular strand and are regarded as leaves. When not visible, their position is indicated by raised or depressed areas of varying shapes. Axes may dichotomize infrequently, some produce much smaller lateral branches, and some exhibit a complex series of branching referred to as H- or K- branching (these probably represent a sequence of bifurcations in the rhizome). Many have a slender vascular strand, 1–1.5 mm in diameter in the center. The genus has long been considered a possible early lycopod, for reasons elaborated later.

Not only is the genus broadly and imprecisely defined, but also species are difficult to distinguish. Several factors contribute to the difficulties: most specimens are vegetative; they are preserved to differing degrees such that major parts of the plant are not always present; branching pattern varies from one specimen to another; and few are extensively enough preserved to determine variation within one plant. Some confusion also has resulted from the historical development of the generic concept.

All evidence is not in, so this constitutes an interim assessment of *Drepanophycus*. We will present the better understood aspects of the genus and state the problems and possibilities in regard to its future characterization. Emphasis is placed on the type species, *D. spinaeformis*, in this discussion.

The genus *Drepanophycus* was established by H. Goeppert in 1852 for specimens from the Spirifer-sandstone near Hachenburg, Nassau, Germany. The specimens consist of stout axes covered with leaf-like appendages which are broadly triangular at the base, often curved and tapered to a distal tip. The axes average 2.5 cm in width and the leaf-like appendages are nearly that long and fairly widely spaced (Fig. 5.4B). Goeppert noted that no veins were visible on the axes or appendages and that the plant was "a very peculiar form different from all fossil fucoids

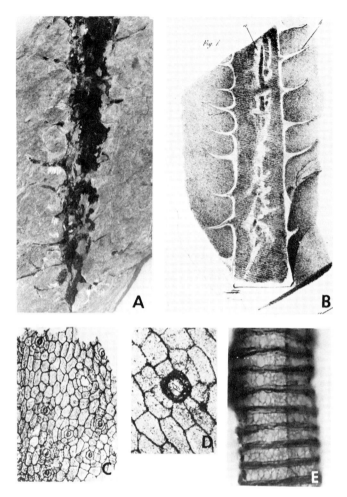

Figure 5.4. *Drepanophycus spinaeformis.* A. Portion of an axis showing the long falcate leaves. × 0.6. B. Illustration of an axis from Goeppert's original 1852 account. Approx. × 1. C. Portion of epidermal cuticle showing numerous stomates. × 58. D. A single stomate enlarged. × 167. E. Portion of a scalariform tracheid enlarged showing the reticulate pattern between bars. × 514. (A. NCUPC Collections. C,D. From Stubblefield and Banks, 1978, E. From Hartman, 1981. Rev. Palaeobot. Palynol.)

known up to now." Because no vascular strand was evident, he considered it to represent an alga rather than a vascular plant.

Few reports of the genus were made for nearly 70 years. However, Dawson in 1871 described nearly similar specimens from the Lower

Devonian of Gaspé, Quebec, as *Arthrostigma gracile*. These plants consisted of axes about 1.5–2.5 cm wide, with rigid leaves departing at nearly right angles in all directions. When broken they leave scars. He made no comparison to Goeppert's species at the time of the above publication, and it is unclear if he knew of it and considered it algal, did not know of it, or merely ignored it.

Specimens from several localities were subsequently assigned to Dawson's taxon from Scotland, Norway, Germany, France, and China. All are vegetative. The similarity between *Arthrostigma* and *Drepanophycus* was first noted by Weiss (1889) and later by others including Kidston (1894), Halle (1916), and Cookson (1926). Since none of the later-described specimens attained the diameter of the original material, they were not combined until 1930, when Kräusel and Weyland described new, smaller specimens from Germany. They made the combination and used the prior name, *Drepanophycus*.

Halle's study of Norwegian specimens in 1916 illustrates the considerable variation in vegetative axis morphology which came to be included in the genus. He described three forms under the name *Arthrostigma gracile*, as follows: 1) forms similar to Dawson's, i.e. the leaves nearly as long as the axis is wide, broad-based, mostly curved, and single-veined; 2) forms with densely arranged leaves, generally shorter than above (6–8 mm long), curved to straight, departing at an acute angle to the axis, and possibly more regularly arranged; and 3) forms with thick, short, distantly spaced thorn-like leaves up to 8 mm long and wide. Halle isolated scalariform tracheids from macerations of the dark central strand in some axes and suggested that stems and leaves may have been soft in texture.

Lang in 1932 illustrated specimens from Scotland from which he showed cuticular patterns, including stomata, and tracheids. He also demonstrated the presence of vascular tissue in the leaves and stomates at the base of the leaves. The leaves were only 5 mm long.

The 1930 and 1935 accounts by R. Kräusel and H. Weyland represent a major reorganization of the data concerning the plants *Arthrostigma* and *Drepanophycus*. They re-examined Goeppert's specimens and studied new ones from areas in Germany. Their 1930 account clarified the use of the name *Drepanophycus* (in which they placed *Arthrostigma*) and illustrated the variability in leaf form and density among the specimens. They also illustrated a possible fertile specimen, a possible shoot apex with very densely spaced leaves, and some aspects of branching. The problems faced by paleobotanists at that time in distinguishing smaller *Drepanophycus* axes from the spiny axes then included in *Psilophyton* (but now placed in *Sawdonia*) was also discussed.

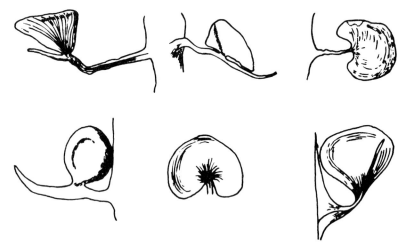

Figure 5.5. *Drepanophycus spinaeformis.* Drawings from Kräusel and Weyland (1935) showing their interpretation of sporangial positions. (Palaeontographica.)

Drepanophycus spinaeformis Goeppert, 1852.

In 1935, Kräusel and Weyland presented information based on new collections which mainly elucidated the nature and location of the sporangia (Fig. 5.5). The latter are described as attached to the upper surface of a leaf in its mid-region by a short stalk. Detached sporangia were observed on or near several other leaves such that some appeared to be located at their tips or in the axils. Film-pulls from some sporangia were macerated and revealed some very crumpled spores ranging from 25–45 μm in diameter. In the text, Kräusel and Weyland reiterate the medial, adaxial position of the sporangia; however, in their reconstruction, the sporangia are shown attached in three ways: as above, terminally, and in the axils. Thus doubt remains concerning the actual mode of sporangial attachment. Unfortunately most of their specimens were destroyed in World War II. *Drepanophycus* has long been considered an early lycopod, and a better understanding of sporangial attachment is necessary in order to better assess its affinities.

A few additional accounts are interesting in this regard: Croft and Lang (1942) illustrated some fertile specimens of *D. spinaeformis* from the Senni Beds of Wales. In these specimens the sporangia "give the impression of being terminal on the leaf." However, they seem to have been strongly influenced by Kräusel and Weyland's account and suggest the terminal appearance is the result of unsatisfactory preservation. Both authors have seen Croft and Lang's fertile specimen and while preservation is not as good as one would wish, the sporangia indeed appear to terminate stalks (or leaves).

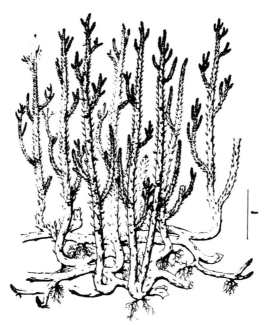

Figure 5.6. Proposed restoration of a representative portion of *Drepanophycus spinaeformis*. (From Schweitzer, 1980b, Bonn. Paleo. Mitt.)

Schweitzer (1980b) discussed specimens assigned to *Drepanophycus spinaeformis* based on collections from several localities in Germany accumulated over many years (Fig. 5.6). He indicates the sporangia are terminal on stalks located amongst the leaves. His specimens consist of axes with relatively short, thorn-like leaves. Few photographs of the specimens are included to document sporangial position or leaf morphology and one wonders if the specimens are correctly designated (or might they be *Kaulangiophyton*?). He also notes (but does not figure) variation in leaf size and morphology on rhizomes vs. aerial shoots. He also illustrates roots.

It is clear that careful and complete investigation of the attachment and location of sporangia in *Drepanophycus spinaeformis*, and also in other species, is still needed.

The anatomy of *Drepanophycus spinaeformis* was first described by Fairon-Demaret in 1971 from Belgian specimens, although Halle, Lang, and Kräusel and Weyland were all able to demonstrate from macerations the presence of annular and scalariform tracheids. Fairon-Demaret confirmed that the vascular strand is an apparently lobed protostele (very crushed in the specimens studied) with the stele 3.3 mm wide. Smaller tracheids occur to the outside, suggesting exarch or possibly mesarch maturation. Leaf traces are present, consisting of about four rows of annular, spiralled, or scalariform tracheids that are mesarch. She, and some other workers, described a distinctive tracheid wall pattern in *Drepanophycus*; the walls have annular and/or helical and scalariform thickenings with the region between them appearing perforated or reticulate (Fig. 5.4E). Hartman (1981) examined these with the SEM and light microscopy and interpreted the pattern to be a result of pyrite degradation of the compound middle lamella; it is however quite consistent on specimens from very different geographic regions and in our opinion a fundamental feature of the tracheids.

Fairon-Demaret (1978) described several very long axes of *D. spinaeformis*, illustrating both regularly and irregularly arranged leaves and the presence of very much smaller lateral branches. She noted that the axes with H- or K-branching usually have only short leaves, while those with long leaves lack that type of branching. She concluded there are two possible interpretations: 1) different parts of the plant exhibited different types of leaves and branching or 2) that two distinct plants (species?) have been included in *D. spinaeformis*.

Stubblefield and Banks (1978) examined cuticles of *D. spinaeformis* using specimens from Gaspé and New York State. They exhibit polygonal and elongate cells and some randomly distributed stomata (Fig. 5.4C, D). The latter consist of two guard cells, a pore, and two reniform subsidiary cells. It is noted that this cuticular pattern, along with other characters, can be used to distinguish *Drepanophycus* from the zosterophyllo-phytes.

Several additional species of *Drepanophycus* have been described, many of which are summarized by Grierson and Banks (1963). Specimens originally described by Dawson from Gaspé as *Lepidodendron gaspianum* were later recombined as *Drepanophycus gaspianus* by Kräusel and Weyland (1948). This species apparently differs from *D. spinaeformis* in having more dense and regularly arranged leaves (Fig. 5.7), but distinc-tions between the two have lessened as additional information has accumulated. Grierson and Hueber (1968) described anatomy of axes as a lobed protostele and also figured cross-sections of leaves. The latter are simple, consisting of epidermis, parenchymatous mesophyll, and a small vascular strand.

Kaulangiophyton (Gensel and Kasper, 1969.
Kaulangiophyton akantha Gensel and Kasper.

Kaulangiophyton is one of several distinctive plants (Gensel et al. 1969) that have been found recently in the Trout Valley Formation of Baxter Park in northern Maine, the age being either upper Lower or lower Middle Devonian. The specimens were found in a horizon only a few

Figure 5.7. *Drepanophycus gaspianus.* A. A specimen from New Brunswick showing broad axis and densely spaced leaves. × 0.87. B. Specimen from New York State. × 0.6. (A. NCUPC Collections. B. From Grierson and Banks, 1963, reproduced with permission of the Paleontological Research Institution, Ithaca, New York.)

centimeters thick along the bank of Trout Brook and are preserved essentially as impressions: however they afford a considerable amount of significant information.

The plants probably attained a height of several decimeters with a general habit quite like that of some of the modern terrestrial species of *Lycopodium*, but with much 'reduced' leaves (Fig. 5.8). The branching axes attain a diameter of 9 mm and display the distinctive H-branching that is also found in certain species of *Zosterophyllum* and *Drepanophycus*. The stout, thorn-like leaves are about 2 mm long and are quite sparsely but evenly distributed over the axes. Some of the upright branches bear quite large sporangia (Fig. 5.9) that measure about 5 × 8 mm, were probably ovoid in life and terminate lateral stalks that are 2–4 mm long. These do not seem to be associated in any regular fashion with the sterile spines. Although spores were not found, there can be little doubt that the sporangia are correctly identified as such.

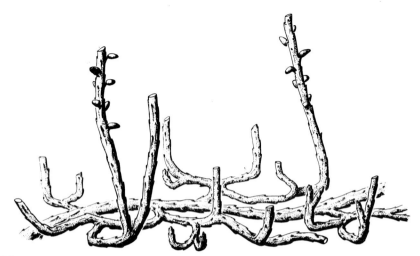

Figure 5.8. *Kaulangiophyton akantha.* Restoration of a portion of the plant. (From Gensel et al. 1969, Bull. Torrey Bot. Club.)

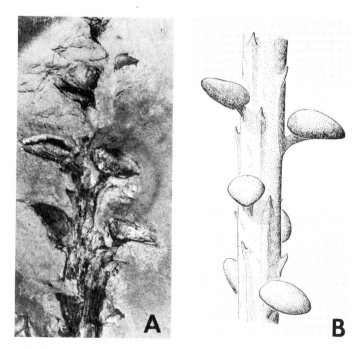

Figure 5.9. *Kaulangiophyton akantha.* A,B. Photograph and restoration drawing, respectively, of a portion of a fertile axis. Approx. × 2. (From Gensel et al. 1969, Bull. Torrey Bot. Club.)

II. LOWER AND MIDDLE DEVONIAN FOSSILS
THAT CLEARLY ESTABLISH THE LYCOPOD LINE

Baragwanathia Lang and Cookson, 1935.
Baragwanathia longifolia Lang and Cookson, 1935.

This lycopod was described as coming from a presumed Upper Silurian horizon in Victoria, Australia. The dating was established largely on the basis of associated graptolites (*Monograptus*) but more recent studies present evidence that suggests a Lower Devonian age. Similar specimens recently collected from another locality (Yea) are, however, dated as Upper Silurian by Garratt (1978). For a more detailed discussion of the age than we can include here, the reader is referred to recent accounts by Banks (1980b) and by Chaloner and Sheerin (1979).

B. longifolia was a relatively large and structurally complex plant for its age, whether late Silurian or early Devonian. It is known from partially petrified compression fossils of densely leafy stems 1 to 2 cm in diameter that bear leaves up to 4 cm long and 1 mm broad (Fig. 5.10). Sporangia that were about 2 mm in diameter are present on some of the stem compressions (Fig. 5.10C) and are described by Lang and Cookson (1935) as follows:

> "The sporangia are arranged in obliquely ascending series across the stem. This arrangement corresponds on the whole to the spiral of the leaf-insertions, but is somewhat confused by both sides of the stem being practically in the same plane owing to the shoot being, as usual, a very thin incrustation. The correspondence between the arrangement of the sporangia and of the leaves is suggestive of the former having been borne on the adaxial surface of the leaf-bases . . . It remains possible that the sporangia were sessile on the stem itself between the leaf-insertions, but they were evidently related in position to the latter." (p. 427).

Spores that are slightly oval in outline and measure about 50 μm in diameter have been extracted from some of the sporangia.

The xylem, which is entirely primary, apparently had a cross-sectional form very similar to that of *Asteroxylon* but being known from partially petrified compressions does not show the pattern as clearly. One of the larger and better preserved specimens is described by Lang and Cookson (1935) as follows:

> "The xylem was stellate in cross-section, the interspaces or bays between the rays of the star extending almost to the centre, and the rays of xylem widening out and subdividing as they extend to the periphery. There appear to have been some 12 or more rays of xylem in this large stele." (p. 434).

Figure 5.10. *Baragwanathia longifolia.* A. Portion of a leafy shoot. × 0.5. B,C. Fertile specimens showing numerous sporangia. Both × 0.7. D. Longitudinal view of tracheids showing annular thickenings. × 150. (From Lang and Cookson, 1935, Phil. Trans. Roy. Soc. London 224B.)

The individual tracheids appear to have been very much like those of *Asteroxylon*, being composed of tubular elements with delicate annular bands (Fig. 5.10D).

When this account was in the final stages of preparation for the publisher, we received a paper by Hueber (1983) describing a new species of *Baragwanathia*, *B. abitibiensis*, from the Devonian Sextant Formation (middle to upper Emsian) of northern Ontario, Canada. The new plant is reported as being very similar to *B. longifolia*. Some information is given on the anatomy of the leaves which were partially permineralized and terete in cross section. The type species was apparently somewhat larger and the presence of cuticle and stomata on the leaves and stems of *B. abitibiensis* are stated to be the most significant differences between the two. Hueber also includes a useful summary of certain aspects of the age and ecology of the *Baragwanathia* flora.

It is appropriate here to mention two other fossils that were included in the Australian *Baragwanathia* flora: *Hedeia corymbosa* Cookson (1935), and *Yarravia oblonga* Lang and Cookson (1935). Both are known from very fragmentary material, the former from a cluster of branchlets, each bearing a terminal sporangium, and the other from a presumed synangial structure. These have been illustrated by several authors and it seems especially significant to note that Hueber (1983), who has examined the type material, suggests trimerophyte affinities for *Hedeia*, and that *Yarravia* may represent a poorly preserved form of *Hedeia*. Further consideration of these rather fragmentary specimens must await additional research and information.

In their treatment of the lycopods (Lycophyta) in the *Traité de Paléobotanique*, Vol. II, 1967, Chaloner and Boureau gathered together data on a considerable number of fossil plants that they included in five families under the order Protolepidodendrales. This is an admirable piece of work but necessarily includes genera that have been described from poorly preserved plants and in many cases nothing is known of the reproductive structures. It is an assemblage that shows an advance over the earlier plants, which we refer to as pre-lycopods, and insofar as known, they present a firm starting point for the group in that the generally accepted morphological and anatomical characters of the lycopods are more clearly established.

A considerable amount of additional information has been brought forth since 1967 through new discoveries and the application of appropriate study techniques, and we have sorted out those genera that seem most significant from a biological standpoint and for stratigraphic useage. Perhaps largely as a matter of expediency we are tentatively including these plants in the Protolepidodendraceae.

Protolepidodendron Krejči, 1880.
Protolepidodendron scharianum Krejči, 1880.

This genus presents another case of a somewhat faltering development to its present concept over a long period of time. As indicated above, the name was first proposed by Krejči more than a century ago but it seems most informative to begin here with a resumé of a description by Kräusel and Weyland in 1932. They give a historical summary, and Grierson and Banks (1963) in their account of the Devonian lycopods of

Figure 5.11. *Protolepidodendron scharianum.* A. Restoration of a portion of the plant. B. Drawings showing sporangial position. (From Kräusel and Weyland 1932. Senckenbergiana.)

New York State review numerous reports of species that have been assigned to the genus.

Kräusel and Weyland's description is based on an examination of the original specimens as well as more recent collections, all or most of which came from Kirberg near Elberfeld, Germany, in the Honsel horizon of the Middle Devonian. The axes (Fig. 5.11A) for the most part attain a diameter of 4–6 mm and are densely covered with spirally arranged, small leaves up to 5 mm long and 0.5 mm broad. They are forked toward the distal end (Fig. 5.11), the two branches being not more than 2 mm long. They mention that some leaves are found apparently unforked but it seems most likely that that is due to breakage during fossilization or the entire leaf was not exposed.

Their specimens included some that were fertile (Fig. 5.11B), and they note they were able to examine at least sixty well-preserved sporangia. The latter are oval shaped, are attached directly to the upper surface of the sporophyll, have a slightly pointed proximal end and display an apparent thickening along the top which may represent a line of dehiscence. Spores were not found but in one case a faint circular configuration is suggestive of poorly preserved ones.

Some of the axes were permineralized with iron pyrite and they reported the stele as being triangular in cross section, with rounded angles; the protoxylem groups are somewhat sunken in a mesarch position, and the tracheids for the most part are scalariform. The vascular tissue is entirely primary, with traces to the leaves emanating from the three points of the triangle (but see below).

In its general habit (Fig. 5.11) the plant is considered to have been a small herbaceous one, with upright axes that branched sparingly and attained a height of 20–30 cm. The fertile leaves were probably borne in groups similar to the manner in which they are arranged in the living *Lycopodium lucidulum.*

Protolepidodendron is a frequently encountered name in the Devonian literature. In their account of 1963, Grierson and Banks note:

> "Over a long period of years numerous specimens from around the world had been attributed to the genus, many of them obviously not closely related. In a review of the genus, Kräusel and Weyland (1940) considered many of these and rejected all except *P. scharianum, P. wahnbachense,* and with reservations, *P. microphyllum. P. karlsteini* they placed in synonymy with *P. scharianum.*" (p. 247).

Other changes have been made in recent years. A reexamination of Belgian and German specimens of *P. wahnbachense* by Fairon-Demaret (1972, 1978, 1979) show them to be distinct from *Protolepidodendron* (see p. 134–137). She also studied specimens of *P. scharianum* from Kräusel and Weyland's original collection, from Elberfeld (Fairon-Demaret, 1980) and found that the leaves, after degaging, are actually five-parted. Also, the apparently triangular vascular strand described by Kräusel and Weyland is, when found better preserved, rounded with numerous peripherally located protoxylem points. Thus these specimens are identical with *Leclercqia* as described by Banks et al. (1972) (see p. 137). This leaves the specimens originally described by Krejči from Czechoslovakia and by Halle (1936b) from Yunnan, China as referable to *P. scharianum;* Fairon-Demaret suggests further preparation and study of those specimens is now needed.

Protolepidodendron gilboense Grierson and Banks, 1963.

This is an especially interesting species as it is based on compression fossils that reveal the leaf morphology very well, and permineralized axes with a well-preserved stele showing a cylindrical vascular strand (Fig. 5.12). It is known from Middle Devonian collections obtained at Gilboa, Schoharie County, New York State.

It was a small herbaceous plant with dichotomizing axes up to 4.5 mm in diameter that bore spirally arranged, persistent leaves. The leaves range from 3–6 mm long, and in reference to the forking, Grierson and Banks (1963) say: "The length of the leaf distal to the bifurcation showed

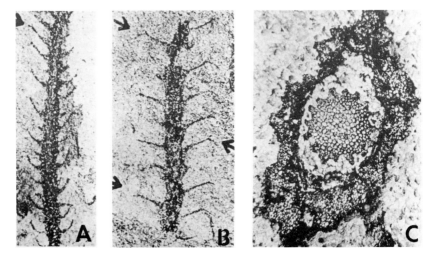

Figure 5.12. *Protolepidodendron gilboense.* A,B. Portions of leafy axes; B shows terminal bifurcation of leaves. A. × 1.2. B. 1.4. C. Transverse section of axis showing shallowly lobed protostele. × 17. (From Grierson and Banks 1963, reproduced with permission of the Paleontological Research Institution, Ithaca, New York.)

considerable variation, never being more than a third and generally one-fourth or one-fifth of the length of the leaf." (p. 249).

The xylem strand is a primary, fluted, cylindrical column about 1.3 mm in diameter composed entirely of tracheids. It is strongly ridged with the protoxylem points (there being sixteen in the specimen available) located at the extremity of the ridges. The protoxylem elements are annular while the metaxylem ones show a transition inwards from scalariform to reticulate to round-bordered pitted.

Estinnophyton Fairon-Demaret, 1978.
Estinnophyton gracile Fairon-Demaret, 1978.

The reader may wonder at our inclusion of the text that follows under this binomial. It deals with fossils that have been described under *Protolepidodendron* but it is now recognized that they cannot be assigned to that genus and, furthermore, it is problematical as to just how they should be classified.

In his comprehensive account of the Lower Devonian plants of Belgium, Stockmans (1940) described some well-preserved specimens of *Protolepidodendron wahnbachense* Kräusel and Weyland, which came from the quarry of Bois de Bescaille at Estinnes-au-Mont, east of the town of Binche. Their age is regarded as being in the upper part of the Lower Siegenian (Fairon-Demaret, 1978).

Figure 5.13. *Estinnophyton gracile.* A. Sterile axes. × 2. B. Portion of a fertile appendage showing two of the sporangia. × 18. (A. From Stockmans, 1940, Mem. Mus. Roy. d'Hist. Nat. Belg., B from Fairon-Demaret, 1972, Bull. Inst. Roy. Sci. Nat. Belg.).

The specimens suggest a small herbaceous plant, probably well under a half meter in height. They consist of axes 2 to 4 mm in diameter bearing leaves in a loose but regular spiral which resemble a small *Drepanophycus* stem, but some of the leaves were noted to be forked. They attain a length of about 4 mm with the division at the distal third of the leaf. It is a good example of how misleading the appearance of a compression fossil may be when freshly exposed and prior to any treatment. It is correspondingly exemplary of what may be revealed through the use of appropriate techniques, and of the ensuing nomenclatorial complications.

On the basis of the forked leaves, some of which are shown in his original illustrations (Fig. 5.13A), Stockmans assigned his specimens to *Protolepidodendron wahnbachense* which had been described by Kräusel and Weyland in 1932 from German specimens, although none of them were found bearing sporangia.

More recently, Muriel Fairon-Demaret (1972) initiated a new investigation of the Estinnes fossils. In this first report she demonstrated what appeared to be pairs of sporangia (Fig. 5.13B) attached singly to the lamina a short distance proximal from the point of dichotomy. The name *P. wahnbachense* was retained at that time. In a later study (1978) resulting from very careful degaging, she reported the fertile leaves as each bearing *two* pairs of sporangia immediately proximal to the point of division (Fig. 5.14). This clearly removed the plant from *Protolepidoden-*

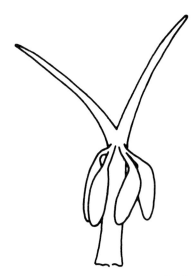

Figure 5.14. *Estinnophyton gracile.* Restoration of a fertile appendage showing the two pairs of sporangia and the bifurcated leaf. About × 8. (From Fairon-Demaret, 1978. Bull. Acad. R. Belg.)

dron and a new binomial, *Estinnophyton gracile* was assigned. The generic name refers to the Belgian locality, Estinnes.

A few more details may be added at this point. The general nature of the small stems, up to 4 mm in diameter, is well shown in one of Stockmans' photos. Fairon-Demaret's notation (1978, p. 601) that they are "up to 4.5 cm" is apparently an error. The leaves are 5 to 7 mm long, with the division at about mid-point. The sporangia, borne on pedicels 0.2–0.3 mm long, are mostly 1.2–1.7 mm long and 0.2–0.4 mm wide; spores have not been found. A specimen that was cut transversely shows what is apparently the vascular cylinder; it is probably an exarch protostele although no cellular structure is preserved.

Space limitations do not allow a detailed discussion of the possible affinities of this interesting fossil, which are given in Fairon-Demaret's account of 1978, but a few comments are appropriate. It does not belong in the genus *Protolepidodendron*; and the way in which the sporangia are borne does not conform closely to typical lycopod morphology.

A point of considerable significance is the early horizon in which *Estinnophyton* occurs; as noted above, Fairon-Demaret dates it as the upper part of the Lower Siegenian, or approximately mid-Lower Devonian. A glance at the charts from Chaloner and Sheerin, 1979, and included in Ch. 12, Figs. 12.4–12.6 reveals that at this time there were few morphological types as far as the manner in which the sporangia are borne, and none of them are closely comparable to the plant of Estinnes. In its general habit, the stems with their persistent leaves suggests a small herbaceous lycopod; the paired sporangia remind one of the trimerophytes and their position on recurved appendages is reminiscent of *Calamophyton*. As Fairon-Demaret notes, some aspects of the leaf morphology and sporangial organization suggest the sphenophylls. But on the overall basis of what is known of the plant, it is our opinion that it stands alone. The unique combination of characters presents a plant that adds to the known complexity of early land plant morphology.

Following this study of the Belgian fossils that were originally named *Protolepidodendron wahnbachense*, Fairon-Demaret has investigated specimens from the type locality where the species was first found, and described by Kräusel and Weyland, in the Wahnbach Valley near Siegburg, Germany. The type specimens were lost during World War II so that a direct comparison cannot be made. However, there seems to be no doubt that the recently collected ones described by Fairon-Demaret (1979) are specifically identical to the originals.

The German fossils consist of stem specimens 2–11 mm in diameter, the longest ones being 10 cm. It is noted that they branch sparingly, but since only short pieces are at hand this feature is not well known. The sterile and fertile leaves are identical in their general form and branching pattern. They average 10 mm long, divide first at mid-point, and a second time shortly thereafter, resulting in a three-dimensional, *twice*-divided appendage (Fig. 5.15) that thus differs from the Estinnes plant. In the case of the fertile leaves, the pedicel of each pair of sporangia is described as being attached just before the second division (Fig. 5.15). The sporangia are cylindrical, measuring 2 mm long and 0.4–0.6 mm in diameter. Spores have not been found.

Fairon-Demaret (1979) renamed the German fossils *Estinnophyton wahnbachense* (Kräusel and Weyland) comb. nov., the only striking distinction being their twice-divided leaves. They are dated as being from the lower part of the Upper Siegenian, slightly younger than the Belgian horizon, but still well down in the Devonian.

Leclercqia Banks, Bonamo, and Grierson, 1972.
Leclercqia complexa Banks, Bonamo, and Grierson, 1972.

This plant is especially well-known from abundant fossil material of excellent preservation; it is appropriately named for Prof. Suzanne Leclercq of Liège who has made so many notable contributions to Devonian paleobotany. Although it possesses the distinctive characteristics of the lycopods these are combined with certain unique features, and it is known to have been widely distributed geographically.

The type species, *L. complexa*, was described in 1972 based on

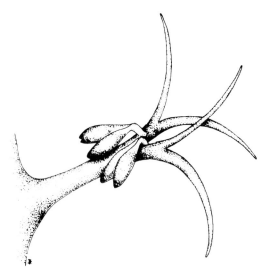

Figure 5.15. *Estinnophyton wahnbachense.* Restoration of a fertile appendage. × 6 (From Fairon-Demaret, 1979. Rev. Palaeobot. Palynol.)

collections from a late Middle Devonian horizon (Panther Mt. Formation) in Schoharie County, New York State. The dichotomizing axes attain a diameter of 7 mm and specimens up to 46 cm long have been found; it may reasonably be assumed that it was a plant that reached a height of at least a meter. The leaves are arranged densely in a characteristic lycopod fashion (Fig. 5.16A). In overall length they range from 4–6.5 mm and are terminated by a sharp tip which is occasionally notched. About midway between the point of attachment and the tip, two lateral branches are given off, each of which almost immediately divides into two parts; a total of five parts per leaf is revealed (Fig. 5.16B, C). Stomata are found generally over the leaves and also have been observed on the stem and sporangia.

The sterile and fertile leaves (sporophylls) are essentially alike; the latter are intercalated between the sterile ones in much the same fashion as is found in the living *Lycopodium lucidulum*. A sporangium about 2 mm in diameter is borne on the upper surface of the sporophyll (Fig. 5.16D) proximal to the lateral segments. It dehisced along a line parallel to the midline of the sporophyll. In their original account, the authors reported the plant as bearing leaves that are eligulate, but in a later report (Grierson and Bonamo, 1979) based on SEM studies, a ligule about 2 mm long with a rounded tip was found on both vegetative and fertile leaves (Fig. 5.16E).

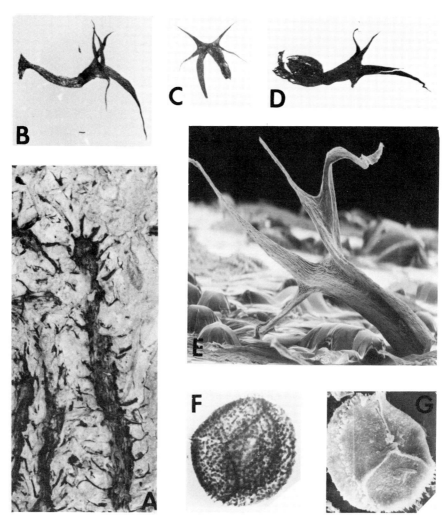

Figure 5.16. *Leclercqia complexa.* A. A stem specimen showing some sporophylls at the lower right portion. × 1.5. B. A single entire macerated leaf with the enlarged base at the left. × 8. C. A single leaf showing clearly all five tips. × 6. D. A single sporophyll and sporangium. × 7. E. An SEM of a vegetative leaf with the ligule near the base. × 35. F,G. Light and scanning electron micrographs of spores. × 345 and × 537. (A–D,F,G from Banks, Bonamo and Grierson, 1972. Rev. Palaeobot. Palynol. E. from Grierson and Bonamo, 1979. Amer. Journ. Bot.)

Well-preserved spores (Fig. 5.16F, G) were found in abundance in many of the sporangia and display a distinctive ornamentation: "The proximal and distal surfaces of the spores are covered with a crowded biform ornament. . . . Each element is composed of a basal enlargement (verruca) at least twice as wide as high (usually 2×1 μm). The rounded apex bears a tiny, non-tapering projection up to 3 μm in length, that is sometimes hooked." (Banks, et al. 1972, p. 34).

There is some variation in spore size, from 60 to 85 μm in diameter, but since this is rather slight and large numbers of spores have been examined the plant is considered to be homosporous.

The presence of a ligule combined with homospory presents an interesting combination of characters and the authors note:

"This suggests that the origin of heterospory and the ligulate condition were not linked as has appeared to be the case in later lycopods."

The center of the stem is occupied by a solid xylary strand that is slightly lobed by fourteen to eighteen longitudinal ridges (Fig. 5.17A). The tracheids range from the peripheral protoxylem cells that are annular to spiral (Fig. 5.17B), to scalariform and oval-bordered pits in the metaxylem (Fig. 5.17C, D). The structure of the tracheid walls deserves special mention, this having been studied in considerable detail by Grierson (1976) with application of the SEM. This is a pioneering and exemplary piece of work and it seems appropriate to include here a selection of his illustrations (Fig. 5.17). Grierson's account includes a detailed discussion of the techniques that he employed that should be useful for future workers.

Leclercqia is widely distributed geographically. Fairon-Demaret (1974) reported it from the Middle Devonian of the Burdekin Basin of Queensland, Australia. *L. complexa* also has been reported by Kasper and Forbes (1979) from the Trout Valley Formation of northern Maine. And Kasper (1977) has briefly reported a new species from the Campbellton Formation of northern New Brunswick, which on the basis of dispersed spore evidence is probably uppermost Lower Devonian. Lastly, as previously noted, Fairon-Demaret (1980, 1981) identified specimens formerly called *Protolepidodendron scharianum* from Elberfeld, Germany, and from Ronquiers and Oe, Belgium, as *Leclercqia*.

Colpodexylon Banks, 1944.
Colpodexylon deatsii Banks, 1944.

Of the two species of this genus that were originally described by Banks, *C. deatsii*, the type species, is the much better known. It was obtained from a roadside quarry near Pond Eddy, Sullivan County, New York State, being of Upper Devonian age (Portage Formation). A

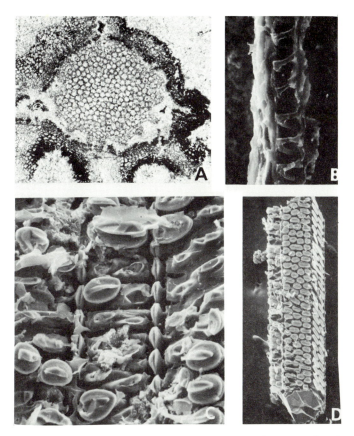

Figure 5.17. *Leclercqia complexa.* A. Cross section of a stem showing exarch protostele. × 20. B. Portion of a carbonized helical element of the xylem. × 633. C. Portions of three tracheid casts showing 1) their lateral interfaces marked by the paired pit cavity casts, and 2) the fractured surface of a pyritic pit canal at about the level of its inner aperture. × 704. D. Portion of a single pyritized tracheid cast revealing more than one wall facet and showing casts of the pit canals and cavities. × 155. (A. Photo supplied by J.D. Grierson. B–D. from Grierson, 1976, Amer. Journ. Bot.)

restoration drawing was given by Banks in 1970 (Fig. 5.18) and additional information and illustrations by Grierson and Banks in 1963 (Fig. 5.19).

The fossils consist of compressions and permineralizations of dichotomizing axes up to 2.5 cm in diameter and sixty cm long. They bear leaves that attain a length of 3 mm and are three-forked with the two lateral branches being half as long as the central one. The fertile leaves (sporophylls) each bear an elliptical sporangium 1.8–3.8 mm long and

Figure 5.18. Drawing of a single fertile leaf of *Colpodexylon deatsii*. (From Banks, 1960, Senckenberg. Leth.)

0.5–1.6 mm wide on the upper side and proximal to the branches. Spores have not been reported. More information is needed concerning the distribution of the sporophylls; in his original account Banks described them as "probably distributed over the plant, not restricted to modified zones."

The xylem is entirely primary and consists of a lobed protostele (Fig. 5.19B) in which the segments are somewhat wedge-shaped and may be peripherally notched. The protoxylem is exarch, or occasionally mesarch and most of the cells composing the stele are scalariform.

The second species, *C. trifurcatum*, described in 1944, differs from the type in minor aspects such as having considerably larger leaf cushions. It extends the known range of the genus, having been obtained from a lower Middle Devonian horizon in Orange County, New York State.

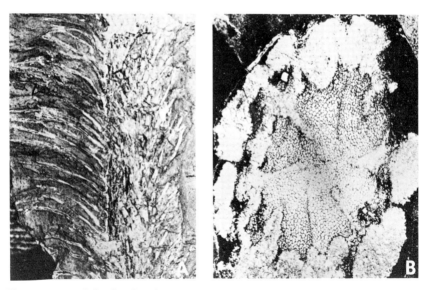

Figure 5.19. *Colpodexylon deatsii.* A. Portion of a leafy axis. × 1. B. Transverse section of the lobed protostele. × 9. (From Grierson and Banks, 1963, reproduced with permission of the Paleontological Research Institution, Ithaca, New York.)

Archaeosigillaria **Kidston, 1901.**
Archaeosigillaria vanuxemii (Goeppert) **Kidston, 1901.**

Of the better known Devonian fossil plants that are regarded as being lycopods yet fall short only in lacking the spore-bearing organs, the genus *Archaeosigillaria*, and especially *A. vanuxemii*, is one of the most important. A great miscellany of specimens have been assigned to the genus over the years and a detailed historical account has been brought together by Grierson and Banks in 1963 that includes much data that we cannot mention here. The following, from their account, seems to make an appropriate introduction to our summary:

> "The genus has steadily evolved in concept and content for more than 120 years despite the fact that the specimen on which the type species is founded has never been formally diagnosed nor described. The lack of a description with quantitative data as well as the absence of a diagnosis for the species have prompted the writers to provide such a diagnosis and to bring up to date information on the species from additional specimens that enlarge and enhance the species concept." (p. 237)

Several hundreds of specimens were available to Grierson and Banks which enabled them to interpret the plant remains in their varied states of preservation (Fig. 5.20). The type locality is cited as Allen's Quarry, Owego, Tioga County, New York State, and the horizon is the Cayuga Formation which lies in the Lower part of the Upper Devonian (Frasnian). The species is known from stems that attain a diameter of 25 mm; they are dichotomously branched and bear persistent leaves that are arranged in spirals. The gross morphology of the leaf is distinctive and has been clearly established only recently in a short paper by Fairon-Demaret and Banks (1978). The leaves as shown in Fig. 5.21, attain a length of about 5 mm and a maximum width of 2 mm and are characterized by the two prominent teeth located slightly proximal to the mid-point, and an acute hair-like tip a portion of which shows in some of their photos.

The leaves are thought to have been persistent. There is no evidence in the genus as a whole that they abscissed and the specimens that have been assigned to *A. vanuxemii* vary greatly in surface pattern resulting from differences in preservation. We have included a photo of part of the type specimen (Fig. 5.20) from the account of Grierson and Banks and they note that ". . . the six-sided leaf bases cannot be regarded as true cushions from which the leaf abscissed, but rather as merely the enlarged base of the leaf."

Some concept of the vascular structure may be gained from a partially permineralized specimen that is illustrated by Grierson and Banks. The xylem is protostelic with a fluted outline of apparently eight

Figure 5.20. *Archaeosigillaria vanuxemii.* A. Portion of an axis showing leaf scars. × 2. B. Portion of an axis showing leaves in side view. × 1.5. (From Grierson and Banks, 1963, reproduced with permission of the Paleontological Research Institution, Ithaca, New York.)

lobes. The maturation is exarch, the metaxylem tracheids being scalariform or reticulate.

In their account of 1963 Grierson and Banks include a chart showing the comparative features of five other species of the genus, of which they recognize three in addition to *A. vanuxemii.* For the most part, the specific differences are those of seemingly minor features of leaf morphology.

III. CHIEFLY UPPER DEVONIAN FOSSILS THAT REVEAL THE ORIGINS OF HETEROSPORY AND THE ARBORESCENT HABIT IN LYCOPODS

A considerable number of presumed lycopods have been described from Upper Devonian and Lower Carboniferous horizons, most of which are known from impressions or compressions, some more or less replaced by mineral matter, and of greatly varied quality in their preservation. A

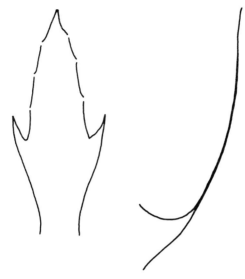

Figure 5.21. *Archaeosigillaria vanuxemii.* Restoration of leaves in front and side views. × 13. (From Fairon-Demaret and Banks, 1978, Amer. Journ. Bot.)

detailed review has been given by W. G. Chaloner and E. Boureau in Volume II of the *Traité de Paléobotanique* (1967) and, in connection with the present account, we acknowledge additional helpful comments from Chaloner. Much of this material, being fragmentary and poorly preserved, conveys but little information. Some of it, however, is interesting and informative and we have selected examples that afford significant information on the origins of heterospory and the arborescent habit in the group.

Among the more conspicuous generic names that one encounters in the literature are *Cyclostigma* and *Bothrodendron*, to both of which there have been assigned numerous species from the Upper Devonian and especially the Lower Carboniferous. Taken as a whole, this assemblage is a very difficult and problematical one due largely to the fragmentary nature of much of the fossil material. The difficulties were expressed quite poignantly, and with a slight feeling of despair by Seward as long ago as 1910, in the second volume of his *Fossil Plants:*

"Without attempting the hopeless task of discriminating between the various Carboniferous and Devonian specimens described under the names *Cyclostigma* and *Bothrodendron*, reference may be made to the following records as illustrating the wide distribution of the genus [*Bothrodendron*]." (p. 257)

We will comment later on the probable distinction between the two genera but we will first present a summary of one of the better known Upper Devonian members of the 'assemblage.'

Cyclostigma Haughton, 1859.
Cyclostigma kiltorkense Haughton, 1859.

The plant has an interesting history as well as contributing significant botanical information. Numerous specimens were discovered in the mid 19th century in the Upper Devonian beds of Kiltorcan Hill, County Kilkenny, Ireland. Our account here is taken chiefly from a detailed summary by T. Johnson in 1913 (he described the plant under the name *Bothrodendron* (*Cyclostigma*) *kiltorkense*, Haughton, sp.), and additional information provided by Chaloner in 1968. Johnson initiated new collecting in 1912 and reported:

> " . . . I was able to spend a week with a collector and two quarrymen in excavating some tons of rock, and in examining closely on both sides each slab as removed. So rich are the rocks that it was, as the quarryman Davis said (as we examined and split open the slabs, showing what the local people called "the drawings on the stone"), "like turning over the leaves of a picture book." (p. 500).

There was sufficient interest in the early days of the discovery of the deposit, especially as to its age, that specimens were sent (December, 1856) to Adolphe Brongniart in Paris for his opinion. As to the specimens of *C. kiltorkense* he replied: he knew "nothing like this fossil, and did not venture to name a family for its reception." Numerous other noted paleobotanists have, over the years, had a hand in passing judgment on the Kiltorkan fossils, including Oswald Heer and Robert Kidston.

C. kiltorkense is known from stem compressions, some of which bear the foliage (Fig. 5.22), and from partially preserved cones with the spores intact in some of the sporangia (Figs. 5.22 and 5.23).

Johnson described stem specimens 30 cm in diameter and two to three meters long, which suggests that the plants attained a height of about eight meters in life. The branching was predominantly dichotomous. It was thus a forest tree of respectable size, at least as compared with the plants we have encountered from earlier Devonian horizons.

The linear leaves are up to 15 cm long and 1 mm broad at the base; there is a single vascular strand and the leaf tapers to a point at the distal end. The foliage was lost by abscission as is indicated by the fact that most of the specimens show only the leaf scars arranged in an ascending spiral, but so slight as to present an almost whorled arrangement. In his more recent study, which deals chiefly with the cones, Chaloner (1968) demonstrated the presence of a pair of parichnos scars, on either side of

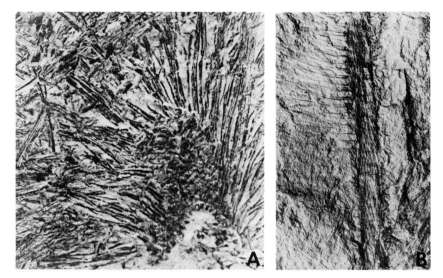

Figure 5.22. *Cyclostigma (Bothrodendron) kiltorkense.* A. Terminal portion of the cone or strobilus. × 1. B. A portion of the stem showing attached foliage, about × 0.6. (From Johnson, 1913, Proceed. Roy. Soc. Dublin.)

the vascular trace scar, on specimens from which the leaves had been shed.

The cones were borne terminally on some of the branches and in life were probably about 6 cm long and 2–3.5 cm in diameter, excluding the free tips of the sporophylls. The latter have been studied from intact cones as well as numerous dispersed sporophylls. They consist of a basal portion about 12 mm long which bears the sporangium, and a linear lanceolate lamina up to 8 cm long. A ligule has not been observed.

The sporangium is up to 4 mm broad and extends the entire length of the basal or proximal part of the sporophyll. As many as thirty-two megaspores have been observed in a sporangium and it is thought that the maximum number was sixty four. They range in size from 760–1520 μm with a mean of 1180 μm. The better preserved specimens bear conical spines up to 40 μm in height. Sporangia containing microspores have not been positively identified and it is therefore not known whether the cones were unisexual or bore both mega- and microsporangia.

Johnson demonstrated that the basal part of the stem passes into a Stigmarian-type rhizome or rootstock:

"I have now several specimens in which the same fossil shows clearly at one end the true *Stigmaria* scars, and at the other end, in continuous

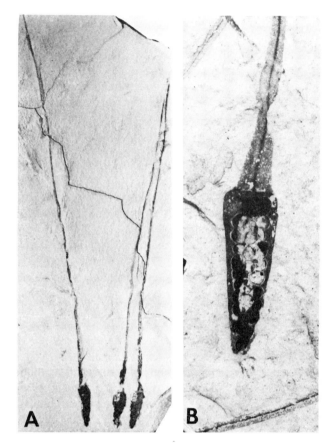

Figure 5.23. *Cyclostigma kiltorkense.* A. Three isolated megasporophylls showing the fertile base (sporangium) and the long sterile lamina. × 0.8. B. Surface view of sporangium with numerous megaspores and sterile lamina with dark central strand. × 3.2. (A. From Johnson, 1913, Proceed. Roy. Soc. Dublin. B. from Chaloner, 1968, Bot. Journ. Linnean Soc. London.)

connection with the *Stigmaria,* a stem impression which shows the usual surface markings and some scattered leaf-scars of *Bothrodendron.* These specimens show that *Bothrodendron* had a normal underground *Stigmaria*-stage." (p. 516)

In 1902 Nathorst described numerous specimens (also under the name *Bothrodendron (Cyclostigma) kiltorkense* Haughton sp.) from Bear Island in the Arctic. Some of those that he figured show the dichotomous branching very nicely and we have included one of his illustrations here (Fig. 5.24). The Bear Island fossils do not include leaves or spore-bearing

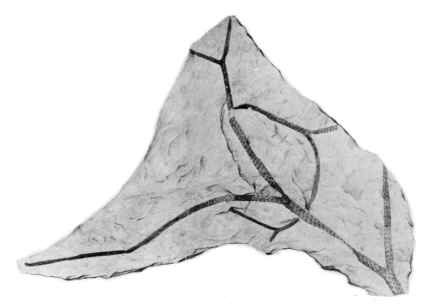

Figure 5.24. *Cyclostigma kiltorkense.* Specimen from Bear Island, showing branching pattern and leaf scars. × 1. (From Nathorst, 1902, Kgl. Svensk. Vet. Handl.)

organs and their presumed identity with the Irish ones lies in the close similarity of the defoliated stems (Fig. 5.25). The age of the two localities is quite close; palynological evidence indicates that the Kiltorcan site is near the top of the Devonian while the Bear Island locality is either at the top of the Devonian or basal Carboniferous.

In view of the usage here of the two generic names, *Cyclostigma* and *Bothrodendron*, a few more words of explanation seem in order. The type species in both is based on stem impressions, thus conveying limited information. As to the sum total of information available on species assigned to the two genera, they are distinguished as follows: *Cyclostigma* is characterized by long linear leaves, no evident ligule, and a large cone with at least six tetrads of megaspores per sporangium. *Bothrodendron* has short leaves, a ligule, and the sporangium includes a single tetrad of spores. Several species of each have been described, in most of which the nature of the spore-bearing organs is unknown. We are indebted to W. G. Chaloner for this summary information and, in a personal communication, he adds a few words which we think are worth including as a dictum that applies rather generally in paleobotanical nomenclature:

"The basic snag is that as we come to more fully understand one species of the genus, the better defined the genus becomes and so the

Figure 5.25. *Cyclostigma kiltorkense.* Specimen from Bear Island showing larger axes and leaf scars. × 1. (From Nathorst, 1902, Kgl. Svensk. Vet. Handl)

more tricky it becomes to use that generic name for poorly preserved specimens. This happened to *Bothrodendron* and *Cyclostigma*, and so the boundary becomes blurred."

Cyclostigma kiltorkense thus presents some significant features which may be summarized as follows:

It indicates a significant increase in size over the Middle Devonian lycopods and in view of the diameter of the stems the presence of a cambium is strongly suggested.

The cones seem primitive in retaining the characteristics of foliage leaves that bore sporangia but they are gathered into a distinct terminal cone.

The leaves were shed by an abscission layer and parichnos strands were present.

Heterospory had been achieved.

Protolepidodendropsis pulchra Høeg, 1942.

In his comprehensive account of the Devonian plants of Spitsbergen, Høeg (1942) described lycopod stem compressions under the binomials

Figure 5.26. *Protolepidodendropsis pulchra*. A stem impression from Spitsbergen described by Høeg, (1942) as *Bergeria mimerensis*. × 0.4 (From Høeg, 1942. Norges Svalbard-og Ishavs-Under.)

Bergeria mimerensis and *Protolepidodendropsis pulchra*; these were obtained from a horizon that is either late Middle Devonian or early Upper Devonian.

    B. mimerensis is represented by stem specimens up to 8 cm in diameter which are described as being preserved in the "bergerioid state." The numerous specimens available to him show considerable variation in the preservation of the outer part of the stem and we have included here one of the seemingly more representative specimens (Fig. 5.26). *P. pulchra* was based on axes ranging up to 1 cm in diameter. Very few specimens had intact foliage and Høeg simply notes that "The leaves are simple and undivided, and may probably attain a length of about 1 cm."

    Høeg suggested that the fossils so described may actually represent one plant, in which *B. mimerensis* constitutes the main stem and *P. pulchra* the smaller branches. Some years later, Schweitzer made additional collections which he described in 1965 and he supported Høeg's concept, stating that it was "highly probable" that the collections represented one plant. He prepared an interesting restoration drawing under the name *Protolepidodendropsis pulchra* Høeg (Fig. 5.27).

    Although neither the basal part of the trunk nor the spore-bearing organs of the plant are known, its general habit which is based on numerous specimens seems well established and gives us some concept of

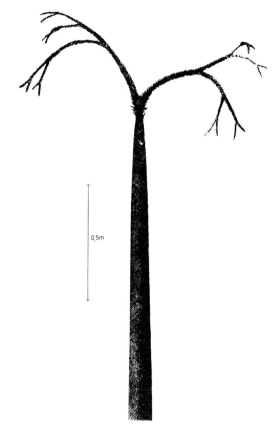

0,5m

Figure 5.27. Reconstruction of the plant *Protolepidodendropsis pulchra* according to Schweitzer, 1965. (Palaeontographica.)

the appearance of a late Devonian lycopod tree of modest dimensions, being probably several meters high.

Prolepidodendron Arnold, 1939.
Prolepidodendron breviinternodium Arnold, 1939.

This generic name was established by Arnold in 1939 on fossils that he had previously (1933) described under the name *Trochophyllum breviinternodium*. They came from the Upper Devonian near Port Allegheny, Pennsylvania.

He figured two specimens in 1939: One is a leafy axis a little less than 1 cm in diameter bearing leaves about 15 mm long; they appear to be in whorls with about twenty leaves per whorl. A distinctive feature of the leaves is the presence of two veins which diverge gradually from the base toward the apex. The second specimen is a defoliated one that shows a low spiral arrangement and the scars are described as resembling those of *Lepidodendron* but lacking the ligule and parichnos scars. Arnold says:

> "*Prolepidodendron* is believed to represent an early lepidodendrid type which is more advanced than *Protolepidodendron* (in which no definite abscission layer had developed), but which is somewhat more generalized than the typical Carboniferous forms. The fructification of *Prolepidodendron* may be of the type recently described as *Lepidostrobus Gallowayi*, a heterosporous cone from the same formation and locality which bears the sporophylls in whorls in a manner similar to the arrangement of the leaves on the *Prolepidodendron* stem." (p. 279–280).

Subsequently other authors have suggested that Arnold's plant should be referred to *Lepidodendropsis* but Grierson and Banks (1963) say that they "consider the assignment of this plant to *Lepidodendropsis* unwarranted in view of the unusual leaf form which serves to distinguish it." (p. 223).

The binomial *Lepidostrobus Gallowayi* was established by Arnold in a short paper of 1935 in which he referred back to his description of the fossil given in 1933. It is based on a compression from "Campbell Hollow, 1.5 miles northeast of Port Allegheny, McKean County, Pennsylvania." The horizon is regarded as the Pocono Sandstone and is stated as "presumably Upper Devonian."

The portion of the strobilus (Fig. 5.28) preserved is one and three-eighths inches wide and four inches long. Arnold isolated megaspores that are about 150 μm in diameter and microspores that are mostly 35 μm in diameter, but some range up to twice that size.

There is some question as to whether these fossils are derived from a horizon that is uppermost Devonian or basal Mississippian, but based on the best evidence that we have the former is correct. And although the vegetative and reproductive structures are not present in organic connection, there is a strong suggestion that they represent a distinctive type of heterosporous lycopod from the time range that we are dealing with.

A Late Devonian Herbaceous Lycopod

Although the great arborescent lycopods that originated in the Late Devonian, becoming conspicuous and abundant in the Carboniferous, claim exceptional attention in almost any general consideration of the fossil plant record, it is evident that small herbaceous forms continued on.

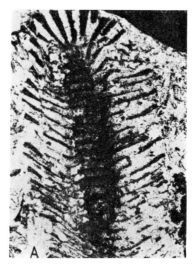

Figure 5.28. *Prolepidodendron breviinternodum.* Portion of a leafy axis. × 0.9 (From Arnold 1939, Contrib. Mus. Paleont. Univ. Michigan.)

They have perhaps played much the same part in the changing floristics of successive geologic ages as they do today—as creeping evergreens on the forest floor or as epiphytes. Numerous species from various geologic ages have been described, chiefly under the generic name *Lycopodites* and *Selaginellites* and they are reviewed in Volume II of the *Traité de Paléobotanique.*

We include a brief description of a representative member of this assemblage from the Upper Devonian.

Lycopodites oosensis Kräusel and Weyland was described in 1937 from specimens of basal Upper Devonian age from Oos near Budesheim, Germany. They were similar in size to some of our modern lycopodiums. The dichotomizing stems were a little over 1 mm in diameter and bore spirally arranged leaves that attained a length of 2 mm (Fig. 5.29). They were spatulate in form, rather thick and rigid and arranged in close spirals. Although a vascular strand was not actually found preserved in the leaves, a central scar (in specimens from which the leaves had fallen) indicates one was present.

Only fragments of sporangia were found in the compressions but these yielded spores that range from 90–120 μm. The plant thus seems to have been homosporous and one in which the sporophylls were not differentiated into a distinct cone.

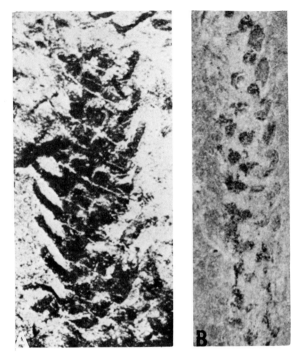

Figure 5.29. *Lycopodites oosensis.* A. Portion of leafy shoot. × 7. B. Portion of sporangium-bearing shoot. × 4.5. (From Kräusel and Weyland, 1937. Senckenbergiana.)

Phytokneme **Andrews, Read and Mamay, 1971.**
Phytokneme rhodona **Andrews, Read and Mamay, 1971.**

The affinities of this fossil, which is thought to be a lycopod stem fragment, are uncertain and its interest lies chiefly in the very distinctive cellular composition (Fig. 5.30A), especially of the cortex and leaf traces. It is known from a stem specimen about 3 cm in diameter and 24 cm long and was found in the Chattanooga Shale (Upper Devonian) in Adair County, Kentucky.

The central core of the stem consists of a very slightly fluted cylinder of primary wood composed of scalariform tracheids. This is surrounded by a band of radially aligned, and presumably secondary, tracheids only 2–4 cells thick. Peripheral to this is a narrow zone that is probably phloem (Fig. 5.30B); this is poorly preserved but in a few places elongate and very thin-walled cells occur that are thought to be sieve cells or their equivalent.

The cortex of the stem presents some very distinctive if not unique features. It includes a narrow inner band 10–12 cells thick of more or less brick-shaped cells. The middle cortex is about 5 mm thick and is composed of a groundwork of large, thin-walled and irregularly shaped cells that is traversed by radially aligned strands of elongate cells. The strands, although essentially radially aligned, tend to be irregular and they divide and anastomose (see Fig. 5.30C).

The outer cortex is 5–6 mm thick and consists of cells that may best be described as short fibers.

The leaf traces depart from the periphery of the primary xylem and consist of a small central strand of a dozen annular tracheids surrounded by thin-walled cells that are probably phloem. They depart from the xylem in a helical pattern but the angle of the spiral is so low as to present a nearly whorled arrangement. They are especially conspicuous in the outer cortex where the trace remains attached to the inner side of an aerenchymatous chamber that is traversed by several parenchyma strands.

Phytokneme is assigned to the lycopods chiefly on the basis of the small spirally aligned leaf traces which probably supplied microphyllous leaves, and on the structure of the xylem cylinder. However, within the many petrified Paleozoic fossils that have been assigned over the years to the lycopods, as well as the living members, there are none, to the best of our knowledge, that have the distinctive features found here.

SUMMARY

The Devonian record of the lycopods is one of the most interesting and well documented for a group of plants that is still extant. It reveals the evolution of heterospory, the development of arborescent members, and suggests how distinctive departures from a 'main line' of evolution may take place.

Paleobotanists working with the very early land vegetation are often asked— "What do the plants look like?", a query that usually implies a comparison with living plants. The vast majority of Devonian plants simply do not look like any living ones. However, one of the oldest, *Baragwanathia longifolia*, (of late Silurian or very early Devonian age) does allow a striking comparison with some living members of the group, such as *Lycopodium lucidulum*. The fossil was larger but is similar in its general habit, leaf morphology, organization of the sporangia, and vascular anatomy.

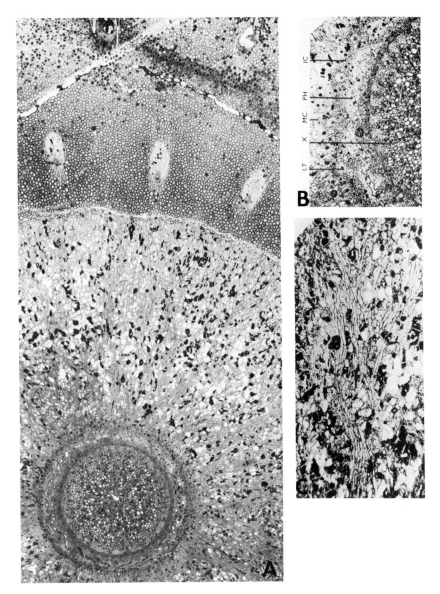

Figure 5.30. *Phytokneme rhodona.* A. A representative sector of the stem in transverse section; showing central xylem strand, inner, middle, and outer cortex. × 13. B. A portion of the central part of the stem showing: lt = leaf trace; x = primary xylem; mc = inner part of the middle cortex; ph = phloem; ic = inner cortex. × 31. C. Part of the middle cortex in transverse section showing distinctive ray-like strands. × 25. (From Andrews, Read and Mamay, 1971, Palaeontology.)

We have included in this chapter certain early Devonian fossils such as *Asteroxylon, Kaulangiophyton*, and the enigmatic *Drepanophycus*, in which the relationship between emergence (sporophyll?) and sporangium does not conform precisely to the lycopod concept but may represent a relationship from which the latter was derived. These also may reflect early stages in the evolution of microphyllous leaves, perhaps being derived from enations. Whether or not this is related to sporangial position is unknown.

In the Middle Devonian, genera such as *Protolepidodendron* and *Leclercqia* reveal the group as well established, becoming diversified, and showing some divergences in leaf morphology. In *Protolepidodendron*, the leaf is forked at or near the tip and in *Leclercqia*, which is homosporous, the leaves divide into five parts in a distinctive manner and bear a ligule, indicating that this puzzling structure is a character that is not necessarily tied to heterospory. The vascular strand in *Leclercqia* leaves may send off short traces to the base of the lateral forks, thus differing from the single-veined microphylls seen in other genera. Is this of significance in considering the origin of these leaves or merely reflecting some of the variation present when lycopod leaves are considered as a whole? (See introductory comments to this chapter).

A discussion of the distinctive genus *Estinnophyton* is also included. It may represent a departure from the lycopod line but its affinities are uncertain. In the two species that have been described, the sporophyll bears two pairs of sporangia, being once-forked in one species and twice-forked in the other.

Numerous genera of late Devonian lycopods have been described from fragmentary stem remains which at least reveal the development of arborescent forms. *Cyclostigma* is one of the better known ones; it was heterosporous with distinct terminal cones and is known from stems up to 12 inches in diameter, thus having been a forest tree of respectable magnitude. Two less well-preserved types not discussed in this chapter offer additional evidence. *Amphidoxodendron*, described from sterile axes from the Middle Devonian of New York State, is regarded by Grierson and Banks (1963) as an early indication of the advent of arborescence. And the numerous sterile axes, with round to oval leaf scars lacking evidence of a ligule or parichnos, placed in the genus *Lepidodendropsis* Lutz, occur in the Late (perhaps Middle) Devonian and especially in the Early Carboniferous. Some of these obtained from the Lower Carboniferous Price Formation in Virginia are up to 30 cm wide. Associated stump casts which may represent basal regions of these stems are at least 2 meters tall and about 45–60 cm in diameter.

In view of the great racial longevity of the lycopods it may be helpful to the general reader to comment briefly on the post-Devonian record. In the Carboniferous, the evolutionary vigor of the group burst forth with the development of the great *Lepidodendron* and *Sigillaria* trees that dominated the forests over vast areas. Quite independently of other groups they developed their own kind of secondary meristems and a unique type of seed.

Apparently, radical changes in the environment brought about a rather abrupt decline in the dominance of these Carboniferous giants toward the end of the Paleozoic. However, smaller herbaceous lines, represented in the Upper Devonian by *Lycopodites oosensis*, continued on to the present. Representatives in successively higher, post-Devonian horizons are described in Chaloner and Boureau's treatment in the *Traité de Paléobotanique*.

Today, although the number of genera is small (*Lycopodium*, *Selaginella, Isoetes,* and *Phylloglossum*), the number of species, their variations in habit and the habitats they occupy, is indeed remarkable. Plants such as the living *Lycopodium lucidulum,* which seems to compare closely with *Baragwanathia longifolia,* indicates an extraordinarily long lineage, and in a New England woodland one may find five or six species of *Lycopodium* in a few minutes of walking, their racial vigor seemingly far from exhausted.

6 / The Sphenopsids (Horsetails)

This distinctive group of plants is represented today by the genus *Equisetum* with about twenty-five species located worldwide, but was more diverse in the past. Two orders are currently accepted, the Equisetales and the Sphenophyllales, the former including extinct plants formerly placed in the Calamitales (Taylor, 1981). Plants unquestionably referable to the sphenopsids date from the Late Devonian to the present, with their height of diversification in the Carboniferous, including the occurrence of large trees and some heterosporous species.

The distinctive features of the sphenopsids are encompassed in the living genus *Equisetum*, commonly referred to as the horsetails or scouring rushes, the latter because of the coarse texture of the stems resulting from the presence of silica. These plants, found along roadsides and in woodlands of the temperate and tropical regions of the world, consist of underground rhizomes and aerial shoots which are ribbed and divided into nodes and internodes (the stems are referred to as being jointed), with the aerial shoots bearing whorls of leaves and sometimes branches. Branches are borne next to leaves instead of in their axils. The leaves are quite small and nonphotosynthetic. Sporangia are borne pendulous just below the expanded tips of specialized axial structures termed sporangiophores, which in turn are borne in whorls within the terminally located cones or strobili. The anatomy of this plant is distinctive in several ways; the central region of the stem consists of a sheet of parenchyma at the nodes, but is hollow in the internodes; the vascular tissue occurs as a continuous ring of xylem and phloem at the nodes but as several discrete strands at the internode. The protoxylem cells break down in internodal regions to form lacunae; parts of the cortex pull apart to form additional spaces. For more details the reader is referred to the descriptions in Bold, Alexopoulous, and Delevoryas (1980) *Morphology of Plants and Fungi* or Foster and Gifford (1974) *Comparative Morphology of Vascular Plants.*

Extinct members of this group correspond morphologically to extant ones, although the cones of some of the former also exhibit whorls of sterile bracts alternating with sporangiophores. Anatomically they either resemble *Equisetum*, but with the development of secondary xylem in some genera, or are protostelic (Sphenophyllales).

One of the greatest puzzles is the way in which these distinctive plants originated. The first plants that can be related to this group with any degree of certainty are from the Upper Devonian, and some appear most closely related to the Sphenophyllales (*Eviostachya, Pseudobornia*). Plants referred to *Archaeocalamites* occur in uppermost Devonian and Lower Carboniferous sediments; these perhaps gave rise to the great calamites of the Upper Carboniferous.

Presumed ancestors of the horsetails have been considered for many years to lie among the group of Middle Devonian plants referred to the Hyeniales (Zimmerman, 1930), which included *Hyenia* and *Calamophyton*, and these have been termed the protoarticulates. The most persuasive evidence in support of this hypothesis is that *Hyenia* and *Calamophyton* exhibit distinctive fertile appendages, consisting of distally extended sterile segments and recurved axial segments bearing sporangia, borne in near-whorls on axes. The impact of Zimmerman's telome theory is especially significant in this regard; following his ideas one could evolve a typical equisetalean sporangiophore via fusion and/or webbing of the terminal sterile segments of *Hyenia* or *Calamophyton*, leaving the sporangia pendant below. Sterile appendages could, via webbing, result in leaves typically seen in sphenophyllums and calamites.

However, in 1965, Leclercq and Schweitzer reported that *Calamophyton* had a complex multistranded vascular system similar to that seen in plants assigned to the Cladoxylales (see ch. 7) and removed it to that order. The "joints" in its stems were shown by Leclercq and Andrews (1960) to be artifacts of preservation. The anatomy of *Hyenia* is poorly understood; a badly preserved specimen described by Kräusel and Weyland showed a V-shaped vascular strand but it is not known if this is the entire amount of vascular tissue or only part of it. The possible role of these plants as sphenopsid ancestors thus has been placed in doubt;

however, we believe that a statement made by Banks in 1968a is worth noting: "It will be a matter of no small interest to learn whether *Hyenia* remains in Sphenopsida or must also be transferred. It will be equally interesting to see if the new anatomical information from *Calamophyton* that seems to make it a cladoxylalean continues to outweigh the morphological evidence that *Calamophyton* does have a primitive sphenopsid sporangiophore." (p. 86).

Further, Skog and Banks (1973) when describing a new genus, *Ibyka* discussed a possible evolutionary sequence leading from trimerophytes via *Ibyka* to Hyeniales and then to undoubted sphenopsids. They deal with the cladoxylalean type of anatomy by suggesting that lobed steles, such as is found in *Ibyka*, could easily evolve into a complex stele as is present in *Calamophyton* and that this in turn, by loss of central parts, could result in the distinctive pattern of many sphenopsids. They state:

> "On the basis of this kind of reasoning it may not be astonishing to find a cladoxylalean type of vascular structure in the Hyeniales nor to suggest that the latter could be converted phylogenetically to a dissected siphonostele like that in *Archaeocalamites* and then in Calamitales." (p. 378).

We have chosen to include discussion of *Calamophyton* and *Hyenia* in our treatment of the Cladoxylales, but do not discount the possibility of their having some role in the evolution of the sphenopsids.

Brief consideration of another protoarticulate must be included here. A vegetative axis resembling *Hyenia* and a fertile shoot resembling *Pseudosporochnus* (see Ch. 7) were described together as a new genus, *Protohyenia*, by Ananiev (1957). Later he described cladoxylalean anatomy for the fertile remains (1964). It seems reasonable, as Bonamo and Banks (1966b) have suggested, to retain these pieces as separate genera; thus *Protohyenia* presently is a doubtful taxon.

We will discuss next those plants whose affinities to the sphenopsida are fairly well established and in so doing will deal with some members whose greatest abundance occurs in the Lower Carboniferous, although their earliest appearance seems to be Upper Devonian. Among these plants there appear to be members of both the Sphenophyllales, known best from the lovely Sphenophyllums of the Upper Carboniferous swamp floras, and the Equisetales (especially the Calamitaceae). Some very fragmentary and poorly understood plants sometimes allied to this group but not treated here are summarized under 'Primitive or doubtful Sphenophytes' in volume III of the *Traité de Paléobotanique*.

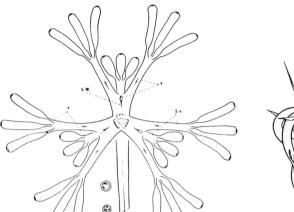
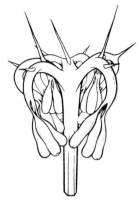

A   **B**

Figure 6.1. *Eviostachya hoegii.* A. A schematic diagram showing the arrangement of the three short and six long sporangiate units of one fertile appendage. B. Restoration of a single fertile appendage. (From Leclercq, 1957, Mem. Acad. Roy. Belgique.)

DEVONIAN PLANTS ALLIED TO THE SPHENOPHYLLALES

Eviostachya Stockmans, 1948.
Eviostachya hoegii Stockmans, 1948.

This genus was first described by Stockmans in 1948 from compressions from several Upper Devonian localities in Belgium where it was "not particularly rare." It consists of strobili about 5 cm long composed of a central axis bearing whorls of sporangiophores with pendant sporangia (Figs. 6.1 and 6.2). A careful study of compression and permineralized remains by Leclercq in 1957 resulted in the addition of considerable detail; this study is impressive in demonstrating what thorough examination using several types of techniques can provide about the construction of a plant.

Leclercq demonstrated that the strobili are borne in groups of at least three on slender, three-lobed axes and with a whorl of bracts located from 6 mm to immediately below each cone. The bracts are about 8 mm long, undivided, bifurcated or many times divided. On some specimens they extend upward for several mm and may cover a good portion of the cone.

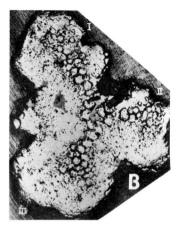

Figure 6.2. *Eviostachya hoegii.* A. General view of a single strobilus. × 4. B. Transverse section through the strobilar axis showing the triradiate vascular strand. × 35. (From Leclercq, 1957, Mem. Roy. Acad. Belgique.)

The cone itself is rather complex for such a small structure. It consists of a central axis with whorls of fertile appendages. Each whorl consists of six fertile appendages (sporangiophores) grouped in pairs; each appendage (Fig. 6.1) trifurcates forming three secondary axes, the central one being shorter than the two laterals. Each of the resultant nine secondaries trichotomize again, with all axes terminating in sporangia. Thus there is a total of 27 sporangia in a complete appendage. Spiny extensions depart from the curved region of the major appendage segments. A major difference between this cone and those of many Upper Carboniferous sphenophyllaleans is the absence of whorls of bracts between the fertile appendages.

Sporangia are 1–2 mm long and 0.3–0.6 mm wide, round-oval on the short segments and more elongate on the long ones. They dehisce through an apical pore. The spores obtained are quite variable in morphology, ranging from ones with coni, to ones with fine spines, to others with fine but longer and occasionally bifurcated spines. Leclercq compared them with Naumova's genus *Acanthotriletes*, the smallest ones with coni being most similar to her *A. famennensis*.

Sectioning of the single permineralized specimen resulted in a detailed understanding of the vasculature of the cone axis and fertile appendage. Axial anatomy consists of a deeply three-lobed vascular strand (Fig. 6.2B) with the ends of the lobes laterally expanded and with two protoxylem areas present at the periphery of each. These areas enlarge and emit a broadly triangular lateral trace. This divides in two, each strand supplying one fertile appendage. Subsequently, each of these traces will divide into 3 terete strands. Each of the terete strands divides into three again resulting in nine traces, each traversing one of the nine distal units of the sporangiophore. Protoxylem tracheids are annular and metaxylem ones are scalariform.

Leclercq interpreted *Eviostachya* as being intermediate in its structure between the Hyeniales (drawing similarity on the basis of the sporangiophore construction) and younger Sphenophyllales (which exhibit a triangular protostele).

Sphenophyllum subtenerrimum Nathorst, 1902.

Further evidence suggesting possible Devonian relatives of the Sphenophyllales is found in the occurrence of fragments of jointed stems, sometimes branched, occurring in the Upper Devonian of Bear Island and Belgium. First described by Nathorst, 1902, these axes typically have swollen nodes and may bear filiform leaves in whorls at the nodes. They closely resemble the Carboniferous species, *Sphenophyllum tenerrimum*, differing mainly in their very slender leaves. The Belgian record is based on specimens collected by Mme. Ledoux-Marcelle in the Upper Devonian of Belgium as noted and figured by Stockmans (1948). While undoubtedly a sphenopsid, much more information is needed to understand the whole plant.

Pseudobornia Nathorst, 1894.
Pseudobornia ursina Nathorst, 1894.

This plant was first described by A. G. Nathorst from Upper Devonian deposits of Bear Island, located between Spitsbergen and Norway in the Arctic. The first collections of *Pseudobornia* were made by Nordenskjold and other explorers on the Swedish Polar Expeditions of the middle 19th century and some specimens were described by O. Heer in 1871 as *Calamites radiatus, Cardiopteris frondosa,* and *C. polymorpha*. More recently, new collections were made and described by H. J. Schweitzer (1967).

The specimens consist of stems (probably aerial shoots and rhizomes) and lateral branches, the former being from 1 to perhaps 58 cm in

Figure 6.3. *Pseudobornia ursina.* A. Part of a presumed primary branch showing the articulations, branch scars and base of a secondary branch. × 0.2. B. Part of a branch showing the whorled arrangement of the leaves. × 0.5. (A. From Schweitzer, 1967, *Palaeontographica,* B. From Nathorst, 1902, Kgl. Svensk. Vet.-Akad. Handl.)

diameter (Fig. 6.3). They exhibit nodal lines or articulations which are curved rather than straight as in other articulates. Occasional round raised areas occur near the nodel lines which may represent the bases of lateral branches. The axes interpreted as stems or rhizomes lack ribbing and the nodal lines are not as pronounced as in other articulates, but the lateral branches often appear strongly ribbed and have more pronounced nodal lines. Leaves are borne in whorls, each with four to six leaves/whorl, on the lateral branches. They are 4–8 cm long, fan-shaped and distinctive in having a long petiole. The latter divides twice and each unit terminates in laminae which consist of long, narrow, wedge-shaped lobes giving the

Figure 6.4. *Pseudobornia ursina.* Part of a branch showing a fertile region. × 0.36. (From Nathorst, 1902, Kgl. Svensk. Vet. Handl.)

margins a dentate or fimbriate outline. Nathorst suggested the leaves were probably shed frequently. An epiphyte called *Codonophyton* is found on many stems.

Nathorst described one specimen with a fertile region terminating a stem, consisting of a spike 32 cm long with closely spaced whorled leaves (Fig. 6.4.). He interpreted the sporangia as being borne on the lower surface of the leaves. However, Leclercq reported in 1964 that each sporophyll consisted of a horizontal portion on the upper surface of which sporangia were borne and an upturned free region divided into two parts. These had a truncated apex and toothed margin. She suggested the sporangia, which are numerous on each sporophyll, were arranged in concentric whorls.

Schweitzer (1967) demonstrated that roots depart from some of the axes interpreted as rhizomes (the occurrence of isolated, but associated roots was noted by Nathorst). He also found some large, smooth stem impressions at one locality which were up to 10 meters long and incomplete, being 58 cm wide at the base and 25 cm at the distal end. These he interpreted as representing major stems of a tree-sized plant, and attributed them to *Pseudobornia* on the basis of the presence of nodal lines and round scars (presumably of branches). Unfortunately, no anatomy has been found in any of the specimens described.

Mamay (1962) reported the occurrence of *Pseudobornia* in Alaska, based on fragmentary specimens of its distinctive leaves, and used it, plus the presence of certain lycophyte remains, to date the sediments as Upper Devonian.

While it is clear that *Pseudobornia* (Fig. 6.5) probably was a shrub or small tree, with distinctive leaves and apparently a large strobilus containing numerous sporangia, it is regarded as of uncertain affinities within the Sphenopsids. It usually is placed in its own family and order and has been considered by some workers tentatively to be closest to the Sphenophyllales.

POSSIBLE EQUISETALEAN DEVONIAN PLANTS

Archaeocalamites Stur, 1875.

This genus is known to occur mostly in the Lower Carboniferous, but some impressions from Bear Island and possibly Germany have been referred to it. The genus is known from the Carboniferous from compressions, pith casts, and permineralized stems. Much of our current knowledge of the plant is summarized in Smoot et al. (1982). The pith casts are characterized by aligned rather than alternating grooves or ribs in adjacent internodes; this also is reflected in the manner in which traces depart from the stems as indicated. The stems bore whorls of several times dichotomized leaves (Fig. 6.6) and strobili, some referred to the genera *Potnocites, Potncuopsis,* and *Protocalamostacnys,* may belong to these stems.

The plant apparently was arborescent, exhibiting secondary growth in which the wood is very dense and regular, with narrow rays. The strobili apparently had superposed whorls of sporangiophores and no sterile bracts. The Devonian records of this genus are in need of clarification, being based on rather fragmentary stem impressions. *Archaeocalamites* apparently is quite a large and well-developed sphenopsid and further

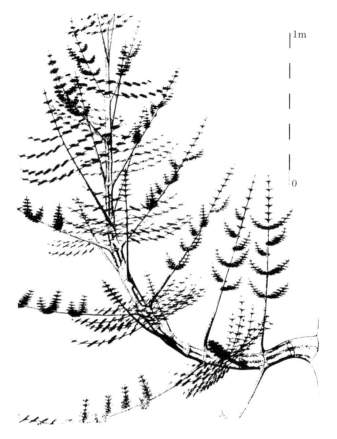

1m

0

Figure 6.5. *Pseudobornia ursina.* Restoration of a primary branch system showing whorled arrangement of the higher order branches and leaves. (From Schweitzer, 1967, Palaeontographica.)

occurrences of it in the Devonian would be of great interest to our understanding of the early history of the group.

SUMMARY

This brief accounting of possible sphenopsids in the Devonian provides a tantalizing glimpse of the early establishment of this unique group of plants, but much remains to be learned about their inception and the changes that led to the various calamites and sphenophyllums of the Carboniferous, the time when the group was at its peak. Certainly by the

Figure 6.6. *Archaeocalamites radiatus.* Drawing of a branch with leaves. × 0.5. (From Stur. Illustrated in Scott, 1923.)

end of the Devonian, both arborescent and herbaceous forms indicative of at least three lineages within the group were well established. These plants are of further interest in that both sterile and fertile appendages are much-divided entities, the sporangia terminating recurved terete axes in *Eviostachya* but apparently being borne in large numbers on specialized sporophylls in *Pseudobornia*. An even more complex strobilus, consisting of sterile and fertile divided segments has been reported from the Lower Carboniferous by D. H. Scott (1897) as *Cheirostrobus*. Most of the calamitean and sphenophyllalean strobili of later Carboniferous times exhibited fusion of parts, forming a peltate sporangiophore with several pendant sporangia, and also exhibited whorls of sterile bracts interspersed between whorls of sporangiophores. Leaves in both are more laminate in younger taxa also.

The origin(s) and early development of the articulates is very obscure. Despite the demonstration of cladoxylalean anatomy in *Cala-mophyton*, the arrangement of its fertile appendages (and those of *Hyenia*) is still the most likely type from which an articulate sporangiophore could have evolved. Perhaps the evolutionary sequence proposed by Skog and Banks (1973), from a trimerophyte ancestor, to intermediates such as Ibykales or Hyeniales, to various early articulates, will be shown to have occurred. Search for well preserved plants in Middle and Upper Devonian sediments is needed in order to learn more about this interesting group.

7 / The Cladoxylopsids, Coenopterids, and Some Possibly Related Forms

A rather detailed introduction to the content of this chapter seems necessary to explain our treatment of the plants that are included. Most of them are difficult, and some impossible, to classify satisfactorily but they add significantly to our knowledge of early land vegetation. The literature dealing with them is replete with well-intended but vague references to their being ferns, fern-like, or fern ancestors. We hope to clarify, at least in some degree, the significance of this comparison.

Plants that are referred to as *ferns* or *fern-like* are abundantly scattered through the fossil record and the literature on fossil plants. However, the term *fern* as applied to living plants is in itself by no means a satisfactory one in view of the diversity of the plants involved. Starting, as a basis for comparison (and mutual understanding) with certain widely distributed living plants such as those included in the genera *Adiantum* (maidenhair fern), *Pteridium* (bracken fern) and the osmundas (Osmundaceae), we are concerned with plants that are probably typical or 'true' ferns to both professional and nonprofessional botanists. Their foliage is representative for ferns in general. But in this context fern-like foliage is found in several other major groups such as some angiosperms (flowering plants), and the fossil pteridosperms. And 'typical' fern-like foliage is by no means a sharply defined entity if we glance over the plant kingdom as a whole.

It will be helpful to start with a few words on the diversity and classification of those living plants that are generally called *ferns*. There are numerous texts and other accounts in which one will find good summary treatments of the living ferns. We are making use of the Bold, Alexopoulos, and Delevoryas text *Morphology of Plants and Fungi* (1980) as a concise, clearly outlined account that we believe meets with favorable acceptance by many botanists. The living ferns (in the broadest sense) are presented under three Divisions: Pteridophyta I, II, and III. We will be concerned here only with their Divisions I and II, since Division III takes up a unique and difficult assemblage of aquatic plants (Marsileales and Salviniales) that are sometimes referred to as 'water ferns', and we deal with them briefly in another chapter.

Thus Pteridophyta I includes the Eusporangiate members with two major classes, Ophioglossopsida and Marattiopsida. Of these the former includes the rather well-known genera *Ophioglossum* (adder's tongue fern) and *Botrychium* (grape fern), while the Marattiopsida includes *Marattia* and several other tropical genera most of which have synangial spore-bearing organs. The problem as to "what is a fern?" begins here as *Ophioglossum* and *Botrychium* are quite distinct from any other living ferns, especially in the nature of their sporangiate organs.

Pteridophyta II includes those plants that seem to be generally regarded as typical ferns. The ordinal name Filicales is used to include such familiar families as the Polypodiaceae, Osmundaceae, Cyatheaceae, and others. There have been many classifications proposed for the 9,000 or so species included in the Filicales but there is little disagreement as to the overall unity of this assemblage, and the application of the term *fern*.

Summarizing very briefly the fossil record of these living groups: very little is known about the Ophioglossales; the Marattiales are clearly defined in the Upper Carboniferous; the Osmundaceae probably originated in the Permian; and others make their appearance at later times. More detailed information may be found in Volume IV of the *Traité de Paléobotanique*.

Because the ferns (Filicales) are a commonly encountered group of interesting plants from the Carboniferous on to the present, it is understandable that their origins have been sought in earlier horizons. Many Devonian plants that have been referred to as fern-like or fern ancestors, perhaps in large part for want of a better guess as to their relationships, have been dealt with as members of the Coenopteridopsida or Cladoxylopsida. Many of these plants share with ferns some degree of anatomical and morphological differentiation in the organization of main axes (stems) and lateral organs (leaves or leaf precursors); this also seems to have led to their being grouped with or near the ferns. The tentative use of these two taxa seems to be the best that is available for the fossils (or most of them) that are described in this chapter. It is very important, Banks (1968a) outlines possible trends from the rhyniophytes to the Filicales, as they are known from the Carboniferous on, is vague and uncertain. The opinions of several authorities who have been concerned with these plants in recent years are included to substantiate our own.

In his comprehensive survey and revision of the Devonian plants Banks (1968) outlines possible trends from the rhyniophytes to the trimerophytes, and from that group to the cladoxylaleans, coenopterids, ferns, and progymnosperms. And in discussing the Sphenophytina and Filicopsida in particular he says: "The origin of these two groups is speculative. The sphenophytes may have been derived from Rhyniophytina. The ferns may have had a similar origin by way of Trimerophytina. Some characters associated with Filicopsida evolved in each of the three groups, Cladoxylopsida, Coenopteridopsida, and Progymnospermopsida, and the various evolutionary series seen among them could equally well have led to ferns. Beck (1962) has cited several reasons for believing that progymnosperms are neither ferns nor progenitors of ferns. Similar reasoning applies to cladoxylaleans and coenopterids. . . . " (p. 98).

In his recent textbook *Paleobotany* (1981) Taylor starts his chapter on the Pteridophyta with a description of essentially the same Devonian plants, although he divides them into three groups: Cladoxylopsida, Rhacophytopsida, and Coenopteridopsida; and in his introductory sentence he says: "The plants described in this chapter may be considered ferns, although some of the Devonian members are better regarded as fern-like."

Scheckler (1974) in a survey of the presumed or possible Devonian 'ferns' states the case in a more positive fashion: "At present no Devonian plant is recognized as unequivocally related to the modern ferns." (p. 462).

It is the opinion of the present authors that, although the ferns (Filicales) may have had their origins somewhere in the general complex of plants described in this chapter, the term *fern*, in the present state of our knowledge, should not be applied to any Devonian fossil.

THE COENOPTERIDS AND *RHACOPHYTON*

The coenopterids or coenopterid ferns as they are usually called originally included a large assemblage of plants that are known in large part from Carboniferous petrifactions. Some possess megaphyllous, laminated leaves while others have three-dimensional, non-laminate 'fronds'; many of them are protostelic, their vascular system being composed of primary wood, there being only one genus with weakly developed secondary xylem present; the sporangia are foliar-borne or terminal and in some (*Botryopteris*) the sporangia are aggregated into massive fructifications; they are mostly homosporous but heterospory is present in one species of *Stauropteris*. In the Carboniferous it is a large, varied, and intriguing assemblage which in part seems to justify being called fern-like, although their relationship to the Filicales is not clear. In some recent studies, some of the genera which possess megaphyllous, laminate leaves, and foliar-borne annulate sporangia have been moved as extinct families to the Filicales. These are summarized in Taylor's text *Paleobotany* (1981). If one accepts these changes, two families remain in the Coenopteridopsida: the Zygopteridaceae (in part) and the Staurop-teridaceae. Earlier detailed summaries may be found in Hirmer (1927) and in Andrews and Boureau (1970), Vol. IV of the *Traité de Paléobotanique*.

Although the coenopterids are known chiefly from Carboniferous fossils, some Devonian plants have been included, the best known genus being *Rhacophyton*. It also has been regarded as a 'pre-fern' (Cornet et al. 1976) of the order Protopteridiales, or placed in its own class or order, the Rhacophytopsida or Rhacophytales. In the latter cases, the distinction is based on the absence of clearly differentiated, laminate megaphyllous leaves in *Rhacophyton*.

Rhacophyton Crépin, 1875.

The type species is *R. condrusorum* Crépin from the Upper Devonian of Belgium. It was originally described by Crépin as *Psilophyton con-drusorum*, later referred to *Sphenopteris* and then to *Rhacophyton*. Further details concerning the history of the genus are found in Stockmans (1948) and in Leclercq (1951) who gave a comprehensive account and a description of *R. zygopteroides*. A third species, *R. ceratangium*, was described by Andrews and Phillips in 1968 and in a later report by Cornet, Phillips, and Andrews in 1976; since it is especially well preserved (including compressions and petrifactions) it forms the basis of our account of the genus.

Rhacophyton ceratangium Andrews and Phillips, 1968.

This is known from abundant specimens taken from a roadside locality near Valley Head, West Virginia (Fig. 7.1). The horizon is in the Hampshire Formation, close to the top of the Devonian (Famennian). The matrix in which the fossils were preserved is a fine-grained, uncon-solidated siltstone which facilitated degaging and the elucidation of the three-dimensional form of the plant.

Figure 7.1. The roadside locality in the Upper Devonian Hampshire Formation near the village of Valley Head, West Virginia, from which the remarkably well-preserved specimens of *Rhacophyton* and *Archaeopteris*

Specimens that are interpreted as the main axis (stem) range up to 2 cm in diameter and it is estimated that the plant attained a height of about 2 meters. The stem bore organs that are referred to as vegetative fronds which are arranged in two vertical and alternating rows on each side. These are essentially two-dimensional, each bearing two rows of primary branches which in turn bore two rows of secondaries that vary considerably in their branching patterns, some being nearly monopodial and others essentially dichotomous (Figs. 7.2 and 7.3).

The fertile fronds or branches are unique and complex structures whose form was revealed only after detailed degaging of numerous specimens. They consist of a central axis bearing *nodal units*, one of which is shown in Fig. 7.4. This consists, typically, of two sterile primary

Figure 7.2. Restoration of *Rhacophyton ceratangium.* (From Andrews and Phillips, 1968, Bot. Journ. Linnean Soc. London.)

branches about 10 cm long and two fertile primary branches. The latter are essentially spherical, rather profusely dividing structures, that bear numerous sporangia (Fig. 7.5A, B) which consist of an ovoid body about 1.3 mm long and 0.3–0.4 mm in diameter with an elongate terminal tip that attains a length of 1 mm. The significance, if any, of this curious tip is not known. The spores measure 50–60 μm in diameter, there being several hundred per sporangium. They are trilete, round, and with a thin outer wall layer which resembles in its morphology the perispore seen on extant fern spores (Fig. 7.5C). It seems reasonably certain that the plant was homosporous.

Although the sterile and fertile fronds all belong to the same plant, the way in which the fertile ones were borne is not entirely clear; some were apparently borne as a special branch of the sterile frond.

Structures that seem to be correctly interpreted as roots have been identified. They are 1.5 mm in diameter and give rise to unbranched laterals that are about 0.1 mm wide. Some of these are petrified and

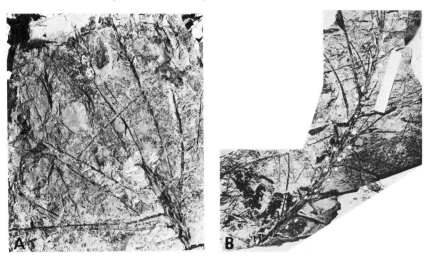

Figure 7.3. *Rhacophyton ceratangium.* A. A main axis bearing several vegetative branches. × 0.2. B. A rather long main axis specimen with portions of lateral branches. × 0.1. (A. From Andrews and Phillips, 1968, Bot. Journ. Linnean Soc. London. B. From Cornet et al. 1976, Palaeontographica.)

Figure 7.4. *Rhacophyton ceratangium.* A single fertile branch showing the two elongate sterile segments and the two densely branched fertile units bearing numerous sporangia. (From Andrews and Phillips, 1968, Bot. Journ. Linnean Soc. London.)

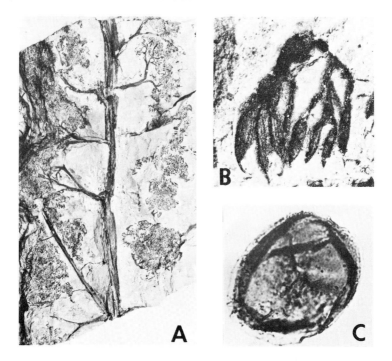

Figure 7.5. *Rhacophyton ceratangium.* A. A main axis bearing parts of several fertile branches—the sporangium-bearing portions are visible. × 0.7. B. Terminal portion of a fertile branch segment showing several sporangia with their elongate tip or 'beak'. × 8.6. C. A single spore. × 700. (From Andrews and Phillips, 1968, Bot. Journ. Linnean Soc. London.)

consist of a single group of primary tracheids surrounded by radially aligned, and presumably secondary, tracheids.

Numerous fragments of petrified axes were found, some of which are identified as the main axis (Fig. 7.6). In all cases they consist of a very slender elongate strand of primary wood which expands at either end. This 'clepsydroid' primary bar is enclosed by a considerable development of radially aligned scalariform tracheids regarded by the authors as being secondary wood. Since wood rays were not actually illustrated some doubt has been voiced as to its origin from a true cambium. A more recent study (Dittrich, Matten and Phillips, 1983) reports and illustrates the presence of rays.

In this connection it is significant to note that Schultka (1978) in a paper describing the anatomy of specimens of *R. condrusorum* from Upper Devonian rocks of Walheim/Aachen says that "Strongly developed secondary wood with many rays is present." The identification of

Figure 7.6. Transverse section of part of a pyritized axis of *Rhacophyton ceratangium.* × 12. (From Andrews and Phillips, 1968, Bot. Journ. Linnean Soc. London.)

'secondary' wood in some of the more primitive vascular plants of the Devonian, on the basis of radial alignment of the tracheids and the presence or absence of rays, seems worthy of discussion relative to its bearing on the origin of cambial activity in general. This seems to be an appropriate point at which to introduce such a discussion.

SOME NOTES ON CAMBIAL ORIGINS AND EVOLUTION

In certain Devonian plants in which radially aligned tracheids are present, and where rays have not been identified, there has been a tendency to regard them as a kind of metaxylem and not the result of a cambium. Leclercq and Bonamo (1971) questioned the validity of calling this tissue 'secondary xylem' in *R. ceratangium*:

> "It is here suggested that the radially aligned tracheids could represent an oriented metaxylem, a kind of xylary mechanical tissue developed in relation to the weight of the heavy pairs of fertile pinnae. Radial disposition of tracheids is not in itself a valid criterion to establish the existence of cambial activity." (p. 110).

Scheckler (1975a) discusses the problem of radially aligned tracheids which may be interpreted as "aligned metaxylem" in some species of

Cladoxylon. He concludes: "In agreement with Leclercq (1970), who also studied this problem, the aligned tracheids seen in some Cladoxylales are interpreted as metaxylem." (p. 36).

In Jackson's *A Glossary of Botanic Terms* (1949) the cambium is defined as follows: ". . . a layer of nascent tissue between the wood and bast [phloem], adding elements to both. . . ." (p. 61).

Dictionary definitions are useful and necessary in establishing order and effective communications but the known activities of this interesting meristem go well beyond the limits of the definition.

Well-preserved secondary phloem is not often present in fossil plants. Its absence in many cases certainly is due to decay but we suggest that, in at least some of the pteridophytes, it is simply due to the fact that it was never produced. Three instances may be cited to substantiate this:

In 1960 Arnold described an especially well preserved Upper Carboniferous lepidodendrid (*Lepidodendron schizostelicum*) stem from Kansas in which cortical tissues are in direct contact with the peripheral cells of the secondary xylem. It would appear that the cambium was a unifacial one that produced only xylem and the parenchymatous ('cortical') cells on the outside were of primary origin. Arnold summarizes as follows:

> "Is it inherent in the cosmic order that a cambial initial must produce both xylem and phloem mother cells? Could a cambium exist that produces only one of these? Then there is the question of the nature of the growth mechanism that produced this secondary xylem. Did it result from a cambium as such that cut off derivatives only on one side, or was there no cambium as such, the radial alignment of the xylem elements having resulted from previous divisions in the procambium? The latter is suggested by the absence of anything resembling the cambium at the periphery of the xylem." (p. 259).

In 1961 Andrews and Mahabale described, rather briefly, some large calamite axes (*Arthropitys communis*) from the Upper Carboniferous of Illinois which have several cm of secondary wood consisting of tracheids and rays, and peripheral to this broad band, and in direct organic connection with it (there being no break or decay), is a rather broad zone of 'typical' cortical tissue. The latter consists of essentially isodiametric cells which compare in no way with the secondary phloem of modern arborescent seed plants. The abundant secondary xylem, and apparent lack of secondary phloem, seems to indicate a unifacial cambium.

Eggert and Kanemoto (1977) described a Middle Pennsylvanian *Lepidodendron* stem from southern Illinois which showed particularly well-preserved primary xylem and phloem; in some of the stems studied, secondary xylem was present but there is no evidence of secondary phloem. Radially seriate, thin-walled cells occurred immediately adjacent to the secondary xylem which they interpret as secondary stelar parenchyma.

One may wonder how the transport of food materials was managed in these lycopod and calamite trees. Lacking specialized longitudinally elongate cells would suggest that it was very slow, and perhaps an important contributing factor that led to the decline and extinction of these plants.

At present we have only fragmentary evidence concerning secondary phloem from plants that are transitional between the pteridophytes and seed plants. In a short account of *Callixylon* axes, Arnold (1930b) described tissues peripheral to the secondary xylem which may represent secondary phloem. They consist of somewhat irregularly shaped parenchymatous cells and he says: "Unlike the secondary xylem of *Callixylon* the phloem (assuming that it is phloem) shows comparatively little differentiation." (p. 430).

Tissue interpreted as secondary phloem has been described in several other progymnosperms, (see Ch. 8) including *Tetraxylopteris* Beck, (1957), *Triloboxylon* (Scheckler and Banks, 1971a), and *Protopitys* (Walton, 1969). However, the presence of sieve cells has not been demonstrated conclusively in these genera. Their first undoubted occurrence is in the Lower Carboniferous genus *Calamopitys* (Galtier and Hébant, 1973), tentatively allied to the pteridosperms. Sieve cells are known to occur in several younger pteridosperms.

The cambium, as it functions in many of the angiosperms, is certainly one of the most remarkable of meristems, from the standpoint of the great variety of cells that it produces to the outside and to the inside, and from the standpoint of the many so-called anomalies (successive cambia, reversal of the usual sequence, etc.) in both dicotyledonous and monocotyledonous angiosperms. Quite clearly this great morphological, and presumably chemical, diversity did not originate *de novo*.

Devonian and Carboniferous plants are beginning to reveal something of the origins of the cambium. We suggest that the sequence has been somewhat as follows: An initial stage in which some continued division of primary tissues resulted in aligned tracheids which are perhaps best called metaxylem; next, ray cells became inserted among the tracheids and meristematic activity became concentrated at the periphery; this resulted in a strong development of secondary wood but with little or no phloem, at least at the outset; this was followed by a bifacial meristematic activity forming (phloem) cells toward the outside and (xylem) cells toward the inside. The continued evolution of complexities in cambial activity in the higher living and fossil vascular plants has resulted in a corresponding diversity in secondary tissue patterns.

THE CLADOXYLS

Cladoxylopsida Pichi-Sermolli, 1959.
Cladoxylales Hirmer, 1927.

Until quite recently a considerable majority of the fossils (i.e. genera that have been assigned to the Cladoxylopsida) were known only from petrified axes. The general nature of the plants, their classification, origins, and what they might have given rise to, present a complex set of problems. At the outset a few introductory notes on some of the 'literature landmarks' may be useful for those wanting a broader treatment than we are able to give here.

In 1927 Hirmer established the order Cladoxylales in which he included the genera *Voelkelia* and *Cladoxylon*, with an emphasis on the Devonian species *C. scoparium*, which is described in some detail below. In 1935, Paul Bertrand brought out a classic and monumental contribution on numerous species of *Cladoxylon* from Saalfeld. These are known from petrified stems and are of mostly Lower Carboniferous age. His study contributed much to our knowledge of the stem anatomy. The most detailed general summary that we have at present is the one given by Leclercq in Vol. IV of the *Traité de Paléobotanique*. The following are cited as general characteristics for the family Cladoxylaceae: The vascular system is a complex polystele composed of anastomosing strands (stelar segments of the whole). The peripheral ones tend to be radially arranged and are I-, U-, V-, or Y-shaped as seen in transverse section; the others are more or less round. The peripheral strands have a 'parenchymatous loop' (bouclé periphérique) developed to a greater or less degree. Genera and species have been delimited on such characters as the number, form and distribution of the stelar segments, the general nature of the 'parenchymatous loop', and the presence or absence of secondary xylem.

Leclercq included the following genera in the Cladoxylaceae: *Cladoxylon* Unger, *Hierogramma* Unger, *Syncardia* Unger, *Arctopodium* Unger, *Voelkelia* Solms-Laubach, *Steloxylon* Solms-Laubach, *Pietzschia* Gothan, *Pseudosporochnus* and *Calamophyton* Kräusel and Weyland. It is pertinent to note, relative to the present account, that the best known members of the Cladoxylopsida are Devonian plants, namely *Pseudosporochnus, Calamophyton*, and *Cladoxylon scoparium*. Two other introductory comments may be appropriate at this point:

Classification. A strong reliance on vascular stem anatomy in distinguishing the fossils included in the cladoxyl group renders the classification tentative. This is discussed at some length by Banks in his 1968a account, and Scheckler (1974) says: "Plants with cladoxylalean anatomy form a distinct group from the Middle Devonian to Lower Carboniferous. Their relationships with other groups of plants are obscure." (p. 470). This problem, which is beginning to be resolved, in no way detracts from the interest of some of the plants that are included.

Peripheral loop. This phrase, or the French equivalent 'bouclé periphérique' has been applied to the more or less bulbous extremity of the primary xylem bar (Fig. 7.8) in the individual xylem strands. Judging from the variation in the descriptions of the structure it may be composed of a solid mass of protoxylem tracheids or a mixture of the latter with parenchyma, preserved in various states of perfection. In his paper dealing with *Cladoxylon hueberi*, Matten (1974) says: ". . . (the term peripheral loop, as used here, refers to the presence of either areas of parenchyma in the position of the protoxylem or protoxylem lacunae)". (p. 312).

Numerous authors have expressed their opinion about this structure and Matten (1974) offers an interesting one, namely that "The 'peripheral loops' may appear only at particular stages in trace formation or in the formation of only certain kinds of traces. . . ." (p. 306). His discussion may be referred to for further information, as well as ones by Scheckler (1974), and Stein et al. (1983).

Cladoxylon Unger, 1856.

As indicated above our treatment of this genus is confined to the Devonian species.

Cladoxylon scoparium Kräusel and Weyland, 1926.

This species is based mainly on one specimen, consisting of compressions and partially permineralized portions of the axes, from the Middle Devonian (Givetian) of Kirberg and Oben, near Holz. They were described in detail in 1926 and a short supplement was added in 1929. Schweitzer and Geisen (1980) re-examined part of the original specimen and their finds will be discussed later.

The general aspect of the plant is shown in the authors' restoration (Fig. 7.7) which serves as a convenient basis for our description.

The largest axis is about 1.5 cm broad; this branches more or less dichotomously and bears sterile and fertile appendages. The sterile appendages ('leaves'), which are reported as being spirally arranged, seem best described as dichotomously divided micro-branches; they range up to 1.5 cm in length in the basal region and to 1.8 cm distally, where they are more abundant.

Just how abundant the fertile organs were seems uncertain but some of the terminal branches bore fan-shaped sporangiate appendages,

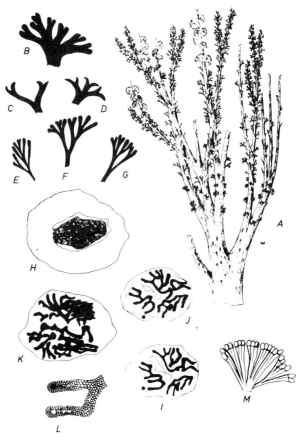

Figure 7.7. *Cladoxylon scoparium.* A. Restoration of a part of the plant. B–G. Sterile (vegetative) leaves. H–K. Transverse sections through various levels of the axial system showing the stelar organization. L. Detail of a distal portion of a stelar segment showing a peripheral loop. M. A fertile appendage. (From the *Tráite de Paléobotanique* Vol. IV, fasc. 1.)

variously and deeply lobed, with several sporangia borne distally in spoon-shaped terminal depressions (but see below).

Some of the axial fragments were found in a pyritized state and, depending on the size, these include numerous anastomosing stelar fragments that are variously shaped, as shown in the figures. These were reported in 1926 as being composed exclusively of primary tracheids, but a later study (1929) revealed sections that show some radially aligned (secondary?) tracheids. These cells are scalariform, reticulate, or rarely bordered pitted.

Schweitzer and Geisen (1980) found a piece of the counterpart of the specimen figured by Kräusel and Weyland as their Plate 13, Fig. 1 in the collection of W. Weyland. They redescribed the fertile appendages, suggesting that overlying matrix, fissures in the rock, and juxtaposition of segments contribute to their fan-shaped appearance and that they are organized instead identically to those of *Calamophyton primaevum*. On this basis they combine the two genera, *Calamophyton* having priority.

The identity of their specimen needs to be clearly established in order to assess the distinctness of *Cladoxylon* and *Calamophyton*.

Kräusel and Weyland, in discussing this material, gave an account of the history of the genus *Cladoxylon* and offered some speculations on affinities. They refer to the cladoxylons as a distinctive Paleozoic group and *C. scoparium* as having a "fern-like appearance", as have other writers since that time. We agree that the genus is a distinctive group insofar as known; the affinities will hopefully be more satisfactorily understood when the general morphology of other species is better known. As to being "fern-like", it is our opinion that there is no close resemblance to plants of the Filicales.

We include a brief mention of two reports of *Cladoxylon* in this country. They are both based on rather fragmentary material but are of interest in extending the known range of fossils assigned to the genus.

Cladoxylon dawsonii Read, 1935.

In 1935, Charles B. Read described *Cladoxylon dawsonii* from the Upper Devonian Genundewa limestone member of the Genesee Shale of New York State. His account is based on a single transverse section in the collections of the U.S. Geological Survey, it having been originally presented to F. H. Knowlton by one John M. Clarke who was a collector for Sir J. W. Dawson.

The section was labelled *Cladoxylon mirabile* Unger apparently by Dawson and was recorded as coming from the "Genesee Shale, Styliola layer (Genundewa limestone), Canandaigua, New York". Read states that it seems probable that the section is from the same block on which Dawson's 1882 report of *C. mirabile* Unger is based, but that there is insufficient evidence to positively assign it to Unger's plant which came from the Upper Devonian of Thuringia. Read therefore assigned the new name, *Cladoxylon dawsonii*.

The stem was in the vicinity of 1.5 cm in diameter and consists of a central stelar system composed of numerous radiating plates of primary xylem some of which are strongly curved and described as forming U- or V-shaped masses. The pitting was probably scalariform and protoxylem loops are present in the distal parts of the xylem strands. Some of the latter have small amounts of secondary xylem at the periphery. The stelar system was surrounded by a broad zone of thin-walled parenchymatous cortex.

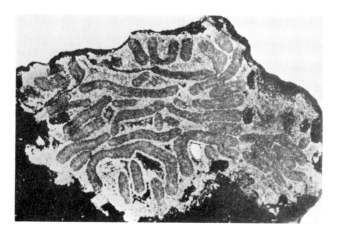

Figure 7.8. *Cladoxylon hueberi.* Cross section of a stem.× 6. (From the *Traité de Paléobotanique* Vol. IV, fasc. 1.)

Cladoxylon hueberi Matten, 1975.

Specimens on which this epithet is based were first described in a doctoral thesis by F. M. Hueber in 1960 (Cornell University); this was followed by a published description by Leclercq in Vol. IV of the *Traité de Paléobotanique* (1970), and the assignment of the species name by Matten in 1974.

The type locality is Cave Mt., Green County, New York and the horizon is the Oneonta Formation, Lower Frasnian (lower Upper Devonian). Matten's specimens are from a locality near Cairo, New York, the horizon being the Kiskatom Formation of the Givetian (upper Middle Devonian).

The plant is well known from an illustration in Leclercq's account which is reproduced in our Fig. 7.8. The part of the axes preserved, showing the vascular system, is about 1.5 cm in diameter and includes up to forty stelar segments that vary considerably in their form in transverse section. They consist of primary xylem only; the protoxylem is mesarch and the metaxylem tracheids are described as being scalaiform, reticulate, or bordered pitted.

The next two plants to be treated, *Calamophyton* and *Hyenia*, are of special interest, as noted in Ch. 6, in that, for many years, they were considered to represent precursors of sphenopsids. Discovery of clado-xylalean anatomy in *Calamophyton* caused its removal from the Proto-articulatae to the Cladoxylales. Although the anatomy of *Hyenia* is poorly understood, it is quite similar morphologically to *Calamophyton* and thus its possible sphenopsid affinities are also doubted. We do not totally exclude the possibility that they may be involved in the evolution of the sphenopsids, but have chosen to treat them here.

Calamophyton Kräusel and Weyland, 1926.
Calamophyton bicephalum Leclercq and Andrews, 1960.

The genus *Calamophyton* was established by Kräusel and Weyland in 1926, being based on specimens of *C. primaevum* from the Middle Devonian of Germany. Other early reports include those of Aderca (1932) and Leclercq (1940).

Our description here deals exclusively with some remarkably well-preserved specimens of *C. bicephalum* which were collected by Prof. Leclercq from the Middle Devonian (lower Givetian) Brandt quarry near Goé, in eastern Belgium. The fossils consist for the most part of compressions which, with careful degaging, revealed the three-dimensional form of the sterile and fertile appendages.

Figure 7.9A shows the largest specimen that was found, and consists of what is referred to as the 'main stem' that divides into two parts which continue to dichotomize. The various branch orders bear sterile appendages or 'leaves' and the distal parts of some branches bear abundant, essentially whorled fertile appendages.

The sterile appendages (Fig. 7.9C and 7.10A) are about 8 to 9 mm long, being somewhat more robust on the main stem than on the distal branches. They were three-dimensional organs with successive branch orders (a maximum of four) being at right angles to each other. It is thought that they were terete and quite rigid since they were not flattened during fossilization. A delicate vascular strand is present in these appendages.

Some of the distal branch parts were predominantly fertile (Figs. 7.9B and 7.10B). The individual appendages, which tend to be whorled, consist of a main stalk, slightly inclined adaxially, that dichotomizes, and each of the two divisions produces three short recurved branches which in turn are terminated by a pair of sporangia. Thus there are twelve sporangia per appendage when all are intact. They are cylindrical, pointed at the distal end, and measure approximately 2.5 × 0.6 mm. It may also be noticed that each of the two major divisions terminates in a rather blunt, equal dichotomy. Spores were not found in these Belgian specimens.

The general distribution and form of the sterile and fertile appendages suggest that the two are homologous. Whether or not the former should be termed leaves is perhaps questionable since they are essentially three-dimensional structures lacking a lamina.

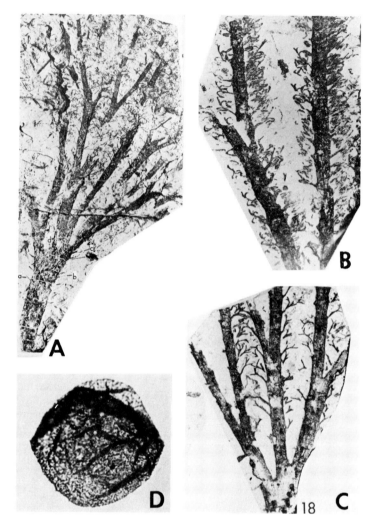

Figure 7.9. *Calamophyton bicephalum.* A. A large specimen showing digitate branching and vegetative leaves. × 0.3. B,C. Portions of fertile and vegetative specimens respectively. × 0.9 and × 0.6. D. A single spore. × 273. (A–C from Leclercq and Andrews, 1960, Ann. Missouri Bot. Garden. D. from Bonamo and Banks, 1966, Amer. Journ. Bot.)

In 1965, Leclercq and Schweitzer described very briefly the vascular anatomy of the Belgian *C. bicephalum* and German *C. primaevum*. They state that the two are probably conspecific and since the Belgian fossils are better preserved we will confine our discussion to *C. bicephalum*.

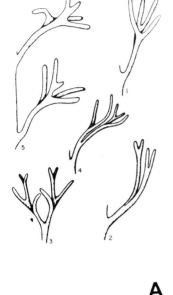

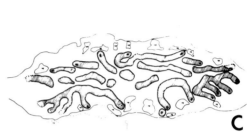

Figure 7.10. *Calamophyton bicephalum.* A. Drawings of several leaves. B. Drawing of an entire fertile appendage. C. Transverse view of an axis from the more distal portion of the plant (somewhat compressed) showing the stelar segments. (A,B. From Leclercq and Andrews, 1960, Annals Missouri Bot. Gard. C. From Leclercq and Schweitzer, 1965, Acad. Roy. Belgique Bull. Cl. des Sci.)

As described for third and fourth order branches that are about 7.5 mm in diameter, the vascular tissue consists of a complex polystelic structure composed of fourteen to sixteen segments varying in size and form as may be noted in Fig. 7.10C. The xylem is entirely primary and the peripheral arms have some protoxylem elements "or a small loop in a mesarch position." This vascular anatomy is regarded as of the cladoxylalean type and the authors note: "The combination of these characters removes *Calamophyton bicephalum* from the Sphenopsids placing it in the fern-like Cladoxyleae." (p. 1398).

Calamophyton bicephalum also has been reported by Bonamo and Banks (1966) from the Marcellus Formation, lower Middle Devonian (Eifelian), of Ulster County, New York State. Their report adds significant information on the spore-bearing organs. They find that the sporangia apparently dehisced longitudinally and spores were present (Fig. 7.9D). They are spherical and range in diameter from 86–166 μm. The exine is ornamented with variously shaped projections ranging in length from less than 1 μm to 4.5 μm; these are compared with the dispersed spore genus *Dibolisporites* of Richardson.

Leclercq (1964) described the fertile organs of *C. primaevum* as being organized identically to those of *C. bicephalum*, further supporting the idea that the two are conspecific, but she did not formally combine them in her discussion in Vol. IV, fasc. 1 of the *Traité de Paléobotanique.*

Arctophyton Schweitzer, 1968.
Arctophyton gracile Schweitzer, 1968.

We make brief reference to this plant here but it should be pointed out that its proper classification is not clearly understood. It is based on numerous fragmentary and associated specimens from Spitsbergen in rocks of lower Middle Devonian age.

The part of the plant that is most satisfactorily understood consists of an axis that may have attained a length of 0.5 m and bore spirally arranged branches that attained a length of not more than 20 cm. These branches in turn bore numerous ultimate branchlets that are non-laminate and divide several times in a dichotomous or slightly pseudomonopodial fashion. Some of them, probably borne toward the distal end of the 'primary' branches, were fertile and consisted of a stalk 2–4 mm long that dichotomized twice at the terminal end bearing a total of four sporangia that are 1.5 mm long and 0.5 mm wide.

The fertile branchlets bear some comparison with the more complex sporangiate organs of *Calamophyton*, and to the genus *Chaleuria* (see Ch. 10). However, it is our opinion that it is not possible to satisfactorily classify *Arctophyton* with the evidence at hand. In critical reviews of possibly related plants, Bonamo (1975) and Beck (1976) voice a similar viewpoint.

Hyenia Nathorst, 1915.
Hyenia sphenophylloides Nathorst, 1915.

The genus was established for plant remains from the Middle Devonian of Hyen in western Norway and was described as consisting of unbranched axes (later shown to depart from a horizontal rhizome) which bear appendages arranged in an alternate to whorled manner. They are narrow, 10–15 mm long and dichotomize distally. Fertile parts were unknown. Nathorst was struck by the resemblance of *Hyenia*, particularly in terms of its apparently whorled leaf arrangement, to the sphenopsids. However, there was no evidence of articulations (nodes and internodes) as in later occurring sphenopsids.

Høeg (1931, 1935, 1945) provided additional information, confirming the rhizomatous habit and he described more completely the morphology of the sterile appendages.

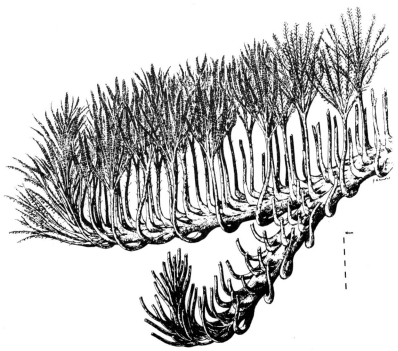

Figure 7.11. A restoration of *Hyenia elegans* by Schweitzer, 1972. (Palaeontographica.)

Hyenia elegans Kräusel and Weyland, 1926.

In 1926 and 1929, Kräusel and Weyland described a second species of *Hyenia, H. elegans,* from the Middle Devonian of the Rhineland, Germany. Their initial collections were of 4 mm wide aerial shoots, these being covered with leaves which are up to 2.5 mm long, 1 mm wide, and one to three times divided. The latter sometimes appear whorled, other times alternate. This species was initially distinguished from *H. spenophylloides* by its smaller size and by the presence of fertile organs.

Fertile axes bear lateral appendages which are two to three times dichotomous and some or all of which terminate in ovoid-fusiform sporangia. Kräusel and Weyland noted Nathorst's comments about similarity of *Hyenia* to sphenopsids, and felt the fertile organs in particular might agree with that affinity.

In 1932, Kräusel and Weyland demonstrated conclusively that the aerial shoots of *Hyenia* were borne on creeping rhizomes which were about 2 cm wide. Bases of branches appear to be enlarged. They also presented

Figure 7.12. Part of a terminal fertile spike of *H. elegans.* × 0.6. (From Schweitzer, 1972, Palaeontographica.)

limited data on anatomy; suggesting that the rhizome contained a V-shaped vascular strand, with the arms of the V directed toward the aerial shoot. Tracheids were scalariform. However, their photos are not very clear and much remains to be learned about the anatomy of this very interesting plant.

Schweitzer (1972) described specimens which he attributed to *H. elegans* that include large rhizomes bearing numerous aerial shoots (Fig. 7.11). The aerial axes may dichotomize one to several times in close succession to produce a more or less digitate pattern as seen in *Calamophyton*. Fertile axes bear very closely spaced lateral appendages which divide in their mid-region and each segment recurves and terminates in a pair of sporangia (Fig. 7.12). The close and regular arrangement of the lateral appendages results in a spicate appearance. Vegetative axes bear more distantly spaced lateral appendages which may be several times divided and are basically spirally arranged. He also suggested *H. sphenophylloides* and *H. elegans* should be combined as one species.

S. Leclercq in 1940 described some specimens from beds at the Lower-Middle Devonian boundary near d'Oe, Belgium, which she initially

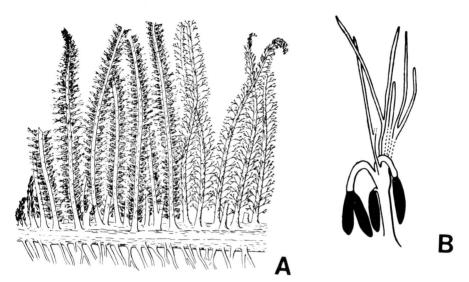

Figure 7.13. A. Restoration of *Hyenia* sp. by Leclercq, 1940. × 0.3. B. Detail of fertile appendage showing long distal extensions. × 4 (After Leclercq, 1940, Acad. Roy. Belgique.)

assigned to *H. elegans*. These specimens were later (Leclercq, 1961) suggested to represent a new species because the morphology of their fertile organs differs from that of *H. elegans* but no new name has been properly published. One specimen was particularly spectacular in showing the basal rhizome at the edge of the rock with thirteen branches departing from it (Figs. 7.13 and 7.14). These are up to 19.5 cm long, all but one are unbranched, and most are fertile. Sporangia-bearing organs are up to 15 mm long and borne in groups of up to six. Each is undivided basally, bifurcating medially into two recurved segments and one extended one (Figs. 7.13 and 7.14). The recurved segments terminate in one to two sporangia each, the extended one divides several more times to form filiform extensions. The sporangia are oval, 2–3 mm long and 0.5 mm wide. Leclercq illustrated globular bodies 5–10 μm in diameter as possible spores but they may instead represent tapetal residue. The fertile organs are the most complex and numerous of any of the species presently attributed to *Hyenia*.

Høeg (1942) described yet another species, *H. vogtii*, from the Middle Devonian of Mimerdalen, Spitsbergen, which differs from the others in its growth habit. He assigned it to a new subgenus, *Hyeniopsis*.

This species consists of axes 5–7 mm wide and up to 20 cm long which bear spirally arranged lateral branches. These are up to 15 cm long and may bear a second order of laterals. All are covered with small thorn-

Figure 7.14. *Hyenia* sp. of Leclercq, 1940. A. Part of a rhizome with several upright fertile branches. × 0.5. B. Portion of an upright branch bearing fertile appendages. × 1.5. (From Leclercq, 1940, Acad. Roy. Belgique.)

like emergences, and bear lateral appendages which are grouped in threes, are spiralled to verticillate, and divide up to five times. There is no evidence of a basal rhizome, and all aspects of axis and branches suggests quite a different type of habit.

This plant actually resembles the genus *Ibyka* in many aspects and closer comparison of the two may yield some interesting results.

Arnold (1941) briefly reported on the occurrence of *Hyenia* from the Middle Devonian of New York State, calling it *H. banksii*. Stems with lateral branches, several departing from one level and all covered with leaves occurring in whorls 8 mm apart, comprise the total data available on this species.

Pseudosporochnus H. Potonié and Bernard, 1903.
Pseudosporochnus nodosus Leclercq and Banks, 1962.

This is another one of the exceptionally informative and well-preserved plants from the Middle Devonian (lower Givetian) Brandt quarry near Goé, Belgium, and is based on a large number of specimens. The genus has a long and somewhat confused history with numerous

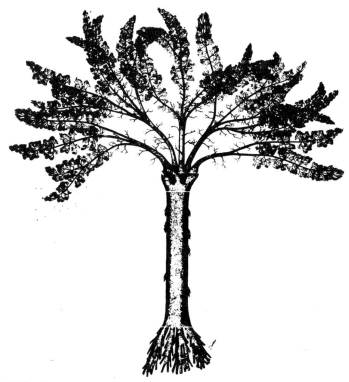

Figure 7.15. *Pseudosporochnus nodosus.* Restoration of the entire plant. (From Leclercq and Banks, 1962, Palaeontographica.)

contributions by different authors. A rather detailed review may be found in the introduction of Leclercq and Banks' 1962 study.

It will be helpful to refer first to their restoration drawing shown in our Fig. 7.15 which depicts the plant as a 'small tree'. It consists of a main stem with a crown-like cluster of primary branches departing from the apex; these first order branches divide dichotomously in a digitate fashion to produce two or more secondary branches which in turn bear the sterile or partially fertile ultimate branchlets or 'leaves.'

As a size index the first order branches vary in diameter from 2.2 to 3.6 cm at their base and 1.1 to 1.9 cm at their apex; some are notably bulbous at the basal end. It is of interest to note that the digitate branching at the distal end of these first order branches seems similar to that present in *Cladoxylon scoparium* and in *Calamophyton*.

The second order branches bear spirally arranged 'fronds' that may be either sterile or partially fertile. The sterile fronds average 4 cm in

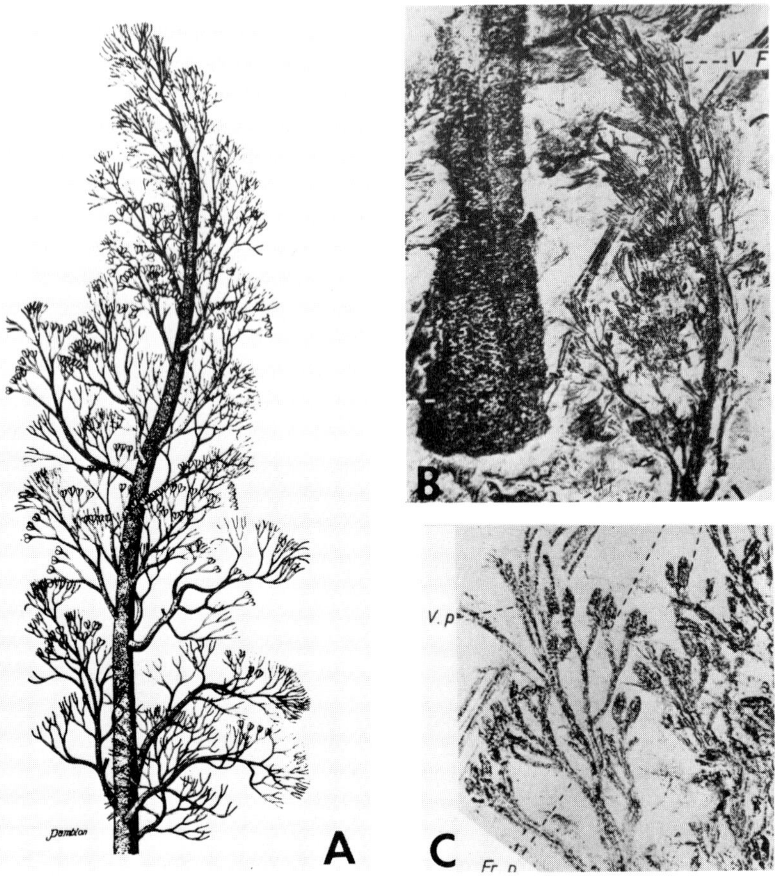

Figure 7.16. *Pseudosporochnus nodosus.* A. Restoration of a distal branch with vegetative and fertile appendages. × 0.8. B. Right, a distal branch segment with sterile and fertile appendages; left, base of a larger branch showing bulbous base and numerous sclereid nests. × 0.68. C. Distal portion of branch system showing sporangia. × 2.6. (From Leclercq and Banks, 1962, Palaeontographica.)

length and consist of a central axis 0.5–1.0 mm in diameter that produces two or more pairs of essentially opposite pinnae. These each dichotomize three times and the terminal part of the central axis dichotomizes (Fig. 7.16A, B).

The fertile fronds, although slightly smaller than the sterile ones, display a similar general architecture. The central branches are fertile and the proximal and distal ones tend to be sterile. A fertile branch dichotomizes twice with each of the four branches bearing a pair of sporangia (Fig. 7.16C). The sporangia are 2 to 3 mm long and 0.8 mm wide. Spores were not observed.

A conspicuous feature of all of the branch orders, including the ultimate branchlets, is the presence of abundant round or elliptical bodies about 0.5 mm broad that are revealed from permineralized specimens to consist of nests of sclereids (Fig. 7.16B). These distinctive sclerotic nests were observed in specimens found among the collections which range up to 8 cm in diameter and are regarded as composing the 'main stem' of *P. nodosus*. They also are found in axial specimens up to 3 cm in diameter which bear downwardly directed branches and have been interpreted as the probable roots of the plant.

In 1968 Leclercq and Lele described specimens (Fig. 7.17) that contained somewhat better preserved vascular tissues, but still do not show the entire stele. This study is based on specimens that have portions of the axes occasionally preserved that are referred to as 'inclusions'. In reference to specimens of the first order of branching they say:

> "The vascular system is polystelic and is distinguishable into an external zone and a central zone of steles. . . . The external zone is composed of more or less straight, separate, clepsydroid arms of different lengths, radially orientated. The central zone consists of circular to oval, separate steles. All the steles are composed of primary xylem and are embedded in a parenchymatous ground tissue, generally not preserved. . . . " (p. 100).

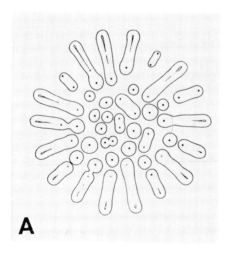
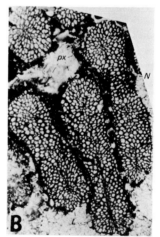

Figure 7.17. *Pseudosporochnus nodosus*. A. Presumed stelar pattern in a major branch. B. Portion of vascular system showing several stelar units. × 28. (From Leclercq and Lele, 1968, Palaeontographica.)

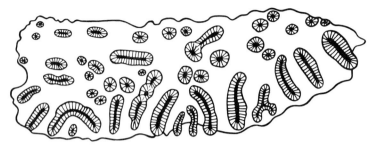

Figure 7.18. *Xenocladia medullosina.* Transverse view of a portion of an axis showing some of the many vascular strands. × 1.5. (From Arnold, 1952, Contrib. Mus. Paleont. Univ. Michigan.)

Xenocladia Arnold, 1940.
Xenocladia medullosina Arnold, 1940.

This plant was first described by Arnold in 1940 on the basis of very fragmentary fossils and in 1952 from more completely preserved ones. The original specimens came from the Middle Devonian Tully pyrite of Erie County, New York State, and the later ones from near Springbrook, Erie County, in the Middle Devonian Ludlowville Formation.

The part of the plant axis that is preserved measures 1 by 5 cm in transverse section and was about 10 cm long (Fig. 7.18). Judging from the curvature of the outer surface it suggests an axial diameter of 10 cm or more.

The transverse surface reveals a peripheral band of radially elongate vascular strands measuring about 2 × 7 mm and these flank an interior group that are circular to oval in cross section and which range down to 1 mm in diameter. They are embedded in a parenchymatous tissue that is partially preserved.

The individual strands are described as being protostelic, with a central primary xylem of tracheids only; the protoxylem areas vary from one in the smaller strands to several in the larger, but the exact number is obscured by incomplete preservation. The surrounding radially aligned xylem is described as secondary and consists of tracheids with one or two rows of bordered pits. Arnold (1940) reported: "There is no visible parenchyma, either as separate cells or in the form of rays." (p. 300).

He notes that *Xenocladia* differs from *Cladoxylon* in the nature of the tracheid wall structure, the absence of a peripheral loop in the primary xylem, and no evidence of the type of foliar trace found in *Cladoxylon*— although this last is a negative character of doubtful significance due to incomplete preservation. He considered that the most closely comparable fossil to *Xenocladia* is *Pietzschia*, the chief difference being the absence of secondary xylem in the latter.

Pietzschia Gothan, 1927.
Pietzschia schulleri Gothan, 1927.

The type species is known from petrified specimens from the Upper Devonian Wildenfels shale of Saxony. The axis is 2–3 cm in diameter and contains a peripheral ring of radially elongate steles consisting of primary xylem with a protoxylem pole at each extremity. The smaller central steles are essentially circular in transverse section and have a central protoxylem. *Pietzschia polyupsilon* Read and Campbell is known from well-preserved specimens of axes described in a preliminary account by Read and Campbell in 1939. These came from the New Albany Shale, of uppermost Devonian or basal Mississippian age. In a chart showing the localities from which petrifactions were obtained the authors list Junction City and Holy Cross, Kentucky, and New Albany, Indiana.

The axis (stem) is about 25 mm in diameter and includes a peripheral ring of about fifty-four radially aligned vascular strands. They are I-, U-, or Y-shaped resulting from frequent branching and anastomosing. Like the type species, each strand consists of primary xylem only with "several protoxylem groups" and are described as being "surrounded by a delicately walled parenchymatous tissue, unquestionably the phloem."

Only the initial stages of trace formation are known. Four strands are involved. A bundle is pinched off from each one and they fuse to form two ∩-shaped strands. The vascular strands are embedded in a predominantly parenchymatous ground tissue with some scattered sclerenchyma strands, the latter becoming more numerous toward the periphery.

Pietzschia is regarded by the authors as being clearly related to *Cladoxylon* but differs from it in the lack of secondary wood, the "reduction" (or absence?) of the peripheral loops, and the arrangement of the vascular strands into a more precisely defined ring.

Rhymokalon Scheckler, 1975.
Rhymokalon trichium Scheckler, 1975.

It is with some hesitation that we include this in the cladoxyl group but regardless of its correct affinities it presents exceptionally interesting stelar structures and adds to the known variety of plants in the Devonian.

It is known from three petrified specimens, consisting of a 'main' or first order axis and two orders of branching (Fig. 7.19), that were obtained from a quarry on Cave Mt. in East Ashland, Green County, New York State. The horizon is the Oneonta Formation of the lower Upper Devonian (Lower Frasnian).

The largest axis that is known, which will be referred to as the first order axis, is 3 cm in diameter. This includes a central xylary mass measuring about 10 × 18 mm in greatest diameter. This mass or 'vascular

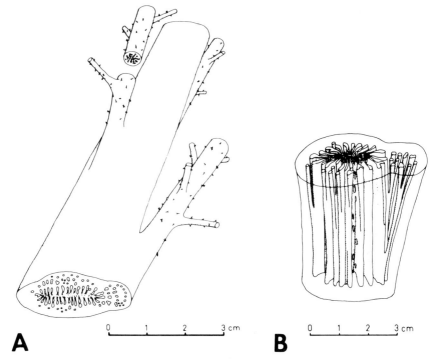

A   **B**

Figure 7.19. *Rhymokalon trichium.* A. Restoration of a part of the plant showing branching pattern. B. Restoration of vascular system of a main axis and primary branch. (From Scheckler 1975, Canadian Journ. Bot.)

cylinder' is composed of plates of xylem, apparently lacking protoxylem, separated by parenchyma (Fig. 7.20A–C). There is some differentiation between the central and peripheral parts and Scheckler says: "The center of the actinostele consists of roughly equal numbers of tracheids and parenchyma while the ribs consist solely of tracheids." A structural comparison is drawn in modern plants with the polyarch roots of some monocots and the fern *Marattia*.

Departing from the peripheral ridges of this central xylem 'core' are many (100–200) more distinct xylem strands which constitute the traces to the second order branches. These strands divide once or twice as they pass out into the base of the second order branches, increase in size, and develop a distinct central protoxylem (Fig. 7.20D). They continue to divide and anastomose in the branch. It is significant to note that "At no time . . . is the actinostele characteristic of branches of the first order reconstituted in those of the second." The protoxylem is described as consisting of tracheids with annular or helical wall thickenings and the

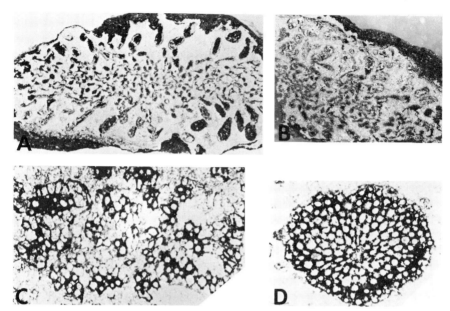

Figure 7.20. *Rhymokalon trichium.* A, B. Transverse section showing general organization of stelar structure. $\times$ 8 and $\times$ 5. C. Detail of inner vascular system. $\times$ 87. D. Detail of single outer vascular strand departing into branch base. $\times$ 100. (From Scheckler, 1975, Canadian Journ. Bot.)

metaxylem of isodiametric or elongate parenchyma and scalariform to circular pitted tracheids.

Scheckler states quite aptly: "Where to place *Rhymokalon* poses a problem". The general vascular anatomy of the first order axis and the development of the vasculature to its branch are distinct from that of any other Devonian plant, or indeed any other plant living or fossil. Its tentative assignment to the Cladoxylales is simply the closest that seems at all possible in view of what we know about it at present.

THE IRIDOPTERIDS AND *REIMANNIA*

The next group of plants we will consider are mostly Middle Devonian in age and are perhaps the least understood of all in terms of their affinities. They are problematical particularly because most are known only on the basis of vegetative anatomy of short lengths of axes with lateral branches. One genus, *Ibyka*, is also known from compressions and includes fertile organs.

Most of these plants were grouped by C. A. Arnold in the suborder Iridopteridinae of the Coenopteridales, since he believed they represented plants more advanced than early Devonian types and showing some similarities to the younger Coenopteridales—thus intermediate in the evolution of the ferns. Wm. Stein recently completed a restudy of those plants included in Arnold's suborder and has revised the concept of the group as follows. He established the order Iridopteridales, to include the genera *Iridopteris, Arachnoxylon, Asteropteris,* and *Ibyka. Reimannia,* originally included in the group, is regarded as a progymnosperm by Stein, but may better be considered as of uncertain affinities. We will return to a discussion of the Iridopteridales as a group at the end of this section.

Iridopteris Arnold, 1940.
Iridopteris eriensis (Arnold) Stein, 1982.

The genus *Iridopteris* was established by C. A. Arnold in 1940 based on a single permineralized axis fragment from the Middle Devonian Leicester Pyrite (Givetian), Erie County, New York State. The axis consisted of a preserved parenchymatous cortex and a five-lobed vascular strand with spacing of the lobes such that four are in opposite pairs and the fifth one is median (Fig. 7.21). Spaces located near the tips of each lobe were compared by Arnold to the parenchyma-filled peripheral loops in zygopterid ferns. Small traces departed from the lobes and one large trace was observed near the end of one lobe. Primary xylem maturation is mesarch.

Recently, Stein (1982a) reëxamined Arnold's specimen of *Iridopteris,* supplemented by a new collection from the early Middle Devonian Purcell Member of the Millboro Shale, Virginia. He added more data on the detailed histology of the plant; one aspect of particular interest is his clarification that the so-called peripheral loops are formed by the breakdown of protoxylem, not parenchyma. This lessens any possible similarity to the zygopterid ferns. Stein also presented details concerning the departure of small and large traces; both depart radially from lobes in nearly the same manner and are whorled with successive whorls being unevenly spaced along the major axis. Within a whorl, three minor traces will depart to the left of the parent rib, and two will depart to the right. Traces in successive whorls alternate. The large trace is elliptical with two protoxylem areas, small traces are terete with a central protoxylem area. The latter bifurcate farther out in the cortex. No evidence of trace structure outside the parent axis was obtained nor is it known what type of structure they supply.

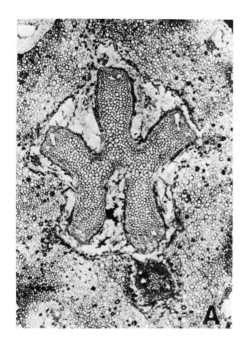
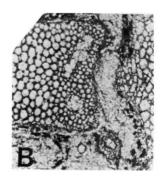

Figure 7.21. *Iridopteris eriensis.* A. Transverse view of the main axis showing two types of traces. × 15.5. B. Distal tip of xylem lobe showing protoxylem lacunae and small trace departing. × 47. (From Stein, 1982, Bot. Gazette.)

Arachnoxylon Read, 1938.
Arachnoxylon kopfii (Arnold) Read, 1938.

Arachnoxylon kopfii was first described as *Asteropteris kopfii* by C. A. Arnold in 1935a from a fragmentary plant axis in which only the multilobed primary xylem strand was well preserved. It was obtained from the late Givetian Windom Shale, of the Moscow Formation, in Erie County, New York State. Read in 1938 renamed the fragment *Arachnoxylon*, adding data from another axis obtained from the Leicester Pyrite in Ontario County, New York State, of the same age. Read considered the plant axis to exhibit characters intermediate between the ferns and psilophytalean ancestors (i.e. plants we today refer to the trimerophytes).

Stein (1981) reëxamined and redescribed the previously studied specimens of *Arachnoxylon*, plus a more recently collected specimen, adding considerably to our knowledge concerning it.

The largest specimen is about 3 cm long and 0.6–1.4 cm wide, consisting of a major axis and lateral branches (Fig. 7.22). The cortex is

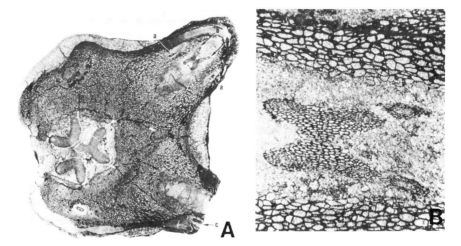

Figure 7.22. *Arachnoxylon kopfii* (Arnold) Read. A. Transverse section showing xylem strand of main axis and parts of two departing branches. × 4.2. B. Lateral branch trace with strands departing to next branch order. × 25. (From Stein 1981, Palaeontographica.)

composed of thick-walled parenchyma cells which are more or less isodiametric in cross section view and somewhat elongate in longitudinal view. The xylem strand consists of primary tissues only, and is six-to-seven lobed; exhibiting an arrangement of three symmetrically disposed lobes departing from a central region, each of which bifurcates or trifurcates. Primary xylem maturation is mesarch. Some poorly preserved, presumed phloem cells, surround the xylem strands.

Two types of lateral traces, termed by Stein major and minor traces, depart in a nearly whorled fashion, with successive whorls distantly spaced. The major trace departs from the tip of a xylem arm and initially appears elliptical in shape: as it traverses the cortex it becomes somewhat hourglass shaped (or clepsydroid) with a pair of protoxylem areas at each end (Fig. 7.22). At, or just before, its departure from the cortex, two lateral traces depart from the ends of the major trace; Stein calls these subsidiary traces and they appear to supply pairs of appendages on the lateral axis. Two subsequent subsidiary traces are produced, this time abaxially from the lateral axes, shortly after the first. The subsidiary traces are terete, with a centrally located protoxylem area.

The minor traces depart from the ends of xylem arms also, but are initially terete with a centrally located protoxylem. These traces divide once, and then are not found in subsequent levels.

Stein suggested that these traces could be interpreted as follows: major traces supply branches, and minor and subsidiary traces supply leaves or structures equivalent to leaves. It is unclear if these remains represent an upright portion of a plant or may represent a rhizome with branches departing from one side of the rhizome at certain intervals. As with most of the other plants included in the Iridopteridales by Stein, much remains to be learned about *Arachnoxylon* including the nature of its reproductive structure and overall branching pattern. It exhibits many anatomical features similar to *Iridopteris* in particular and it seems reasonable to ally it with that genus.

Ibyka Skog and Banks, 1973.
Ibyka amphikoma Skog and Banks, 1973.

This genus is based on compressions, impressions, and segments of permineralized axes from the late Middle Devonian Moscow Formation (Givetian) of Schoharie County, New York State and was interpreted by Skog and Banks as a possible sphenopsid precursor. It shares characters of anatomy with *Iridopteris* and *Arachnoxylon* and on that basis has been included in the order Iridopteridales by Stein (1982a). Until further evidence is obtained concerning the evolution of the horsetails and their precursors, both possibilities remain open, as do others.

Ibyka consists of axes on which are borne spirally arranged lateral branches, these in turn bearing spirally arranged ultimate appendages (Fig. 7.23). The latter dichotomize several times and terminate in recurved tips or in small presumed sporangia (no spores have been obtained). Some structures identical to ultimate appendages also occur on the main axes.

The anatomical features of *Ibyka* include the following: a lobed protostele consisting of three main arms, each of which bifurcate to produce a total of six lobes (Fig. 7.24); protoxylem areas located near the tips of each arm which may break down forming lacunae; poorly preserved phloem in which are found some thin-walled cells and some tanniniferous cells; and a cortex composed of parenchyma, sclereids, and tanniniferous cells. No secondary tissues are produced. Trace production to lateral branches is unclear but it appears that the end of a xylem arm enlarges, separates, and a five-to-six lobed trace re-forms as it traverses the cortex. Traces to the ultimate branches are terete; the protoxylem lacunae elongate radially, split to form two regions, the trace separates and both trace and xylem arm retain one lacuna.

Asteropteris Dawson, 1881.
Asteropteris noveboracensis Dawson, 1881.

This genus was established by Dawson in 1881 for permineralized axes 2.5 cm in diameter which were obtained from the Upper Devonian Portage Group at Milo, New York State. He found the vascular tissue to

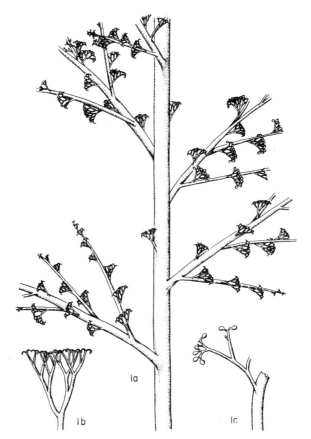

Figure 7.23. *Ibyka amphikoma.* 1a. General restoration of a portion of the plant. 1b,c. Details of the sterile and fertile ultimate appendages. (From Skog and Banks, 1973, Amer. Journ. Bot.)

consist of radiating plates of sclariform tracheids surrounded by parenchyma. Lateral traces were produced in close succession and resembled those of the Zygopteridaceae.

Paul Bertrand in 1913 presented more information on *Asteropteris*, describing the vascular strand as an actinostele with approximately four major arms which each bifurcate up to three times (Fig. 7.25). Individual segments of these arms are of unequal length, and many exhibit a peripheral loop. The scalariform tracheids are 12–25 μm in diameter. It was not possible to determine if the peripheral loops were represented in this case by parenchyma or are protoxylem lacunae.

Several arms had produced lateral traces which are clepsydroid in shape but with two peripheral loops of unequal size at each end. The

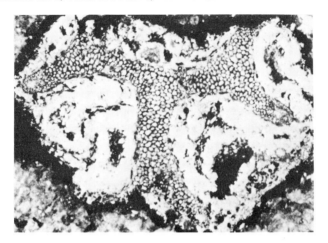

Figure 7.24. *Ibyka amphikoma.* Transverse section showing vascular structure of the main axis. × 24. (From Skog and Banks, 1973, Amer. Journ. Bot.)

cortex where present just outside the xylem consists of thick-walled, possibly secretory cells.

The nature of the lateral traces is closely comparable to that of many plants included in the Zygopteridaceae and because of that similarity, many workers have placed *Asteropteris* in that group (Andrews and Boureau, 1970; Barnard and Long, 1975). Much more information is needed to better assess the affinities of this plant.

Reimannia Arnold, 1935.
Reimannia aldenense (Arnold) Stein, 1982.

Reimannia aldenense was established by Arnold (1935a) for per-mineralized axes from the Middle Devonian Ledyard Shale Member of

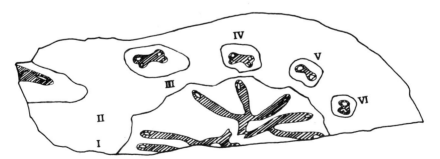

Figure 7.25. *Asteropteris noveboracensis.* Transverse view of portion of major axis with vascular strands departing to branches. (From Bertrand 1913.)

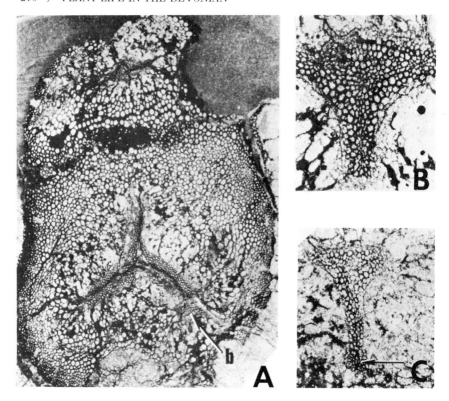

Figure 7.26. *Reimannia aldenense.* A. Transverse section of major axis with a departing branch at top of photo. × 9.4. B. Vascular strand in lateral branch. × 38. C. Same, with trace departing at bottom. × 31.6. (From Stein, 1982b, Palaeontology.)

the Ludlowville Formation in Erie County, New York State. The specimens consisted of axes with a three-angled xylem strand composed of primary tissues only. Protoxylem strands occurred near the tips of each angle, often having disintegrated and represented by a lacuna. Arnold termed these peripheral loops.

Little additional information was obtained concerning *Reimannia* until Stein's re-examination of Arnold's specimens in which he (1982b) described the anatomy of three levels of branching for *Reimannia*. The first order axis consists of a three-lobed vascular strand; the number of protoxylem areas is unknown although Stein suggested several were present (Fig. 7.26A). Primary xylem maturation is mesarch, and metaxylem tracheids possess elliptical pits. Second order axes are initiated as a single strand which is diamond-shaped or four-angled; after departing the parent axis it gradually becomes three-angled with protoxylem strands near the tips of angles and one in the middle (Fig. 7.26B). This axis produces a subopposite pair of third-order laterals, each with a small terete stele. These bifurcate shortly after their departure from the axis. Another third-order lateral is produced by elongation of the remaining portion of the second order stele (Fig. 7.26C) and departure of a trace. The pattern of branching in second and third orders is reminiscent of that of the presumed progymnosperm *Proteokalon*, although the latter also produces secondary tissues.

Stein allied *Reimannia* with the progymnosperms despite its apparent lack of secondary tissues, mainly on the basis of its apparent predictable pattern of branching and angled vascular strand. While the material present might represent distal parts of a plant with secondary growth, no evidence to support that possibility is presently at hand and we feel it equally possible that *Reimannia* might be allied with an as yet unrecognized group of plants derived from the trimerophytes.

Returning to the concept of the Iridopteridales as established by Stein (1982a), all the genera included share these features: a vascular strand consisting of deeply lobed primary xylem; the lobes are undivided or bifurcate or trifurcate; one to several protoxylem strands located at the tips of lobes, often disintegrating leaving lacunae; traces depart radially in whorls. In two genera, *Arachnoxylon* and *Iridopteris*, two types of traces are formed. Primary xylem maturation is mesarch, wall patterns on tracheids are similar and those aspects of cortical anatomy presently known are similar.

Stein considered these plants as more evolved than trimerophytes, but less evolved than ferns and probably not related to them. He also discounts possible protoarticulate affinities. It seems very likely that several pteridophytic plant groups existed in Early-Middle Devonian times which had little or nothing to do with any presently existing ones. Especially crucial in assessing the evolutionary relationships of the plants included in the Iridopteridales is learning more about their reproductive structures. Also of interest would be more complete knowledge concerning the nature of lateral appendages in these plants, especially in view of the two types of lateral trace production.

SUMMARY

The various plants dealt with in this chapter are among the least understood Devonian plants in terms of their overall morphology and affinities. While more complex than trimerophytes, they lack distinct leaves and annulate sporangia which characterize filicalean ferns. At present, our best assessment is that they may represent several extinct vascular plant lineages, probably derived from a trimerophyte ancestor, with possibly only *some* of them being among the lineages leading to one or more of the major groups of ferns (Ophioglossales, Marattiales, Filicales). It is also possible some of these plants may represent precursors to the horsetails.

These plants offer evidence, as do progymnosperms, on the possible early stages in the differentiation of branching systems (stems) leading to megaphyllous leaves, by exhibiting a change in the shape of their vascular strand at some level of branching. Later stages, involving planation and lamination of some portion of the branching systems, are exhibited by a variety of plants, including both progymnosperms and the early seed plants discussed in the next chapters. Galtier (1981) offers evidence on laminate leaves in Lower Carboniferous filicalean ferns and possible pteridosperms, and surveys the occurrence of these structures and the possible ways they evolved. He notes that laminate leaves occur earlier in progymnosperms (Upper Devonian) than do fronds with laminate pinnules in filicalean ferns (Lower Carboniferous).

It also is apparent that plants of probably very different affinities exhibited vascular strands with peripheral loops, that these are sometimes regions of parenchyma and sometimes areas where protoxylem cells originally occurred and were destroyed and that they may be present at some levels in a plant and absent at others.

The polystelic organization of vascular tissue characteristic of the plants included in the Cladoxylales is quite variable and as stated by Fairon-Demaret (1979): "The systematic weight accorded to the 'polystelar complex' of the two Middle Devonian genera in which external morphology is known [and to other genera as well—authors] probably has been overestimated." (p. 157) (as indicated translation).

One of the greatest challenges in future Devonian plant studies is the further elucidation of the overall morphology and systematic relationships of these enigmatic but very intriguing plants.

8 / The Progymnosperms

One of the most significant and fascinating of recent contributions to our knowledge of evolution in the plant world has been the recognition of a group of plants which to some degree bridge the gap between pteridophytic and seed-bearing plants. It is based on Charles Beck's (1960) demonstration that certain fossils of fern-like foliage known under the generic name *Archaeopteris* were borne on petrified stems of gymnospermous anatomy known under the generic name *Callixylon*.

A great deal of study by many people has been devoted to the plants that are now included in this group, both before and after Beck's observations were reported. Consequently some aspects of the historical development involved seem appropriate to a clear understanding of the present status of our knowledge.

When he announced this discovery Beck gave the following synopsis of the new group of plants as he envisaged it at the time:

Progymnospermopsida Beck, 1960.

Free sporing plants with a pteridophytic type of reproduction and foliage combined with gymnospermous stem anatomy showing secondary xylem:

Aneurophytales: *Aneurophyton, Eospermatopteris, Tetraxylopteris*
Protopityales: *Protopitys*

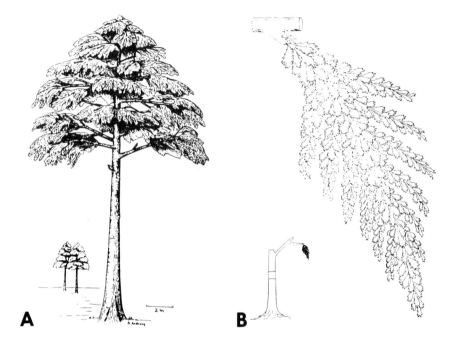

A   **B**

Figure 8.1. *Archaeopteris.* A. Generalized habit reconstruction. B. Restoration of the lateral branch system of *Archaeopteris macilenta.* (A. From Beck, 1962, B, 1971, Amer. Journ. Bot.)

Pityales: *Pitys, Callixylon, Archaeopteris*

Since 1960 a considerable number of generic entities have been assigned to the Progymnospermopsida. Some of these, as will be evident, are based on very fragmentary fossils that represent but a small portion of the whole plant. Our emphasis will be on the better known members of the group, that is, those that satisfy the basic features: pteridophytic reproduction and gymnospermous anatomy.

THE CONCEPT OF PROGYMNOSPERMS AND *ARCHAEOPTERIS* DAWSON, 1871.

By far the best known progymnosperm 'plant assemblage' includes those fossils described under *Archaeopteris* and *Callixylon*. As an introduction: *Archaeopteris* is based on fossils that superficially suggest large fern fronds that attained a maximum length of about one meter, and at least forty species have been described. *Callixylon* is known from petrified stems and trunks and most of the specimens that have been found are less than a half meter in diameter; however some have been reported in excess of 1.5 meters in diameter: they have abundant secondary wood of a gymnospermous (coniferous) type in which the radial walls of the tracheids are characterized by grouped bordered pits (Fig. 8.10).

In his account of 1962 ("Reconstructions of *Archaeopteris*, etc. . . . ") Beck included a reconstruction drawing of an *Archaeopteris-Callixylon* tree, which we have reproduced here (Fig. 8.1A). In the case of such a rather large forest tree, some of which may have attained a height of 100 feet, it is not possible to show significant detail but this will serve as a convenient starting point for a more detailed treatment. Hereafter when referring to the plant as a whole we will simply use the name *Archaeopteris*.

In addition to the known attachment of an *Archaeopteris* 'frond' to a *Callixylon* stem as reported by Beck in 1960 it may be noted that both have been found (frequently associated) in numerous Upper Devonian localities including Ireland, USSR, Bear Island, northern USA, China, Ellesmere Land and other Canadian localities.

We will deal with several of the better known species of *Archaeopteris* but at this point a brief summary of the history of the type species *Archaeopteris hibernica* (Forbes) Dawson seems appropriate. So far as we are aware, the first reference to the plant was under the name *Cyclopteris hibernica* Ed. Forbes in the Proceedings of the British Association for 1852. In his *Traité de Paléontologie Végétale,* Vol. I, 1869, Schimper transferred it to *Paleopteris hibernica* (Ed. Forbes) and included some classic illustrations in his plate 36. The final designation, *Archaeopteris hibernica* (Forbes) Dawson was established by Dawson in 1871.

Archaeopteris macilenta (Lesq.) Carluccio et al., 1966.

Fossils attributed to this species have been the source of several detailed studies, three of which are especially informative: Compression fossils reported by Phillips, et al. 1972, from the Hampshire Formation near Valley Head, West Virginia; similar fossils reported by Beck, 1971, from the Katsberg Formation near Cannonsville, Delaware County, New York; and petrified specimens described by Carluccio, et al. 1966, from a lower Frasnian horizon of Schoharie County, New York. All of these are of Upper Devonian age.

The portion of the plant shown in Figs. 8.1B and 8.2 was referred to in the literature prior to about 1966 as a compound frond, consisting of a rachis (main axis), two rows of pinnae (primary branches), which in turn bore the pinnules (leaves); the pinnules were also borne along the main axis.

The studies of Carluccio, et al. 1966, and Beck 1971 noted above, based on petrified pieces of the 'frond', showed that the rachis and pinna axes actually contained a eustele (Figs. 8.2C & 8.3), conforming to typical stem anatomy. In the rachis or main axis, two kinds of vascular strands depart from the lobes, all in the same ontogenetic spiral: small ones which

Figure 8.2. *Archaeopteris macilenta.* A. Portion of lateral branch including central axis bearing leaves and fertile primary branches. × 2.6. B. Detail showing fimbriate nature of the leaves. × 2.3. C. Transverse view of the central axis of a lateral branch showing the radial symmetry of the xylem; the latter is partially distorted by crushing during fossilization. × 22.3. (A,B from Phillips et al. 1972 Palaeontographica. B. from Carluccio et al. 1966, Amer. Journ. Bot.)

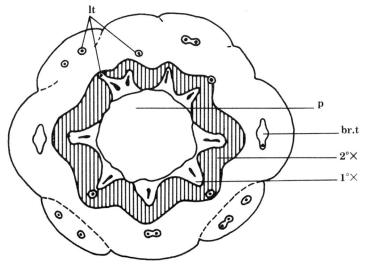

Figure 8.3. *Archaeopteris macilenta.* Drawing of the vascular pattern of the central axis of a lateral branch. lt = leaf trace; br.t. = branch trace; p = pith, $1°\times$ = primary xylem; $2°\times$ = secondary xylem.

supply the leaves borne along the main axis, and somewhat larger ones that become four-lobed and supply the primary branches. The latter depart regularly from the same two lobes thus accounting for the two-rowed arrangement of the primary branches. There is also some secondary xylem present in the main axis and the radial walls of the tracheids show the characteristic *Callixylon* type of grouped pits. Therefore, most paleobotanists since that time have regarded the 'frond' as a modified branch system in which the 'pinnules' (borne on both the main axis and the primary branches) are actually the leaves. We accept this terminology but we will comment on it again below.

The primary branches bear the leaves in a very dense spiral and they may be either sterile (vegetative) or fertile (bearing sporangia). Fig. 8.2A & 8.4 show a portion of the main axis and the proximal parts of two primary branches; the latter bear essentially flabelliform sterile leaves at their base, becoming partially to wholly fertile distally. The wholly fertile leaves are non-laminate, or only partially so, dichotomize one to several times, and bear numerous sporangia. The sporangia are fusiform, 1.5 to 3.4 mm long, dehisced longitudinally and contained either very large numbers of microspores that range from 33 to 68 μm in diameter or fewer megaspores that are up to 420 μm in diameter.

We will introduce other species of *Archaeopteris* more briefly but it will be helpful at this point to comment further on the 'frond' or modified branch system as a whole. This is a unique structure, a good deal is known about it but the story is not complete. Different species of *Archaeopteris*

Figure 8.4. Restoration of part of branch system of *Archaeopteris macilenta.* (From Phillips et al. 1972, Palaeontographica.)

have been identified chiefly on the gross morphology of the vegetative leaf, which differs strikingly in different species from nearly entire to very deeply dissected. There is also some variation within a single species. Also the various ways in which a branch system may be preserved will reveal seemingly different ways in which the leaves are borne, either on the main axis or the primary branch. Beck (1971) has discussed and illustrated this in some detail. He has also found that there may be one or two more rows of leaves on one side of the main axis than on the other side.

It is especially important to add that the modified branch may vary within a species relative to its apparent function. Some are entirely sterile and apparently photosynthetic, others may be entirely fertile (Fig. 8.5) while others, as described above, perform both functions. If nothing else is brought out in this book, it will certainly be made clear that Devonian plants have contributed much to our knowledge of the origin and evolution of leaves and have shown that to be dogmatic about a definition of this morphological entity is precarious.

Figure 8.5. A nearly complete fertile branch of *Archaeopteris fimbriata* from Bear Island. Approx. 0.3. (From Nathorst 1902, Kgl. Svensk Vet. Handl.)

Archaeopteris halliana (Goeppert) Dawson, 1871.

This species was found mixed with *A. macilenta* at the West Virginia locality described by Phillips, et al. 1972; we introduce it in order to draw some comparisons between species with reference to sterile leaves, sporangia, and spores.

The sterile leaves of the two species are quite distinct. Those of *A. macilenta* tend to be strongly dissected with the divisions ending in delicate, almost hairlike, tips, while those of *A. halliana* are broadly fan-

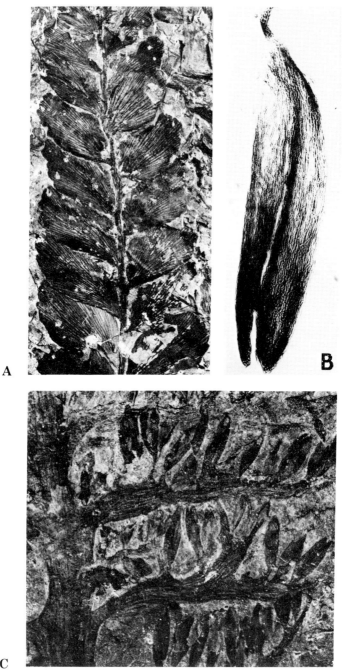

Figure 8.6. *Archaeopteris halliana.* A. Primary branch showing the nearly entire leaves. × 1.5. B. Sporangium of *Archaeopteris* sp. showing longitudinal dehiscence. × 30. C. Primary branch bearing abundantly fertile leaves with sporangia attached on their upper surface. × 0.8. (From Phillips et al. 1972, Palaeontographica.)

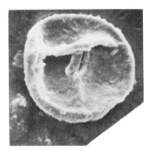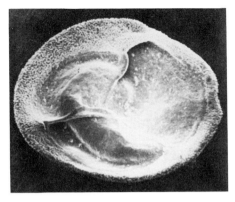

Figure 8.7. Micro- and megaspore (left and right) of *Archaeopteris*. × 206 and × 1045. (B. From Phillips et al. 1972, Palaeontographica.)

shaped, and nearly entire or slightly serrate (Figs. 8.6, 8.8). In both species the sporangia are borne in large numbers on the fertile leaves. Some concept of the abundance, mode of attachment, and some variation in size of the sporangia may be noted in Fig. 8.6C which shows a portion of a primary branch with several fertile leaves of *A. halliana*. Both species are heterosporous but the striking difference between the two in sterile leaf morphology is not matched with similar differences in the megasporangia and megaspores (Fig. 8.7). Some measurements for the two seem worth recording in this context:

	Diameter and length of sporangia in μm	Diameter of megaspores in μm
A. macilenta	320 × 1570	150–420
	450 × 1310	
	575 × 1470	
A. halliana	450 × 2240	170–460
	545 × 2400	
	640 × 1930	

It will be noted that the magasporangium length in *A. halliana* is somewhat greater than that for *A. macilenta*; there is rather little difference in the size of the megaspores or in their wall ornamentation. The microspores generally vary from 33–64 μm in diameter with no significant difference between the two species. Hopefully future investigations will shed more light on the correlation of differences between presumed species.

Figure 8.8. Restoration of *Archaeopteris halliana* branch system. (From Phillips et al. 1972, Palaeontographica.)

Callixylon Zalessky, 1911.

In order to complete this general picture of an *Archaeopteris* plant we will bring in a description and supplementary notes here on the petrified stems (Fig. 8.9) usually called *Callixylon*. The genus was founded by Zalessky on fossils from the Upper Devonian of the Donetz Basin in south Russia. Chester Arnold in his doctoral thesis published in 1930 gave a detailed and informative summary of the several species that had been described at that time. He also prepared the more recent account (1970) that is contained in Volume IV of the *Traité de Paléobotanique*. Several other paleontologists in this country have also had a hand in the earlier history of collecting and studying *Callixylon* including J. S. Newberry, J. W. Dawson, D. P. Penhallow, G. R. Wieland and Loren Petry.

The chief distinguishing features of the petrified stems are the ring of numerous mesarch primary xylem strands that are present around the periphery of the pith and in contact with the secondary wood, and the grouped arrangement of the bordered pits in the radial walls of the tracheids composing the secondary wood (Fig. 8.10). The rays of the

Figure 8.9. *Callixylon.* A partially reconstructed stump on the campus of East Oklahoma State College. (From Beck, 1964, Torrey Bot. Club Mem.)

secondary wood are usually uniseriate but occasionally biseriate. As to growth rings Arnold stated in 1934:

"Vague growth rings appear to be present in all forms of *Callixylon*, and they can be seen in almost any specimen provided the block is large enough to include the very widely spaced ones."

In an account of 1931 describing *C. Newberryi* (Dawson) Elkins and Wieland, Arnold gave a detailed account of a log in which he states:

"The Indiana specimens are quite variable in size. The largest specimen encountered . . . is nine feet long, three feet broad at the base, and tapers to about eighteen inches at the top." (p. 210). And: "In the exhibition hall of the United States National Museum there is a similar trunk from the New Albany shale of Lebanon, Kentucky, which was identified as *C. Newberryi* by its surface features. It is twenty feet high and tapers from about eighteen inches at the base to about four inches at the top." (p. 227)

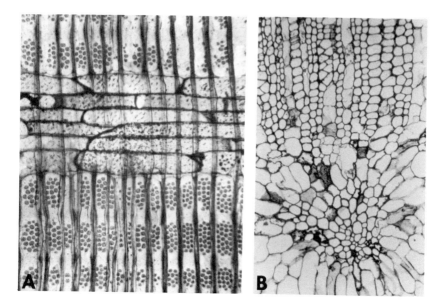

Figure 8.10. *Callixylon newberryi.* A. Longitudinal section of stem showing characteristic grouping of bordered pits in radial walls of secondary tracheids, rays. × 121. B. Transverse section of stem showing a primary bundle and its relation to the secondary wood. × 51. (From Beck, 1970, Biol. Rev.)

In 1934 Arnold described *Callixylon whiteanum* from the Woodford chert of Oklahoma (probably lower Mississippian) and although his report was actually based on studies of log fragments he stated that a stump specimen had been found that was 1.5 meters in diameter. Whatever the maximum diameter and height of the main axis may have been, it is evident that this was a forest tree of respectable magnitude.

Just what constitutes the generic limits of both *Archaeopteris* and *Callixylon* is still not clear; this in no way detracts from their interest and we will mention certain facets of the problem as we continue. Accordingly, at this point we will mention briefly another species of *Callixylon* in order to add to the known variation in the several species that have been described.

Callixylon arnoldii Beck, 1962.

The specimens on which this species was based were collected from the Falling Run member of the Sanderson Formation which is regarded as being of early Mississippian age; it thus extends the range into the base of the Carboniferous. The secondary wood is distinctive in having predominantly uniseriate bordered pits that are circular and more widely separated than in other species. Beck notes the resemblance of the pits to those of the abietinean conifers, although he does not infer a close relationship.

In view of its general importance, we will include brief descriptions of a few more of the better known *Archaeopteris* species (the most recent taxonomic treatment we know of being by Kräusel and Weyland in 1941).

Archaeopteris latifolia Arnold, 1939.

This is of special historical importance as it is the first case in which heterospory was reported. Arnold recorded the species from several localities in the Upper Devonian of Pennsylvania and New York State. He regarded it as being closely related to *A. hibernica*, the type species on which Dawson's 1871 account was based. The leaves are 1 to 2 cm long, rounded or broadly obovate with a serrate margin. Specimens were found bearing sporangia and large numbers of them were scattered through the rock matrix. The attached sporangia had lost their spores but they were found intact in many of the dispersed ones. The sporangia are of two kinds: One which averages 2 mm long and 0.3 mm in diameter, and contains numerous microspores 35 μm in diameter. Arnold stated that the number per sporangium was "probably a hundred or more", but his photographs suggest that 400–500 would be a more likely number. The megasporangia are about the same length but average 0.5 mm in diameter and contain about eight to sixteen spores that are about 300 μm in diameter.

Although the attached sporangia had shed their spores, and the above observations are based on the isolated ones, the two are identical in form and no other plant was found associated with *Archaeopteris*. Thus, with the evidence that we now have from other localities and other species, there can be little doubt that *A. latifolia* was heterosporous.

Archaeopteris jacksoni Dawson, 1871.

In 1965 Pettitt described some fertile specimens which he attributed tentatively to this species. They are clearly referable to *Archaeopteris* but the species is somewhat in doubt. These had been collected by W. Graham-Smith from the Upper Devonian deposits of Scaumenac Bay, Quebec, in 1937, and are deposited in the British Museum of Natural History. The megasporangia are 1.2 to 2.6 mm long and 0.3 to 0.5 mm in diameter and contain fifteen to twenty-five megaspores that range from 110–370 μm in diameter; the microsporangia are slightly longer (1.7 to 2.8 mm) and contain several hundred microspores 45 to 70 μm in diameter.

Archaeopteris from Ellesmere Island

In 1904 Nathorst briefly described *Archaeopteris* specimens that had been collected on the Sverdrup Expedition of 1898–1902 to Ellesmere Island. In the summer of 1962 one of us visited the area that Nathorst's fossils came from and obtained additional material (Andrews, Phillips, Radforth, 1965). Two species were found in the Upper Devonian rocks near the head of Goose Fjord in southwestern Ellesmere. *A. obtusa* Lesquereux is characterized by broad, wedge-shaped, nearly entire leaves which may be up to 7 cm long and *A. fissilis* Schmalhausen by sterile leaves that are best described as filamentous and dichotomize one to three times.

The primary branches of *A. fissilis* bear a mixture of sterile and fertile leaves. The sporangia are 1.6 to 2.0 mm long, dehisce longitudinally and in a few of them spores were found of about 60 μm in diameter. Some larger spores were found in the rock matrix which may be the megaspores but this can be noted only as a possibility.

Two additional items of information may be worth recording. At one of the localities, where we found *A. obtusa*, a well-preserved piece of a petrified root was found that displayed the characteristic pitting of *Calllixylon*. And nearby we found stem impressions varying from 5–20 centimeters broad which, since no other fossils were found in the general area, it is reasonable to suppose were the *Archaeopteris* trunks. These were exposed in great abundance on ledges only a few yards from the ice cap to the east at the head of Goose Fjord. One could readily envisage an *Archaeopteris* forest of trees like the one shown in Beck's restoration (Fig. 8.1), in striking contrast to the present-day high Arctic landscape, with no plants more than a few inches tall.

Archaeopteris is at present the key plant to our understanding of the Progymnospermopsida. It was certainly a dominant element of the Upper Devonian floras of what is now the north-temperate-to-Arctic areas. It may have played an important role in the origins of later groups such as the Cordaites, the Lebachian conifers, and the Noeggerathiopsida, as suggested by Charles Beck (1981). As a generic concept, a considerable amount of information has accumulated but there is much that remains to be known. We would especially like to know more about reproduction in other species. Were some of them homosporous? And did any of them achieve the seed-plant level? We would emphasize that the identity of a species in the present state of our knowledge is not clear. It is worth noting the opinion of a leading authority on the group:

"It is possible that "*Archaeopteris*" represents a large plexus of forms which in time will be separated into several distinct taxa. Certainly the large variation in the genus of webbed and unwebbed leaves, leaves which are spiral or opposite and decussate or even planated on ultimate branches, homospory to heterospory and emerging differences in anatomical structure would indicate that more than one genus is encompassed." (Bonamo, 1975. p. 574–5)

Archaeopteris has had an interesting and somewhat checkered career especially in its earlier days. William Carruthers and W. C. Williamson, in the 19th century, suggested that it might belong in the fern family Hymenophyllaceae; A. G. Nathorst suggested an assignment to the Marattiaceae, and Newell Arber was frankly puzzled about its natural affinities and once wrote to D. H. Scott, saying, "The man who gets out the real nature of *Archaeopteris* deserves the VC [Victoria Cross]" (Andrews, 1980. p. 129–130).

OTHER PROGYMNOSPERMS

When a new and distinctive group of fossil plants is recognized and proves to be of considerable general interest there is a tendency on the part of paleobotanists to use it as a reception center for fossils that are difficult to classify otherwise and seem to offer some features that justify this action. This has been notable in the case of the 'psilophytes' after the discovery of the Rhynie plants and seems to have been the case in the early years following the identification of the pteridosperms in 1904. The progymnosperms now present another example.

Since 1960 a considerable number of generic entities have been assigned to the Progymnospermopsida; a few of these reveal a significant portion of the plant that is represented, others are very fragmentary and their assignment to the group is questionable, and still others probably are parts of plants previously reported under a different generic name.

Two very useful analyses of the status of the group have been drawn upon in preparing the present account: "The Progymnospermopsida: Building a Concept" (1975) by Patricia Bonamo, and "Current Status of the Progymnospermopsida" (1976) by Charles Beck. Beck deals with the numerous cases where two or more generic names probably refer to different parts of one and the same plant; this is a highly condensed source of information and includes an extensive bibliography. Bonamo supplies a revised classification which recognizes only a very few genera as being positively referable to the group.

As a working basis for what follows in this chapter we include here a table (Table 8.1) from Bonamo's study which we will follow, with certain modifications. At the time this was published some twenty-seven genera

TABLE 8.1. Progymnospermopsida (*sensu stricto*)*

Incertae sedis (anatomy)		Incertae sedis (morphology)
	Aneurophytales	
Eospermatopteris	*Aneurophyton*	
Sphenoxylon	*Tetraxylopteris*	
Paleopitys	*Rellimia*	
Triradioxylon	*Triloboxylon*	
Proteokalon	Protopityales	
Cairoa	*Protopitys*	
	Archaeopteridales	
Callixylon	*Archaeopteris*	*Svalbardia*
Siderella		*Eddya*
Actinopodium		
Actinoxylon		

*Modified From Bonamo 1975.

had been assigned to the progymnosperms, the identifying characters of which are: Free sporing plants with a pteridophytic type of foliage and reproduction, and a woody habit with gymnospermous stem anatomy including cambial activity and secondary wood.

Only six of the genera fully conform to these requirements:

> *Archaeopteris*
> *Protopitys*
> *Aneurophyton*
> *Tetraxylopteris*
> *Rellimia*
> *Triloboxylon*

Many of the presumed progymnosperm genera, as presently recorded, represent fragmentary, and apparently terminal portions of plants that attained a considerable size. It is also very likely that some of the reports deal with fossils that will ultimately prove to be congeneric with previously described ones.

Archaeopteridales, continued

The generic entities that follow are included tentatively in this order. The available evidence indicates that this is a correct assignment but all of the distinctive features of the progymnosperms are not present.

Svalbardia Høeg, 1942.
Svalbardia polymorpha Høeg, 1942.

This genus was established by Ove Arbo Høeg on the basis of specimens he collected from a locality known as Planteryggen (Plant Ridge) in the Mimer Valley of Spitsbergen (upper Middle Devonian or lowermost Upper Devonian). Devonian plants were first discovered in Spitsbergen in 1882 by Nathorst, and reported by him in 1884. More recently Høeg has participated in expeditions (in 1924, 1928, and 1939) to the area (Fig. 8.11) and his classic account *The Downtonian and Devonian Flora of Spitsbergen* (1942) draws upon previous collecting and studies, gives a good historical account of previous activity, and contains descriptions of many new fossil plants. One of us (HNA) had an opportunity several years ago to examine some of the Spitsbergen collections in Oslo. As Høeg notes, the preservation in many cases is not especially good and it is greatly to his credit that he was able to extract as much information as he did.

There is considerable discussion in the literature relative to *Svalbardia's* status as a valid genus—centering chiefly on whether or not it should be included in *Archaeopteris*. The specimens of *S. polymorpha* described by Høeg consist of a main axis that attained a length of at least

Figure 8.11. Professor Ove Arbo Høeg at work on one of his several fossil plant hunting expeditions in Spitsbergen.

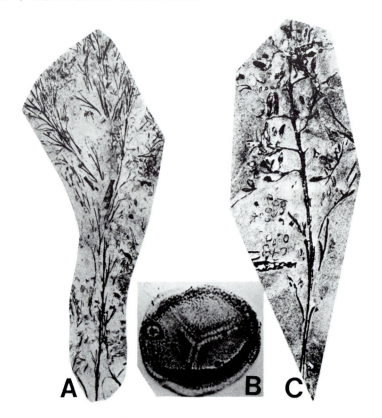

Figure 8.12. *Svalbardia polymorpha.* A. Part of a vegetative branch showing filiform leaves. × 0.7. B. A spore from *Svalbardia.* × 446. C. A fertile branch showing sporangia. × 2.6. (A,C from Høeg, 1942, Norges Svalbard-og Ishavs-Under. B, from Vigran, 1964, Norsk Polarinst. Skr.)

one meter, and bore primary branches whose arrangement is not clear; these in turn bore "pinnule-like organs about 2.5 cm long" which are several times divided to the point of being filiform (Fig. 8.12A). The ultimate fertile branchlets (leaves) are also divided and bear up to a dozen sporangia in their middle region (Fig. 8.12C); the sporangia are "generally erect, pear-shaped, or cylindrical with rounded tops, about 1.5–2 mm long and 0.5–0.7 mm in diameter". (p. 193). Although spores were not clearly identified in the sporangia, Høeg reported that he found in his macerations a predominance of nearly smooth-walled, oblong spores 60–70 µm long. Vigran (1964) described these spores as *Lycospora svalbardiae*, and noted that they also were obtained from sporangia of *Svalbardia*. They are trilete, cingulate(?), 65–83 µm in diameter, and all but the contact areas are covered with small coni (Fig. 8.12B). Other workers have suggested they resemble more closely the dispersed spore genus *Geminospora*. Høeg has the following to say about the plant:

"Together with *Enigmophyton* and some lepidophytes this plant [*S. polymorpha*] is the dominant component of the flora of Planteryggen. The various plant organs, stems, foliar parts, and sporangia, which are here regarded as belonging to one and the same species, were found in great abundance on slabs of a yellowish shaly sandstone. The impressions are mostly indistinct when dry, but when moistened, preferably with xylol, they often show up excellently against the matrix, the remains of organic matter making the outlines appear as distinct black lines." (p. 70)

He also offered a very significant comment on the probable relationship between *Svalbardia* and *Archaeopteris*:

"As to *Archaeopteris*, there is certainly a very great difference between our plant and most of the species of that genus. But *A. fissilis* (Schmalh.) em. Nath., from the Upper Devonian of Donetz (Schmalhausen 1894) and of Ellesmere Land (Nathorst 1904), forms a connecting link, and in reality some of our specimens bear such a resemblance to this species that they scarcely can be distinguished from it, at least if one has to judge by the illustrations published . . . the resemblance also goes very far in the matter of the fructifications, which in their general morphology are identical with those of all species of *Archaeopteris*." (p. 80)

Several other species of *Svalbardia* have been described which we will comment on briefly. The most recent that we are aware of is *S. banksii* Matten (1981) which comes from the Delaware River Flags, Oneonta Formation (Frasnian) of Sullivan County, New York. The unwebbed leaves are borne spirally on the ultimate branches, are up to 3.2 cm long and they dichotomize in more than one plane two to four times. Matten draws a comparison with *Actinoxylon* which is of interest in possibly correlating compression and petrified fossils. He notes:

"Similarities between *Actinoxylon* and *Svalbardia* include: helically arranged ultimate branches; three-dimensionally disposed leaves, and unwebbed, much divided leaves. In addition, *Svalbardia banksii* and *Actinoxylon banksii* are the only two species in the Archaeopteridales for which dichotomies of the leaves in more than one plane can be demonstrated." (p. 1388)

Chaloner (1972) described *S. scotica* from a Middle Devonian horizon on Fair Isle, off the north coast of Scotland; it has somewhat shorter leaves than those of *S. banksii*. Other fossils that have been assigned to the genus include *S. avelinesiana* Stockmans (1968) from the Middle Devonian of Belgium and *S. osmanica* Petrosian and Radczenko (1960) from the Devonian of Russia.

Various authorities differ concerning the most appropriate treatment of the fossils that have been assigned to *Svalbardia*. In his study of 1971, "On the Anatomy and Morphology of Lateral Branch Systems of *Archaeopteris*", Beck has the following to say: "Several genera, e.g., *Svalbardia* Høeg (1952) and *Actinoxylon* Matten (1968) have been considered to strongly resemble *Archaeopteris* but for one reason or another, have been separated from it. In the light of this study, however, I can no longer accept *Svalbardia* as a valid genus." (p. 777)

Since we are not attempting to produce a definitive monographic analysis of the Devonian plants, we leave the matter of the distinction between the two genera to the future when, hopefully, additional information will enable one to more clearly resolve the problem.

Eddya Beck, 1967.
Eddya sullivanensis Beck, 1967.

This is an interesting fossil plant with unique features and quite good preservation although its relationships within the Progymnospermopsida are questionable.

The collections on which Beck's account is based were obtained from the Delaware River Flags near Pond Eddy, Sullivan County, New York. It is of lower Upper Devonian age. The plant was first described briefly under the name *Ginkgophyton* by Beck in 1963.

Beck's restoration (Fig. 8.13) shows the plant as one that may have attained a height of a half meter. The main axis attained a diameter of not more than 5 mm and its anatomy is of particular interest. The vascular tissues consist of four to five primary bundles arranged around the periphery of what was apparently a parenchymatous pith. The bundles were probably mesarch and in contact with the secondary xylem which is a band about 1.4 mm thick. It consists of tracheids and rays that are mostly uniseriate. The tracheids are pitted on both tangential and radial walls and on the radial walls they are grouped in the manner of *Callixylon*.

In the formation of the foliar vascular system, a trace departs from one of the primary bundles, divides in two planes and then continues to divide in a tangential plane to form a dichotomous system in the wedge-shaped, entire (flabelliform) leaves similar to those of *Archaeopteris obtusa*. The latter are arranged alternately on the stem and range in length from 4 to 6 cm (Fig. 8.14). The main axis branches as shown in the restoration but the exact mechanism of branching is not clear and only the basal portions of the branches are preserved. Beck notes: "It should be

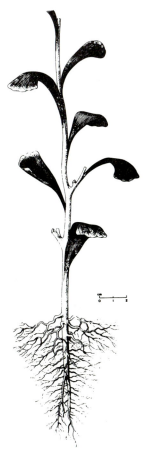

Figure 8.13. *Eddya sullivanensis.* A restoration of the plant. (From Beck, 1967, Palaeontographica.)

emphasized that there is no evidence of the development of lateral branches from buds. From a developmental standpoint, therefore, the apical meristem must have divided (segmented) longitudinally and unequally whenever branches were produced." (p. 4) The stem with its regularly arranged leaves gives the impression of having a branching mechanism somewhat advanced over that of the pseudomonopodial main axes of many of the earlier Devonian plants. Beck comments on this as follows:

"Considerable variation between obvious pseudomonopodial and monopodial branching must have existed among the several groups of

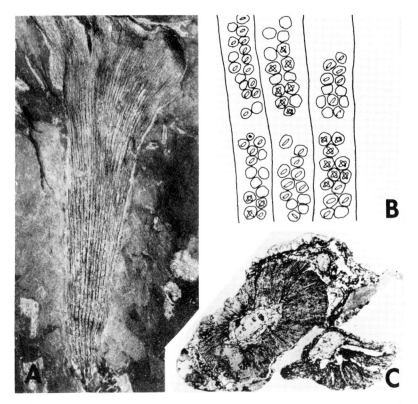

Figure 8.14. *Eddya sullivanensis.* A. typical leaf. × 1.8. B. Drawing of secondary tracheids showing pits on radial wall. C. Transverse section of two stems, one with a leaf base. × 8.6. (From Beck, 1967, Palaeontographica.)

early seed plants and their precursors. The presence in *Eddya* of a relatively straight stem suggests that it might have evolved to a stage at which, in branching pattern, it was closer to a typical monopodial type than to a typical pseudomonopodial type. It is not surprising to find in *Eddya* a developmental pattern that, apparently, is similar to that of some primitive seed plants since other evidence as well suggests that it is probably closely related to some of these groups." (p. 13)

The root system consists of a strange assortment of vascular strands dominated by four irregularly shaped masses consisting chiefly of secondary xylem.

As to its systematic relationships, Beck tentatively assigns it to the progymnosperms chiefly on the basis of the *Callixylon*-like wood.

Perhaps the most important bit of information that *Eddya* affords is the relationship between the leaf morphology and xylem structure. There are numerous reports of leaves of this type and Beck sums this up very well as follows:

> "Over the years a large number of interesting fossils consisting of large flabelliform leaves, usually detached, but occasionally spirally arranged on a compressed axis, have been reported from Devonian and other Paleozoic rocks. These fossils have provided important and largely unresolved problems of taxonomic affinity as well as problems of nomenclature. The majority of the Devonian specimens of this type have been assigned to the genus *Psygmophyllum*. Among other genera to which such specimens have been assigned in the past are *Ginkgophyllum*, *Ginkgophyton*, *Platyphyllum*, *Enigmophyton* and *Germanophyton*." (p. 2).

The account of *Eddya* presents one of the few instances when a flabelliform leaf of this kind has been found attached to an axis with significant structure preserved. A rather extensive review of the literature concerning fossils of this kind is given by Beck in the *Eddya* paper.

Actinopodium Høeg, 1942.
Actinopodium nathorstii Høeg, 1942.

The specimens on which this was founded were collected by Nathorst in 1882 from a Middle Devonian horizon of Mimerdalen, Spitzbergen. The material is petrified and described by Høeg as follows:

> "Leafless axis consisting of a broad parenchymatous cortex and a star-shaped stele, composed of a mixed pith with irregular metaxylem, probably mesarch protoxylem, and an outer zone of secondary xylem, surrounded by phloem." (p. 194).

The description and illustrations seem to conform closely to that of the main axis of the 'modified branch' of *Archaeopteris*.

Høeg did not report *Archaeopteris* from Spitsbergen but we have noted the very close similarity (if not actual identity) between *Svalbardia polymorpha* and certain species of *Archaeopteris*, such as *A. fissilis*. It is therefore very likely in our opinion that *Actinopodium* represents the main axis of a *Svalbardia* branch. This of course cannot be taken as fact until specimens are found that include the features of both.

Actinoxylon Matten, 1968.
Actinoxylon banksii Matten 1968.

The description of this plant is based on pyritic petrifactions found in a quarry near Cairo, Greene County, New York. The horizon is the Kiskatom Formation of the upper Middle Devonian (Givetian).

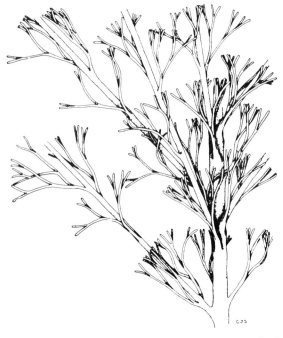

Figure 8.15. *Actinoxylon banksii.* Restoration of a portion of the branch system. (From Matten, 1968, Amer. Journ. Bot.)

The general aspect of the part of the plant that is preserved may be best understood by reference to the author's restoration (Fig. 8.15). It is known from a 'main' axis, referred to as the penultimate branch which bears the ultimate branches in a spiral arrangement and the latter bear leaves in a subopposite and decussate arrangement. The primary xylem of the penultimate branch is a six-pointed actinostele of about 2.5 mm in diameter (Fig. 8.16); there is usually one protoxylem near the tip of each arm and another just within it on the same radius. The preservation of the primary stele is not quite as good as might be wished; the wall ornamentation of the tracheids is described as including elements that are helical-reticulate, scalariform-reticulate, scalariform, and circular-pitted. It is thought to be protostelic, that is, composed of tracheids only, but the preservation in the center is imperfect. Some of the specimens show a small amount of secondary xylem, and the cortex is composed of an inner paraenchymatous zone and an outer sclerenchymatous one. The xylem strand in the ultimate branch is a three- or four-lobed protostele.

The leaves are borne on both the penultimate and ultimate branches and in the latter the phyllotaxy is opposite to subopposite and decussate.

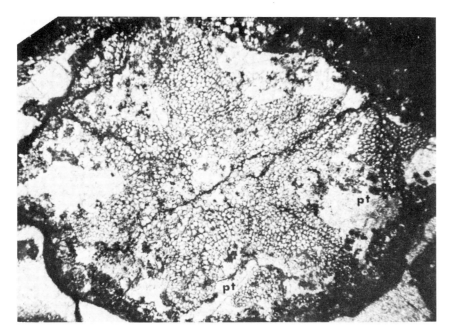

Figure 8.16. *Actinoxylon banksii.* A transverse section of a penultimate branch showing the six-lobed actinostele. × 30. (From Matten, 1968, Amer. Journ. Bot.)

The leaves dichotomized three or four times, attained a length of at least 5 cm and were terete in cross section.

Aneurophytales

Aneurophyton **Kräusel and Weyland, 1923.**
Aneurophyton germanicum **Kräusel and Weyland, 1923.**

Eospermatopteris **Goldring, 1924.**
Eospermatopteris erianus **(Dawson) Goldring, 1924.**

It is not our intent or desire to overemphasize the nomenclatural and taxonomic problems that some of our Devonian plants present. However, the two generic names cited here have appeared many times in the literature more or less linked together, with arguments pro and con concerning their relationships. We will try to present as brief yet clear an explanation as we can concerning the present status of these fossils. In 1978 Serlin and Banks brought out a detailed study of the plants and their related problems—"Morphology and Anatomy of *Aneurophyton*"—and

Figure 8.17. *Eospermatopteris erianus.* Cast of a representative stump specimen. Diameter at base is approx. 60 cm. × 0.06. (From Goldring, 1924, New York State Mus. Bull.)

we can do no better as an introduction here than to quote from their account.

Concerning *Eospermatopteris*: "Dawson (1871) described two new species of fossil tree ferns from New York State solely on the basis of short lengths of trunks which he thought were covered by aerial roots. *Psaronius erianus* was invested by parallel roots, *P. textilis* by roots that anastomosed. Goldring (1924) found many similar stumps [Fig. 8.17] *in situ* at Gilboa, New York, during quarrying operations associated with the building of a dam on the Schoharie River. Associated with the stumps, but never in organic connection, she found isolated seeds, microsporangia, and tripinnate fronds up to six feet long bearing twice bilobed, recurved ultimate appendages. Impressed by the possibility that all these isolated organs comprised a single plant, she described the genus *Eospermatopteris* and renamed Dawson's two species *E. eriana* and *E. textilis*. She regarded the aerial roots described by Dawson as sclerenchymatous cortex like that found in seed ferns. This and the mega- and microsporangia convinced her that she had found a forest of pteridosperms which she referred to as Dawn Seed Ferns." (p. 343)

Concerning *Aneurophyton*: "Kräusel and Weyland (1923) named the genus *Aneurophyton* for fernlike foliage that was two or three times pinnate, bearing bilobed, recurved ultimate appendages. Fertile fronds were less branched but otherwise comparable to the sterile. Sporangia were stalked and oblong-elliptic. Vascular strands were present in all parts except the ultimate appendages, and included secondary xylem. In 1926 and 1929 Kräusel and Weyland added to the original description such features as variability in the form of fertile pinnules, the presence of a triangular protostele with three mesarch protoxylem strands, and repetition of the anatomical pattern in all orders of branching except the ultimate appendages. The last two characteristics led them to suggest that the frond of *Aneurophyton* was not a typical frond (now called a branch system) and that it might be a three-dimensional rather than a planated structure." (p. 343).

In their account of 1978, Serlin and Banks have described sterile and fertile branch systems as well as petrified specimens showing the structure of the xylem (Fig. 8.18). This is based on fossils collected from an abandoned quarry one mile southeast of Pond Eddy, New York, the age being lower Upper Devonian. The most complete sterile branch system that was found is 115 mm long and displays three orders of branching in all of which the arrangement is spiral. The ultimate appendages are one to three times dichotomized and are described as being unwebbed and probably three dimensional (Fig. 8.18A). The fertile specimens terminate in 'fructifications' that are 10 mm long and consist of a "central stalk that dichotomizes once to form two arms that bear short stalks 0.5 mm in length and 0.5 mm in width." The stalks are inserted in two rows on each arm, each terminating in one exannulate sporangium (Fig. 8.18B, D). The sporangia average 2 mm in length, have a blunt tip and dehisce longitudinally. Up to eighteen sporangia have been found composing one fructification.

The xylary anatomy is described as follows for one of their figures that we have reproduced here (Fig. 8.18C):

> "The primary xylem is triangular in transverse section. The sides of the triangle are slightly concave, the apices rounded. . . . Four protoxylem strands are present between branchings. One is central, three occur near the apices of the lobes. . . . Thus protoxylem is mesarch". (p. 347)

The secondary xylem consists of tracheids and rays, the latter being uniseriate and ranging from two to twenty-seven cells high. The radial walls of the tracheids are covered with one to five rows of elliptical bordered pits.

In view of the importance of Kräusel and Weyland's classic series of three papers (1923–1929) on the German Devonian plants, a few additional notes on *Aneurophyton* are included here. Their original description appeared in the first (1923) report in which *Aneurophyton*

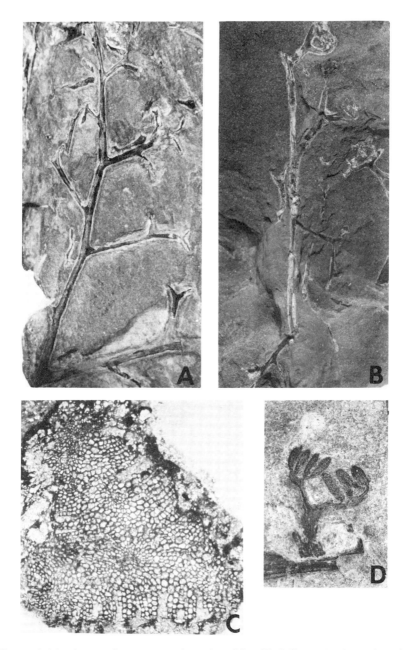

Figure 8.18. *Aneurophyton germanicum* from New York State. A. A sterile n + 2 branch bearing penultimate branches in spiral order. × 1.3. B. An n + 2 branch showing lyre-shaped fructification borne terminally on the uppermost branchlet. × 1.5. C. A transverse section of the n + 2 axis showing the three-lobed primary xylem and some secondary xylem. × 33. D. A lyre-shaped fructification of a specimen collected at or near the type locality in Germany showing several sporangia. × 5. (From Serlin and Banks, 1978, reproduced with permission of the Paleontological Research Institution, Ithaca, New York.)

was presented as a plant with somewhat fern-like foliage on which sporangia, which lacked an annulus, were borne in clusters, and with secondary growth present in the axes. They included an historical sketch of plants that might be described as leading up to this, including reports by previous paleobotanists such as Count Solms-Laubach and A. Gilkinet. In the second part (1926) they gave a further description of the fertile and sterile (vegetative) frond specimens with a restoration drawing. As a summary relating to the various parts described they said:

> "If all these specimens should belong together, they show that *Aneurophyton germanicum* was a stately plant of a tree-like growth. Accordingly, the previously considered possibility that one is dealing with a liana is incorrect." (transl from Part II, 1926, p. 130).

At this time they also considered the possible relationship of their fossils with *Eospermatopteris*. While pointing out that the several parts described by Miss Goldring were not found in organic connection, they note:

> "It is clear that this cannot be considered as a definite proof. But at the same time we cannot dispute the possibility of Goldring's reconstruction. Here it is to be emphasized that there is a great similarity between *Eospermatopteris* and *Aneurophyton*."

They seem to have had some confidence in Goldring's restoration (her Plate 1, 1924 and shown here in Fig. 8.19), as they add: "If one wants to get an idea of the appearance of the *Aneurophyton*-plant, it is advisable to start with Goldring's reconstruction for *Eospermatopteris*." (transl, Part II, 1926. p. 131). In their part III, 1929, they describe the sporangia in further detail, noting the presence of spores and, based on quite well-preserved petrifactions, describe the triradiate form of the primary xylem and development of the secondary xylem.

As this account was approaching the final stages of preparation, we received the recent detailed study of *Aneurophyton germanicum* by Schweitzer and Matten (1982) based on abundant new collections, as well as some original counterparts, from the type localities near Wüppertal. Their study confirms much that has been reported by previous investigators and adds to an understanding of the branch complex. As shown in Fig. 8.20A, the branch displays an initial dichotomy followed by apparent trichotomies, with the distal parts ending in nonlaminate, sterile (vegetative) dichotomies or fertile 'sporangiophores' (Fig. 8.20C). The latter are described as lyre-shaped and bear fourteen or more sporangia. Some additional information is also given on the vascular anatomy of the distal portions (Fig. 8.20B).

Figure 8.19. Restoration of *Eospermatopteris erianus* by Goldring, 1924. (New York State Mus. Bull.)

Finally, a few more comments may be in order relative to the history of *Eospermatopteris* and its relationship, if any, to *Aneurophyton*. Whatever this relationship may eventually prove to be, the plant fossils described by Winifred Goldring under the name *Eospermatopteris* are both significant and interesting.

A few extracts from her history of the discovery of *Eospermatopteris* are helpful in understanding the story. She notes that:

"In the fall of this year [1869] a great freshet swept the upper valley of the Schoharie in the vicinity of the village of Gilboa tearing out bridges, culverts and roadbeds, but greatly benefiting science by exposing in the bedrock standing stumps of trees." (1924, p. 50).

The fossil stumps, chiefly casts, that were exposed were turned over to Sir J. W. Dawson who named them *Psaronius erianus* and P. *textilis*. It

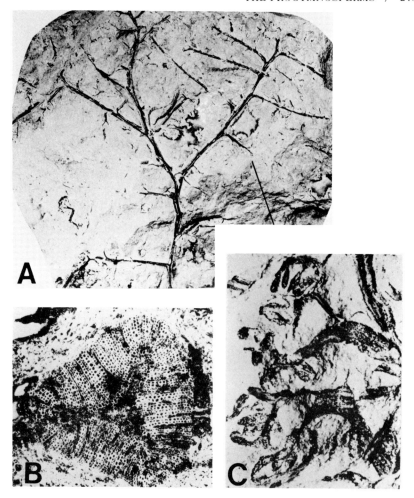

Figure 8.20. *Aneurophyton germanicum.* A. Portion of the branch system, × 0.5. B. Transverse section of an axis with three-angled primary xylem surrounded by secondary xylem, × 17. C. Terminal axes bearing sporangia, × 5.7. (From Schweitzer and Matten, 1982, Palaeontographica.)

was not until the 1920s that a renewed interest in the fossils led to attempts to relocate the original horizon; this effort resulted in the discovery of two or more horizons where the stumps were found. Miss Goldring records the following concerning the fossiliferous horizons:

> "At all three tree horizons the stumps were found with their bases resting in and upon shale and in every case in an upright position with the trunk extending into the coarse sandstone above. . . . The coasts of

those times were unstable, and it was along the borders of such a coast line that at least three successive forests of these trees reared themselves to great heights, were submerged, destroyed and buried. ... Later in that same season [1920 ?] and again in the summer of 1922 that and other localities were worked and reworked until a fairsized collection has been obtained containing seeds [megasporangia], male fructifications, pieces of foliage and roots. ... The above facts leave no doubt that the fruiting bodies, foliage and roots, occurring so frequently wherever the shale layers in question can be worked, belong to the Gilboa trees." (p. 51–54).

Two corrections may be inserted at this point. The fossil stumps were erroneously referred to the Carboniferous tree fern genus *Psaronius* and the various plant parts were not found in organic connection to the stumps, and later investigators have expressed "considerable doubt", rather than "no doubt", as to the significance of the association.

The stump casts, including the proximal parts of their spreading roots, range in size up to specimens with a stem diameter of two feet at a point two feet above the base, and their total height was estimated at about 13 meters. The presumed foliage consisted of fronds at least 2 meters long, the largest fragment that is illustrated being 62 cm long. Structures described as seeds and borne in pairs on frond fragments were later shown by Kräusel and Weyland to contain spores, and the presumed microsporangia are problematical. Thus there is no good evidence to classify *Eospermatopteris* as a seed plant.

Miss Goldring's account of 1924 contains a restoration figure of *Eospermatopteris* as the trees may have looked in life (Fig. 8.19) and her 1927 popular account includes a photo of the landscape restoration in the New York State Museum in Albany.

As we have indicated, the most recent and very informative summary is that of Serlin and Banks (1978) and as a final follow-up to that we are indebted to Bruce Serlin for supplying the following information in a recent letter:

"Now to answer your question on Miss Goldring's material. I went to the state museum to examine her material during the research. From what I saw, it appeared that her sterile material *could* be called *Aneurophyton* though I say this with some hesitation for the following reason. When I was at the Museum, the only sterile specimens they could locate were very small branch fragments such as are illustrated on her plate 10, Figs. 1–4. The larger specimens such as the ones from which text Figs. 3 and 4 were drawn and the large slab illustrated in Plate 9 had been apparently mislaid. Without these larger specimens it is difficult to rule out the possibility that the sterile fragments belong to another progymnosperm in the *Aneurophyton* line. Further doubt is added by the nature of the fertile specimens she illustrated and I examined. They bear no resemblance to those reported by either Kräusel and Weyland or myself, subsequently. If nothing else, this points to the fact that Miss Goldring was dealing with an assemblage of parts from more than one plant source." (Nov. 4, 1981).

Aneurophyton olnense Stockmans, 1948.

This was reported by Stockmans based on compression and impression fossils from the Upper Devonian of Belgium. His description is brief but his illustrations show numerous small sporangial aggregates borne on short side branches along the penultimate axis. The plant is an interesting one and it seems likely that more information might be forthcoming from further study. It may be added that Schweitzer and Matten (1982) have examined the fossils and note that in their opinion it " . . . is not a species of *Aneruophyton* but belongs more probably in the Coenopteridales."

Rellimia Leclercq and Bonamo 1973.
Rellimia thomsonii (Dawson) Leclercq and Bonamo 1973.

The plant described under this name has a history in the literature that is perhaps even more complex and tangled (and still controversial) than that of *Psilophyton* which we related in Chapter 4. It has been described under numerous generic names by numerous authors; it is essential to record some of this here but we think it will be more meaningful to the reader if a description of the plant, as it is now known, is presented first.

The most complete account that we have is that of Leclercq and Bonamo (1971) which is based on well-preserved specimens from the Brandt quarry of Goé, Belgium; this is a Middle Devonian (Givetian) horizon. The account deals primarily with the fertile (spore bearing) parts.

As to its general habit the authors note that large axes 30 to 50 cm long and 10 to 20 cm wide have been found associated with the other fossils but they conclude: ". . . the available information so far gathered does not support the view that *Milleria* [*Rellimia*] *thomsonii* had reached a tree habit, but points to the view that it was a bushy, woody plant of rather large size." (p. 102).

Vegetative and fertile specimens have been found closely associated and there seems to be little doubt as to their specific identity. Two such specimens are shown in Fig. 8.21, aligned as they probably were in life. The parts revealed include a main axis with three orders of branching, the branches being spirally arranged. The proximal ones are vegetative, bearing simple leaves.

The most conspicuous feature of the specimen lies in the dense and distinctive recurved terminal sporangial clusters. Fig. 8.22A is a schemat-

Figure 8.21. *Rellimia thomsonii.* Two specimens showing the vegetative branching system (below) and terminal fertile branches (above) as they may have appeared in life. See description in text. × 0.5. (From Leclercq and Bonamo, 1971, Palaeontographica.)

ic reconstruction of the branching pattern of this terminal unit, lacking the sporangia. It will be noted that the fertile terminal unit as a whole is composed of two "first order pinnae" resulting from an initial dichotomy, and each of these adaxially curved pinnae bear in turn second and third order pinnae in alternate arrangement. The sporangia are borne on the ultimate divisions in such abundance as to obscure the branching pattern in most specimens until they are carefully degaged.

The sporangia are slender, nearly cylindrical organs with an apiculate tip, and are usually borne in pairs, but are occasionally found in groups of

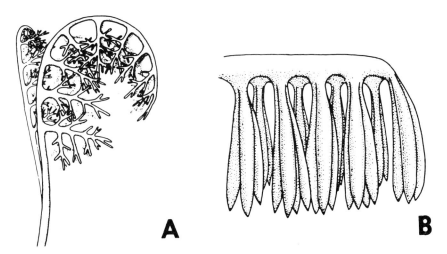

Figure 8.22. *Rellimia thomsonii.* A. Reconstruction of a fertile organ. B. Detail of terminally borne sporangia. (From Schweitzer and Matten, 1982, Palaeontographica.)

three or four. They attain a length of 3.5 mm and a diameter of 0.5 mm. The mature spores are spherical, pseudosaccate and trilete; they range from 77–140 μm in diameter, including the bladder (Fig. 8.23B), the body of the spore ranging from 55–87 μm. They are referable to the dispersed spore genus *Rhabdosporites*. The plant is considered to be homosporous.

In 1977 Bonamo added significantly to our knowledge of this species with a description of fertile specimens (Fig. 8.23) from the vicinity of Gilboa, New York. The particularly significant aspect of this account is her description of portions of petrified axes. In cross-section the primary xylem appears as a three-armed mesarch protostele (Fig. 8.23C), this being the first description of well-preserved primary wood in the plant.

In 1938 Kräusel and Weyland described fossils from the Rhenish Middle Devonian under the binomial *Protopteridium thomsonii* which are apparently conspecific with the Belgian ones described above. They demonstrated axes with an irregular triradiate primary xylem surrounded by secondary xylem consisting of rays and tracheids, the latter being of a scalariform or reticulate type. Most recently, Schweitzer and Matten (1982) have described a considerable number of specimens (also under the binomial *Protopteridium thomsonii*) from the Middle Devonian fossil deposits of Wüppertal-Elberfeld and Lindlar in Germany. These add to an understanding of the fertile parts of the plant. Their term 'fertile organ' apparently identifies the terminal dichotomy with its subsequent divisions as described by Leclercq and Bonamo. The ultimate terminations are

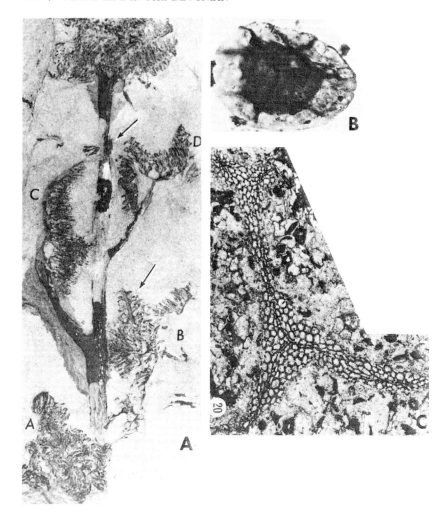

Figure 8.23 *Rellimia thomsonii.* A. Fertile branch showing spirally arranged sporangia-bearing branches. × 1.3. B. Proximal view of a spore of *Rellimia.* × 264. C. Cross section of an N + 3 branch showing three-lobed primary xylem and different stages of trace formation at the extremities of the arms. × 63. (From Bonamo, 1977, Amer. Journ. Bot.)

shown to bear marginal clusters of sporangia as shown in Fig. 8.22B. The entire structure, that is the two branches of the 'fertile organ' and its components, is strongly recurved.

Rellimia thomsonii is important as one of the best known representatives of the Progymnospermopsida and was apparently a significant element of the mid-Devonian landscapes since fossils that can be assigned to it with some confidence have been found in Bohemia, Scotland, Germany, Belgium, Russia, and New York State. However, the naming of fossils that are assigned to *R. thomsonii*, or presumably closely related plants, can at best be described as 'unfortunate.' It is not appropriate nor does space allow a detailed review but at least a brief summary seems required for those who may be especially concerned with the plant in the future.

In 1871 J. W. Dawson described a fossil from northern Scotland under the name *Ptilophyton thomsonii* which is now accepted as the holotype of *Rellimia*. Some of the specimens described as *Hostinella hostinensis* Stur (1882) and *Spriopteris hostinensis* Potonié and Bernard (1904) are among others now referred to *Rellimia*. The generic name *Protopteridium* was introduced by Krejči in 1880 and several species were assigned to it, including *P. thomsonii*, *P. hostinense* and *P. pinnatum*.

Leclercq and Bonamo regard *Protopteridium* as a *nomen nudum* and the different species that have been assigned to it to represent varying modes of preservation. The generic name *Milleria* was introduced by W. H. Lang in 1926, and in their account of 1971, Leclercq and Bonamo chose to follow Lang's designation. It was later called to their attention that *Milleria* had been used previously for a living angiosperm and in a brief notation (1973), they proposed a new name (anagram) *Rellimia*. As noted above, Schweitzer and Matten (1982) in their description of the German fossils retain the generic name *Protopteridium* which they consider to be valid.

Some rather voluminous summary reviews of the publications bearing on this problem may be found in the works of Leclercq and Bonamo (1971, 1973), and in Schweitzer and Matten (1982). Also, Matten and Schweitzer (1982) and Bonamo (1983) have written detailed accounts supporting the validity of *Protopteridium* and *Rellimia* respectively. We have chosen to use the name *Rellimia*. But regardless of the name, the relevant fossils reveal a considerble amount of information about a very significant Devonian plant.

Tetraxylopteris Beck, 1957.
Tetraxylopteris schmiditii Beck, 1957.

This was described initially by Beck on the basis of petrified remains from the Upper Devonian of Delaware and Sullivan Counties, New York.

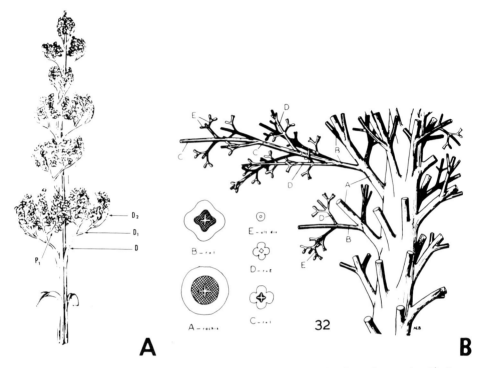

Figure 8.24. *Tetraxylopteris schmidtii.* A. Diagrammatic view of an entire fertile branch. B. Restoration of a part of the plant showing the stem bearing sterile branches. (A. From Bonamo and Banks, 1967. Amer. Journ. Bot. B. From Beck, 1957, Amer. Journ. Bot.)

The vegetative anatomy is quite well preserved and although the fertile parts are present they are not well preserved. Clearer knowledge of the sporangiate organs comes from a later report by Bonamo and Banks (1967) based on specimens from Greene County, New York, which occur in the Oneonta Formation of Frasnian (lower Upper Devonian) age. If reference is made by the reader to these original accounts it should be remembered that Beck's paper appeared prior to the recognition of the Progymnospemopsida (1960) and the Bonamo-Banks account appeared after that date.

The stems of *Tetraxylopteris* attain a diameter of 2.5 cm and bear "branch systems" in a rather dense spiral. We have here, as in *Archaeopteris*, an unusual morphological entity in the "branch system" or frond. Beck refers to the main axis of the branch system as the rachis and this bears three orders of axes that are arranged in an opposite or subopposite and decussate manner, the ultimate divisions being once or

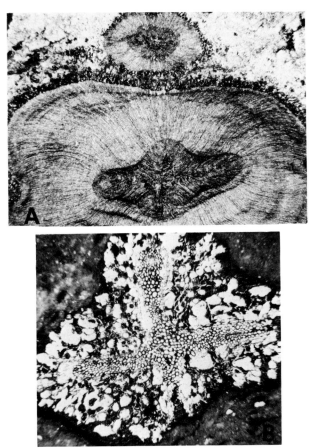

Figure 8.25. *Tetraxylopteris schmidtii.* A. Transverse section of base of the primary branch showing secondary xylem. × 5. B. Transverse section of primary branch showing cruciform primary xylem. × 33. (A. From Scheckler, 1971, Amer. Journ. Bot. B. From Beck, 1957, Amer. Journ. Bot.)

twice lobed (Fig. 8.24B). Although the primary xylem is not well preserved in the stem itself, it is in the several orders of branching in the frond, and there is some evidence to indicate that the stem had the same kind of structure. In the frond axes the primary xylem strand is a cruciform protostele (Fig. 8.25). Conspicuous protoxylem points are usually evident near the ends of the arms of primary xylem and another one is found toward the center, making a total of eight, plus a ninth one in the center of the stele. Branch traces originate at the ends of the arms and are supplied directly by the outer protoxylem which in turn is replaced by the inner one in that arm.

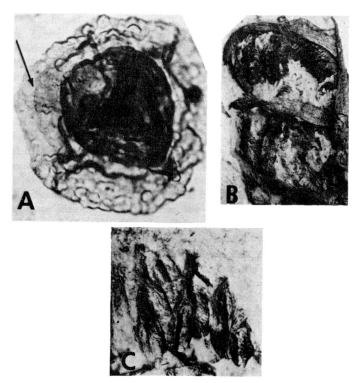

Figure 8.26. *Tetraxylopteris schmidtii.* A. Distal view of a spore. × 395.
B. Lateral view of one of the four major branches of the sporangial complex.
× 4.6. C. Detail of terminally borne sporangia. × 8.5. (From Bonamo and Banks,
1967, Amer. Journ. Bot.)

Secondary xylem (Fig. 8.25A) is present in both the stem and the
rachis. It consists of tracheids and rays that are uni- or multiseriate. The
tracheids bear circular bordered pits on all walls although they tend to be
somewhat more dense on the radial walls. Definite growth lawyers are
evident in the secondary xylem. Secondary phloem is present, thick-
walled fiber cells being especially conspicuous. Cells identified as sieve
cells, as well as parenchyma are present.

Our knowledge of the fertile parts is derived chiefly from the later
report of Bonamo and Banks. The fertile branches or 'sporangial
complexes' are arranged on the stem in opposite, decussate pairs. The
main stalk of each sporangial complex dichotomizes twice and then each
of the four resultant branches divides pinnately (Fig. 8.24A). The
sporangia are densely clustered on the ultimate branchlets (Fig. 8.26B)
with an estimated 2,800 to 4,700 composing each sporangial complex.

The sporangia are elongate-oval (Fig. 8.26C) with an acute apex; they are 2–5 mm long and 0.4–0.8 mm in diameter and open by means of a longitudinal slit. The spores range in diameter from 73–176 μm and are spherical, trilete and pseudosaccate (Fig. 8.26A), i.e. having an exoexine that is expanded into a partial bladder.

Tetraxylopteris like several other genera of Devonian plants has been the subject of a good deal of discussion and controversy as to its classification.

It is evident that there is a close relationship between *Tetraxylopteris* and *Rellimia*, the chief differences between the two being the three-lobed vs four-lobed primary xylem, the generally spiral vs. decussate manner of branching and the double vs. single primary branch (pinna). In their discussion of the relationship between the two, Leclercq and Bonamo (1971) indicate that the retention of separate generic names is justified but they note: "Taken together these characteristics may well indicate a later stage of evolution reached by one species of a single genus during the time available between the appearance of *Milleria* [*Rellimia*] in Middle Old Red Sandstone strata and that of *Tetraxylopteris* in the Upper Devonian." (p. 107).

Triloboxylon Matten and Banks, 1966.
Triloboxylon ashlandicum Matten and Banks, 1966.

This member of the progymnosperm group was first described on the basis of pyritized axes by Matten and Banks in 1966. Additional information was presented on the anatomy by Scheckler and Banks in 1971 and a problematical fertile specimen was reported by Scheckler in 1975. It presents a unique combination of features, including axes with a three-armed protoxylem surrounded by secondary wood, small dichotomizing flattened vegetative branchlets, and sporangia presumably borne on short, twice-dichotomized branches arranged in three longitudinal rows along the 'main' axis. The various specimens were obtained from Greene County, New York, in a lower Frasnian (lower Upper Devonian) horizon.

The main axis as shown here (Fig. 8.27A) attains a diameter of about 13 mm; it contains a three-armed primary xylem strand and in each arm there are two to four protoxylem groups in a median position. Some parenchyma cells are admixed with the metaxylem tracheids which may be scalariform or circular-bordered. There is a fairly strong development of secondary xylem (Fig. 8.27B) which consists of tracheids with crowded, multiseriate bordered pits on all walls, and uniseriate rays that are one-to-thirty cells high. The secondary phloem is partially preserved and includes fibers, rays and parenchyma. Some specimens show conspicuous groups of fibers alternating with parenchyma in the outer cortex, a feature that cannot help reminding one of the outer cortex of the Carboniferous pteridosperms.

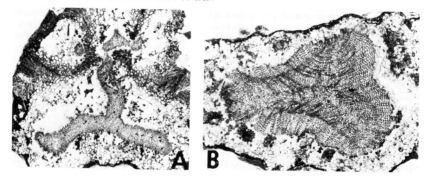

Figure 8.27. *Triloboxylon ashlandicum.* A. Transverse section of an axis showing only primary growth with lateral traces in various stages of departure. ✕ 13. B. Axis showing some secondary xylem. ✕ 13. (From Scheckler, 1975b. Amer. Journ. Bot.)

The main axis bears spirally arranged ultimate branches with a similar but somewhat more delicate xylem strand and these in turn bear the spirally arranged ultimate appendages ('leaves') which are apparently flattened and three times divided. According to Scheckler (1975), the main axis also may bear, in between certain ultimate sterile branches, three rows of short fertile branches (or 'fertile organs') as shown in his restoration. Each of these dichotomizes twice in one plane, the four terminal branch endings in turn bearing the sporangia. The latter are attached by short stalks laterally along the terminal division, and occasionally in a terminal position. The sporangia are 2–3.5 mm long and 0.5–0.8 mm in diameter and taper to an apiculate tip; some have been observed split down one side, suggesting a longitudinal dehiscence. The number of sporangia per fertile organ is estimated to be twenty to forty. Spores have not been observed.

It seems appropriate to add that the preservation of the fertile parts of the plant is not as good as one would wish and in this connection Scheckler notes:

> "This reconstruction is based mainly upon a single specimen . . . and it seems reasonable that other fertile axes of *T. ashlandicum* might differ somewhat in the distribution of fertile organs along the main axis." (1965b, p. 931)

Incertae Sedis—*Triradioxylon*

Triradioxylon **Barnard and Long, 1975.**
Triradioxylon primaevum **Barnard and Long, 1975.**

Our knowledge of this plant is based on petrified specimens of Cemenstone Group age (Lower Carboniferous), from the volcanic ash

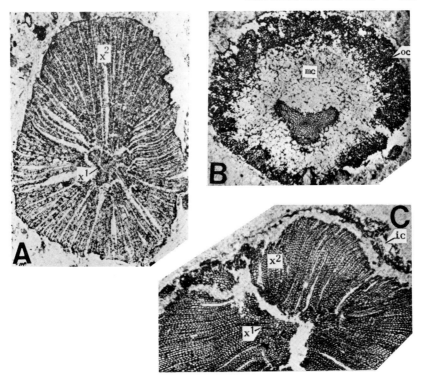

Figure 8.28. *Triradioxylon primaevum.* A. Transverse section of stem showing primary and secondary xylem. × 6.6. B. Transverse section of attached petiole. × 13. C. Transverse section of axis with secondary wood near distal end of stem. × 20. (From Barnard and Long, 1975, Trans. Roy. Soc. Edinburgh 69:231–49.)

deposits of Oxroad Bay, East Lothian, along the southeast coast of Scotland. We noted initially that our overall objective was to present an account of early vascular plant evolution without rigidly keeping within the confines of the Devonian. This chapter is followed by one that deals with the earliest evidences of seed plants and since important developments in plant evolution do not necessarily respect our geologic time boundaries we will be concerned there with fossils from late Devonian and early Carboniferous horizons. *Triradioxylon* is tentatively considered by Barnard and Long as *incertae sedis* in the pteridosperms but they note: "In stelar anatomy *Triradioxylon* shows a greater resemblance to certain Middle Devonian plants than to most contemporaneous Lower Carboniferous forms." A brief description will make this comparison evident.

It is known from well-preserved specimens of the stem and petiole. The stem is about 1 cm in diameter with a triradiate protostele composed entirely of tracheids with one or more protoxylem groups in the central region and near the ends of the arms (Fig. 8.28A, C). The rather strongly developed secondary xylem consists of one to two seriate rays and tracheids which have multiseriate bordered pits on all walls. The petiole trace is also three-armed (Fig. 8.28B) and may have some secondary xylem.

In both stem and petiole the outer cortex is composed of well-developed fiber strands alternating with parenchyma areas, a feature that seems to have been developed first in the progymnosperms and is particularly distinctive of the pteridosperms. The authors draw comparisons with several progymnosperms, including *Aneurophyton*, where the differences lie in the location of protoxylem groups and possibly the way in which the pinna traces depart.

As yet no reproductive structures are known for *Triradioxylon* and on the basis of what we have it seems to be a close decision as to whether it should be classified as an advanced progymnosperm or a primitive pteridosperm.

Periderm Formation in Early Land Vascular Plants

In a short but significant paper of 1972, Scheckler and Banks report finding periderm formation in petrified specimens of *Triloboxylon*, *Tetraxylopteris* and *Proteokalon*, which range in age from mid-Givetian to Frasnian. They note: "It is perhaps no surprise that we have found in this group what we believe to be the earliest occurrence of periderm in a fossil plant." (p. 58).

Their illustrations (Fig. 8.29) show rather typical periderm in *Proteokalon* and *Tetraxylopteris*. It is initiated just within the outer cortex and starts as a nearly continuous circumferential band. Cells identified as phellogen, phellem and phelloderm are described. The phellem cells are arranged in radial rows without intercellular spaces, some being thin-walled and others thick-walled. The phelloderm is described as being parenchymatous, similar to the cortical parenchyma cells but distinguished by their radial alignment. They add: "The presence of a well-developed periderm provides yet another feature linking the progymnosperms to the gymnosperms." (p. 62)

More recently Banks (1981) has reported periderm activity as a wound reaction in a stem fragment of *Psilophyton dawsonii* from the Lower Devonian (Emsian) of Gaspé Bay, Quebec.

Protopityales
Protopitys Goeppert, 1850.
Protopitys buchiana Goeppert, 1850.

This is another interesting example of the development of a generic concept that is the result of the work of several investigators over the

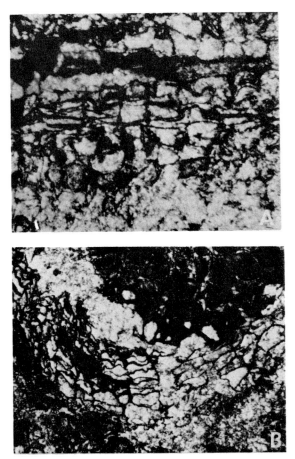

Figure 8.29. Periderm. A. *Tetraxylopteris schmidtii.* Radial section of cortex with early divisions of the phellogen. × 140. B. *Proteokalon petryi.* A transverse section of the cortex with early divisions of phellogen. × 97. (From Scheckler and Banks, 1972).

years. A summary of the history involved will help in understanding the plants and we will include this in the description that follows. The latter is based on accounts of two species, *P. buchiana* and *P. scotica* Walton; we deal with them together since there is some doubt as to the real distinction between the two.

Protopitys was a forest tree of some magnitude, being known from stem fragments up to 1.5 feet in diameter, according to D. H. Scott (1923). The central area is particularly distinctive: The pith region is oval in transverse section and consists of parenchyma with a thin band of primary xylem around the periphery (Fig. 8.30A). At either end of the oval there are two primary xylem strands: these are slightly mesarch, the

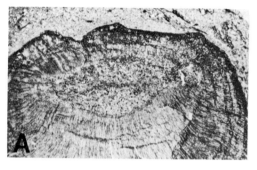

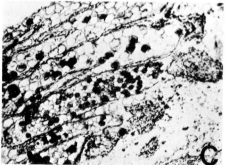

Figure 8.30. *Protopitys buchiana.* A. Transverse section of stem showing pith, primary, and secondary wood. × 4. B. Stem fragment with portions of fertile branch system. × 1.2. C. Section through sporangia showing spores of different sizes. × 18. (A. From Scott, 1923, Vol. II. B,C. From Walton, 1957, Trans. Roy. Soc. Edinburgh 63:333–40.)

protoxylem being located near the outer part of the primary xylem strand. These primary strands divide to form a pair that pass out and then fuse to form a leaf trace. The latter presumably supplied leaves that were distichously arranged. The metaxylem elements are scalariform.

The secondary wood is distinctive, consisting of tracheids that have elongated bordered pits, and rays that are uniseriate and mostly one cell high, but occasionally they are two or three cells high.

The vegetative leaves are not known but fertile ones (referred to as sporophylls) are present which seem best called fertile appendages in the light of our present knowledge. They are short branches that dichotomize several times and bear sporangia apparently in small clusters (Fig. 8.30B). The sporangia attain a length of nearly 3 mm and probably were fusiform. They contained numerous spores that varied considerably in size (Fig. 8.30C). At this point some historical information will be appropriate.

The type species *P. buchiana* was first described by Goeppert in 1845 based on specimens from the Upper Devonian of Falkenberg in

Silesia. Later studies of these fossils were contributed by Kraus and by Solms-Laubach, and these are adequately reviewed by Seward in his *Fossil Plants* (Vol. III, 1917). In 1957 Walton described *Protopitys scotica* from specimens collected from the Calciferous Sandstone Series (Lower Carboniferous) of Dunbartonshire, Scotland. The close similarity in the stem structure of the two species leaves no reasonable doubt as to their generic identity (if not specific identity), and this was the first time that the sporangiate organs were described. More recently D. L. Smith (1962) has described additional specimens of *P. scotica* also from Dunbartonshire.

Walton described the spores as falling into three size categories; Smith's somewhat more detailed studies describe them as ranging from 75 to 355 μm in diameter but with about 70% of them in the range of 90 to 150 μm, followed by a sharp increase to a larger size (about 160 to 360 μm). One cannot be sure that we are dealing with a heterosporous plant but there is reasonable support for Walton's view: ". . . we have here a preliminary stage in the evolution of definite megasporangia and microsporangia."

In 1969 Walton described a stem fragment with associated roots from the Lower Carboniferous of Yorkshire. This was mentioned briefly by Scott in his *Studies* in 1923 under a manuscript name *Protopitys radicans* Kidston. This name was not validly established and Walton states that the specimen is clearly referable to *P. buchiana.*

The roots that are associated with the stem have an oval diarch xylem strand that is typically fern-like. Walton notes: "Scott and Kidston had the opportunity, according to their correspondence with one another, of seeing connections between these roots with a stem." With the material that remained for his study Walton was unable to confirm this observation but seemed favorably inclined toward it. It would add a most interesting feature to this plant, a presumed progymnosperm, if the roots were of the pteridophytic nature that is indicated.

Walton's photos show another feature of the plant that is also of interest but he did not elaborate on it in his text. The portion of the stem that he described shows a well-preserved zone at the periphery of the secondary xylem which includes secondary phloem and some inner cortex. The cells composing the phloem are not as regularly aligned in radial rows as the tracheids, and seem to be only slightly elongated. Also, in one part of the transverse section there is an "anomalous" zone consisting of several narrow rows of alternating secondary xylem and phloem. Hopefully, more abundant material may be found in the future to clarify these interesting anatomical features.

Finally a few comments on the classification of *Protopitys* are interesting. On the basis of its distinctive anatomy Solms-Laubach instituted a new family name, Protopityaceae. Seward regarded it as a "... generalised type exhibiting in the structure of its stem both Filicinean and Coniferous features." D. H. Scott (1923) included it in his treatment of the pteridosperms but said: "The general result of our brief survey is to confirm the impression that *Protopitys* is an isolated type." (p. 155). Walton proposed "... to institute a new group of Pteridophyta, the Protopityales, on account of the fact that it had pteridophytic reproduction." (1957, p. 338).

Most recently, *Protopitys* finds a place in the Progymnospermopsida on the basis of its pteridophytic reproduction and gymnospermous anatomy.

SUMMARY

In 1960 Charles Beck established the new group of fossil plants, the Progymnospermopsida, on the basis of organic connection of *Archaeopteris* and *Callixylon*. He wrote: "There can be no doubt of the determination of the leaves of the specimen of *Archaeopteris* and the axis to which these are attached as *Callixylon*. (p. 356)

Archaeopteris is known from foliage fossils which display pteridophytic reproduction, several species of which have been demonstrated to be heterosporous. *Callixylon* is known from petrified axes with gymnospermous anatomy. This stands as one of the most important advances in our knowledge of vascular plant evolution in bridging, to some degree, the gap between pteridophytes and seed plants.

We are still in the early stages of understanding the limits of the group. Since 1960 numerous generic entities have been established in addition to *Archaeopteris*. Of these we consider *Tetraxylopteris*, *Rellimia*, and *Aneurophyton*, which present the combination of identifying features of the group, to be the most significant. Of the others, or at least those that we have described on the preceding pages, some may represent valid genera while others are probably parts of the four genera that are known from more complete fossils. A few comments on these somewhat doubtful entities will serve to explain the kind of uncertainties and controversies that prevail at present.

The distinction between *Archaeopteris* and *Svalbardia* is questionable and representative. When he established the genus *Svalbardia* in 1942, Høeg, one of our most competent and knowledgeable paleobotanists, was unsure of its validity as a distinct generic entity. He wrote:

"As to *Archaeopteris*, there is certainly a very great difference between our plant [i.e. *Svalbardia polymorpha*] and most of the species of that genus. But *A. fissilis* (Schmalh.) em. Nath., from the Upper Devonian of Donetz (Schmalhausen 1894) and of Ellesmere Land (Nathorst 1904), forms a connecting link, and in reality some of our specimens bear such a resemblance to this species that they can scarcely be distinguished from it. . . ." (p. 80)

It is questionable as to whether there is any difference in the gross morphology of the 'modified branch' in the two, and the ultimate foliar divisions (leaves) of certain species of *Archaeopteris* grade into the type found in *Svalbardia*. At the present time some authorities accept a generic distinction between the two and others do not. The present writers incline toward the latter opinion.

Correlating with this problematical distinction based on gross foliar morphology, it is especially pertinent to add that three fossils based on rather fragmentary petrified remains, *Actinopodium* Høeg from Spitsbergen, *Actinoxylon* Matten from New York State, and *Siderella* Read from Kentucky, may very well be axes of the *Archaeopteris-Svalbardia* assemblage.

The urge to describe even fragmentary remains of plants that apparently belong to so interesting a group as the progymnosperms is understandable, but some of the genera that have been established almost certainly will not stand the test of time and we suggest that more restraint would be appropriate.

Devonian plants have contributed much to our understanding of the evolution of leaf morphology while bringing with it frustrations in the use of acceptable terminology. A botanist concerned with living flowering plants in a temperate woodland generally has little difficulty in distinguishing between leaf, stem, root and reproductive organs, but with Devonian plants, where these structures were in the early stages of evolution, the case is quite different. In the case of *Archaeopteris*, the shift in our concept of the structure (see Fig. 8.1B) that has been regarded as a compound 'fern-like frond' to that of a 'modified branch bearing laminate leaves' is based on anatomy. Because in most vascular plants the xylary structure is radially symmetrical, while the foliar structure is a fragment of it (bilaterally symmetrical), we have applied this to the progymnosperms. But not all paleobotanists accept this without an argument. In her keenly critical review "The Progymnospermopsida: Building a Concept" (1975), Bonamo says: "In reality, what constitutes a leaf in the Progymnospermopsida is speculative for we have no way of knowing just what portion functioned as a photosynthetic unit. Thus arguments and theories of what portion of a "frond" or "branching system" constitutes a leaf are without real meaning." (p. 573)

We do have to have some kind of terminology in order to communicate with one another but the fascination and perhaps the most important lesson that we learn from Devonian paleobotany is that taxonomic boundaries and evolutionary lines of development do not necessarily coincide.

9 / The Early Seed Plants

One of the most important phases in the evolution of plant life on the Earth was the development of the seed and its early stages of diversification. Before going further it will be useful to define what we mean by a *seed*—a strict botanical definition is that a seed is a *fertilized ovule,* and an *ovule* refers to an *integumented megasporangium.* (i.e. an enclosed megasporangium). At times in paleobotany the word *seed* is used interchangeably with *ovule,* even when it is not known if the ovule in question had actually been fertilized.

The decades of the 1960s and '70s have been an especially productive period with an almost startling number of previously unknown seed plants, especially those assigned to the pteridosperms, having been discovered and described. Much of this progress involves pteridosperms of Upper Carboniferous age which lie outside the scope of our study, but of special interest here are the records of seed plants in the late Devonian and numerous discoveries of apparently primitive seed plants, in significant diversity, in the early part of the Carboniferous. We therefore extend slightly beyond the title of our account to bring in some of the Lower Carboniferous seed plants in order to develop a sequence from the earliest land vascular plants of the late Silurian to the beginnings of a significant seed plant assemblage in the early Carboniferous.

At least four quite distinct lines of seed plants seem to have come into existence in the Carboniferous and these may be defined briefly as a matter of reference:

1. Lyginopteridalean pteridosperms, in which the seeds are small (measured in mm), the integument and nucellus are fused except for the distal part of the latter which may form an elaborate pollen collecting mechanism.

2. The medullosan pteridosperms in which the seeds are quite large (measured in cm), the integument and nucellus are joined only at the base, and the nucellus is vascularized.

3. The Callistophytalean pteridosperms in which the seeds are generally small, slightly flattened, winged (platyspermic), and non-cupulate, and with the integument and nucellus joined only at the base.

4. Platyspermic winged seed of intermediate size also occur, found chiefly in the Upper Carboniferous; in these the integument and nucellus are joined only at the base, and they were presumably borne by the cordaites.

These are approximate categories. Fossil seeds are found in many states of preservation which contribute varying amounts of information and to some extent different interpretations. There are many problems involved in arriving at natural relationships. As one example, platyspermic (bilaterally symmetrical) seeds are also found in the lagenostomalean pteridosperms as will be described on a later page.

Most, if not all, of the late Devonian and early Carboniferous ovules seem to be referable to the lyginopteridalean pteridosperms. In the present state of our knowledge we cannot be at all sure as to whether the other groups defined above originated independently or evolved from the lyginopteridalean pteridosperms, and any detailed consideration of these groups lies outside our presentation.

Very little of what we are reporting in this chapter was known two decades ago; we now have some fragmentary and fleeting glimpses of the earliest seed plants which are, however, significant and interesting. We are concerned chiefly with the following: The nature of the earliest seeds (in the Upper Devonian); certain unique structural features that characterized the pollination mechanism of the early Carboniferous seeds; and the great size that some of the early seed plants achieved.

THE EARLIEST FOSSIL SEEDS

Archaeosperma Pettitt and Beck, 1968.
Archaeosperma arnoldii Pettitt and Beck, 1968.

The existence of seed plants in the late Devonian has been suspected by paleobotanists for some time but one of the most convincing pieces of

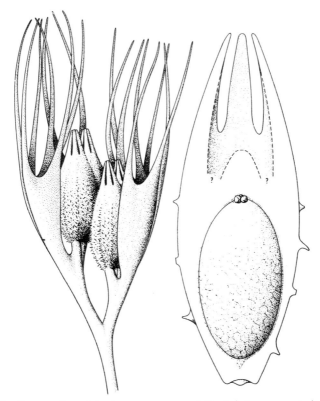

Figure 9.1. Restoration of *Archaeosperma arnoldii*, showing general organization (left) and detail of seed (right). (From Pettitt and Beck, 1968, Contrib. Mus. Paleont. Univ. Michigan.)

evidence that this is indeed true comes from cupulate seeds described under this name. The fossils were first reported, rather briefly and not named, by Chester Arnold in 1935, having been obtained from a roadside exposure six miles west of Port Allegany, Pennsylvania, the horizon being in the Oswayo Formation (Famennian) of the Upper Devonian. Pettitt and Beck continued the investigation using appropriate maceration techniques and the following description is taken from their account.

The fossils consist of compressions representing a terminal part of the plant that bore them. It branches twice as shown (Figs. 9.1, 9.2) and includes two pairs of seeds which are partially enclosed by telomic branch endings described as a cupule, the proximal parts of which are partially fused (Figs. 9.1 and 9.2).

The individual seeds are about 4.2 mm long and about 1.5 mm broad. The integument is divided into several lobes at the micropylar end, and it

Figure 9.2. *Archaeosperma arnoldii.* This is the most complete specimen reported, showing a dichotomy of the common axis producing two cupular axes. Two seed pedicels are visible also. × 8. (From Pettitt and Beck, 1968, Contrib. Mus. Paleont. Univ. Michigan.)

Figure 9.3. *Archaeosperma arnoldii.* A megaspore with three small abortive spores of the tetrad at the apex. × 9. (From Pettitt and Beck, 1968, Contrib. Mus. Paleont. Univ. Michigan.)

is covered with small spines that are especially dense toward the attached end of the seed. By careful maceration the authors were able to insolate the megaspores intact (Fig. 9.3). They vary somewhat in size from 1.0 mm × 0.5 mm to 3.8 mm × 1.7 mm. The megaspore tetrad actually consists of one very large and apparently functional spore and three aborted ones at the apical end. The entire fossil or fructification (Fig. 9.1) is interpreted as being composed of two cupules each of which partially enclose two seeds.

Several other seeds included in our account here are enclosed to a greater or less degree in sterile filaments or branch endings that are described as cupules.

In 1964 Chaloner and Pettitt reported isolated spore tetrads (Fig. 9.4) consisting of one large megaspore (about 2 mm long) and three aborted ones. These were collected from the Upper Devonian Escuminac Formation of Escuminac (Scaumenac) Bay, Quebec, and given the name *Cystosporites devonicus.* At that time it was suggested that they might have come from a lepidocarp seed. However, Pettitt and Beck (1968) regard

Figure 9.4. *Cystosporites devonicus,* showing one functional megaspore and the abortive spores at the apex. From Escuminac Bay, Quebec. × 55. (From Chaloner and Pettitt, 1964, Palaeontology.)

them as very similar to the megaspores of *Archaeosperma* and Chaloner in a recent letter tends to confirm that opinion.

Gillespie, Rothwell, and Scheckler (1981) have reported fossils that are similar in their general organization to *Archaeosperma arnoldii* but from a somewhat older horizon in the Hampshire Formation of Randolph County, West Virginia. The fructification or cupule consists of a four-parted terminal branch system bearing four seeds that are rather loosely enclosed by cupular lobes, there being four of these peripheral to each seed. The latter are 5–6 mm long with a smooth integument that is fused to the nucellus in the basal third of the seed. The lobing of the integument seems to be somewhat deeper than in *Archaeosperma* and is described as being: "... only slightly greater than that of *Genomosperma latens*." [see page 274] and "The top of the nucellus (megasporangium) bulges conspicuously into the space encircled by the free tips of the integument lobes ... and its apex is elongated into an apparent salpinx-like tube." (p. 462).

These are rather tantalizing reports. They leave no doubt that seed plants existed in late Devonian times but we still have no knowledge of what the plants as a whole looked like. It may be significant that species of *Archaeopteris* occur in all three areas associated with the seed remains. Several species of *Archaeopteris* as well as *Rhacophyton* are reported by Gillespie, et al. from the horizon where their West Virginia seeds come from; *Archaeopteris* is a well-known fossil in the cliffs of Escuminac Bay where *Cystosporites* was found; and in reference to *Archaeosperma*, Chester Arnold wrote: "The intimate association of these structures with *Archaeopteris* introduces the possibility that they are the seeds of this plant." (p. 284)

Others have speculated on the possibility that some species of *Archaeopteris* were seed bearing. Hopefully, continued searching will give us an answer.

There is also some evidence that indicates the presence of another group of seed plants in the Devonian. Chaloner, Hill, and Lacey (1977) have described a platyspermic seed from late Devonian rocks of Kiltorcan, Ireland. This was actually described very briefly by T. Johnson in 1917 who assigned the name *Spermolithus devonicus*. They are a little over 3 mm long, definitely platyspermic, with a slightly eccentrically placed inner body that is interpreted as the megaspore, and an apical cleft that apparently represents the micropylar region (Fig. 9.5).

These seem comparable to certain Carboniferous seeds assigned to the genus *Samaropsis*, some of which are of cordaitalean affinities. We can only note that they bring in a new type of seed in the Devonian.

Several other fossils have been described from the Devonian rocks which are suggestive of being referable to seed plants. We mention one of them which has received considerable attention and which is perhaps indicative of a more extensive seed-plant flora as yet little known. *Moresnetia zalesskyi* was described by F. Stockmans in 1946 and a more detailed account appeared in his summary of the Upper Devonian floras of Belgium in 1948. The fossils are delicate terminal branch systems, the branches being about 1 to 2 mm in diameter which dichotomize at rather acute angles and terminate in a cluster of free segments that suggest a cupule (Fig. 9.6). However, seeds have not been found in them. Several of Stockmans' specimens were further investigated, using balsam transfers, by Pettitt and Beck (1968) and we quote from their results:

"There is obvious similarity between some features of *Moresnetia zalesskyi* and *Archaeosperma*. In the former ... a dichotomously branched axis gives rise in the terminal region to a number of free, more or less flattened segments. Each segment is deeply divided into two "lobes" which extend distally as long narrow projections. In these

Figure 9.5. *Spermolithus devonicus.* Seed preserved as a chloritized compression. × 22. (From Chaloner, Hill, and Lacey, 1977, Nature. Reprinted by permission from Nature, Vol. 265, No. 5591, p. 233. Copyright.)

respects they are exactly like the cupule segments in *Archaeosperma.*" (p. 145)

SOME APPARENTLY PRIMITIVE SEEDS FROM THE LOWER CARBONIFEROUS

In 1960 Albert Long intiated a series of studies in which he described a unique and numerous assemblage of early pteridosperm plants, based in large part on petrified seeds from the Lower Carboniferous rocks of Berwickshire, southern Scotland. Others, of course, have contributed to this great development in our knowledge of the origins and evolution of seed plants and some of their work is included in our discussion.

Our focus is on a representative sampling of the seeds and the distinctive structural features that characterize them.

Hydrasperma tenuis Long, 1961.

We have selected this to introduce features that, with some modifications, are typical of many of the Lower Carboniferous seeds. It is

Figure 9.6. *Moresnetia zalesskyi.* A specimen from Stockmans' collection in Brussels (Upper Devonian, Belgium). × 3. (From a photograph in Pettitt and Beck, 1968, Contrib. Mus. Paleont. Univ. Michigan.)

known from some ten specimens that were found in a block of calciferous sandstone from the River Whitadder near Hutton Bridge, Berwickshire. Unless otherwise indicated, the other seeds described below are from the same general area.

The seeds are about 3.8 mm long (Fig. 9.7) and have a maximum diameter of 1.0 mm; they are radiospermic (radially symmetrical in cross section) and taper toward the base. The integument forms several lobes at the distal end where they are free from the nucellus and a single vascular strand is present in each one.

It may be helpful here to combine this account of *Hydrasperma* with an introduction of the special terms that have been used in describing pteridosperm seeds in general. In reference to the figure shown here, the outer layer or covering is the *integument*; it is fused to the nucellus except at the distal end where it becomes lobed for the distal one-third of the seed and the lobes tend to flare outward. In other seeds, this distal part of the integument may be continuous, forming a *canopy* over the terminal

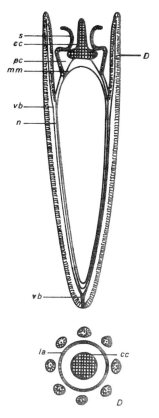

Figure 9.7. *Hydraspermis tenuis.* Diagrammatic median longitudinal section and transverse section at level D. s = salpinx; cc = central column; mm = megaspore membrane; pc = pollen chamber; vb = vascular bundle; n = nucellus. × 21. (From Long, 1961, Trans. Roy. Soc. Edinburgh 64:401–19.)

part of the nucellus, and entirely enclosing the latter with the exception of a small opening, the micropyle. Within the integument are the variously modified parts of the *nucellus* (megasporangium wall) including a distal outer flask-shaped part called the *lagenostome*. In *Hydrasperma*, and notably in many of the Lower Carboniferous seeds, the terminal part of the lagenostome is elaborated as a *salpinx* which served as a specialized pollen-collecting device. The distal part of the nucellus immediately within the base of the lagenostome is sometimes referred to as the *plinth*, and the central part of this extends up as a *central column*. Between the basal part of the central column and the lagenostome is a cavity known as the *pollen chamber*. Within the main body of the nucellus is the *megaspore membrane* (wall of the functional megaspore) and in some seeds this

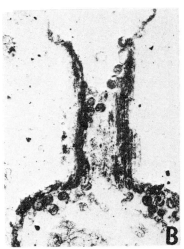

Figure 9.8. *Hydrasperma tenuis.* A. A plexiglass reconstruction of the cupular organ. × 5. B. Apical part of a seed showing sporomorphs in the salpinx and lagenstome. × 53. (From Matten et al. 1980, Biol. Journ. Linnean Soc. London.)

contains the *prothallus* (female gametophyte) with several *archegonia* at the micropylar end.

In *Hydrasperma tenuis* the number of free lobes of the integument varies in different seeds from eight to ten; they each contain a single vascular strand and tend to flare outward slightly so that no micropyle, as such, is formed. Pollen grains measuring 50 μm in diameter were found in most of the seeds, the maximum number in any one being forty-one. Only fragmentary remains of the female gametophyte were found.

In 1980, Matten, Lacey, and Lucas added significantly to our knowledge of the *H. tenuis* plant with a description of the cupulate structure that contained the seeds (Fig. 9.8). Their fossils consist of petrified remains from an intertidal outcrop at Ballyheigue, County Kerry, Ireland, which comes within the Coomhola Formation of uppermost Devonian age.

The cupules are small campanulate organs, with overall measurements of about 9.6 mm wide and 8.7 mm high. The cupule seems best described as an open, three-dimensional terminal branch unit composed of as many as twenty-four ultimate divisions. Their illustrations include an especially interesting one (Fig. 9.8B) that shows numerous sporomorphs (presumably pollen) within the pollen chamber.

The seeds that are described in the remaining part of this section have been selected to show the variations in seemingly critical or distinctive features such as the nature of the terminal part of the integument (whether or not a distinct micropyle is developed), the form of the salpinx, and the gross symmetry of the seed.

Genomosperma Long, 1960.

Long described two species of *Genomosperma* (Long, 1960b), *G. kidstoni* and *G. latens*, which are of particular interest in showing apparently early stages in the evolution of the integument and the salpinx.

Genomosperma kidstoni (Calder) Long, 1960.

This species is known from seeds (Fig. 9.9A) that are 10 to 15 mm long which are distinguished by having an integument composed of lobes that are separate for almost their entire length. In most specimens there

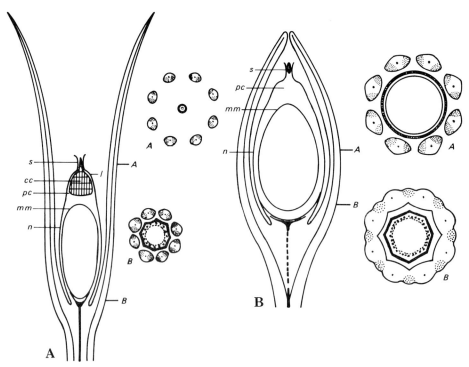

Figure 9.9. A. *Genomosperma kidstoni* (Calder) Long. Diagrammatic reconstruction of median longitudinal section, and transverse sections at levels A and B. s, salpinx; cc, central column; pc, pollen chamber; mm, megaspore membrane; n, nucellus. × 6. B. *Genomosperma latens* Long. Diagrammatic reconstruction of median longitudinal section, and transverse sections at levels A and B. s = salpinx; pc = pollen chamber; mm = megaspore membrane; n = nucellus. × 6. (From Long, 1960b, Trans. Roy. Soc. Edinburgh 64:29–44.)

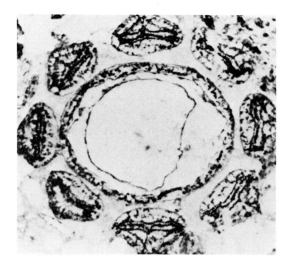

Figure 9.10. *Genomosperma latens.* Transverse section through the median part of the seed showing eight integument lobes surrounding nucellus. × 16. (From Long, 1960b, Trans. Roy. Soc. Edinburgh 64:29–44.)

are eight lobes but the number varies from six to eleven, and they flare out distally so that there is no formation of a micropyle. The distal end of the lagenostome forms a very feebly developed salpinx. The vascular strand in the rather long pedicel divides to form a central one that ends in the form of shallow funnel at the base of the nucellus, and a ring of strands which individually supply each of the integumental lobes. Pollen found in the pollen chamber measures 56–67 μm in diameter.

Genomosperma latens Long, 1960.

This species (Fig. 9.9B) is founded on seeds that are about 8 mm long and are borne on pedicels that are shorter (about 3 mm long) than those of *G. kidstoni* where they measure 10 mm. In most of the specimens examined there were eight lobes which are described as being adnate to one another near the base but very shortly above this point (0.9 mm) they become separate but quite closely appressed (Fig. 9.10) and at the distal extremity they curve inward to form a nearly continuous sheath over the apex of the nucellus. The general organization of the nucellar apex in the two seems to be similar, although no central column was found in *G. latens,* and the pollen in the two are of essentially the same size and form.

One might wonder whether the differences here are stages in development or actually constitute distinct species. Long felt quite sure about the latter choice. In any event, we seem to have a significant view of how a closely enclosing envelope so characteristic of the integument of most seed plants came into existence.

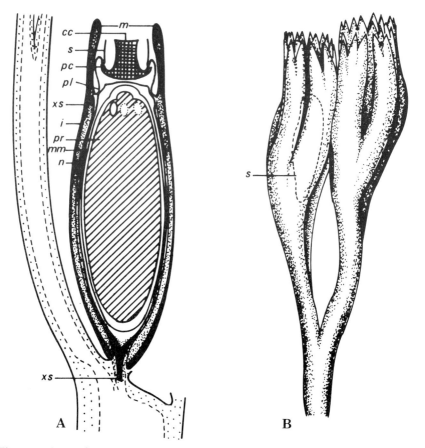

Figure 9.11. *Stamnostoma huttonense.* A. Diagrammatic reconstruction of longitudinal section of seed. m = "micropyle" opening; cc = central column; s = salpinx; pc = pollen chamber; pl = plinth; xs = vascular strand; pr = pro- thallus; mm = megaspore membrane; n = nucellus. × 26. B. Restoration of a pair of cupules showing position (s, dotted line) of one seed in left-hand cupule. × 5. (From Long, 1960a, Trans. Roy. Soc. Edinburgh 64:201–15.)

Stamnostoma huttonense Long, 1960.

The seeds here are quite small, ovoid, and not more than 3.7 mm long (Long, 1960a). The integument forms a continuous sheath throughout its length and contracts slightly at the apex but hardly enough to merit calling the opening a micropyle. A short but distinct cylindrical salpinx and a well-developed central column (Fig. 9.11 & 9.12) are present. Pollen grains with a maximum diameter of 56 μm are present in the pollen chamber and occasionally between the integument and nucellus. The

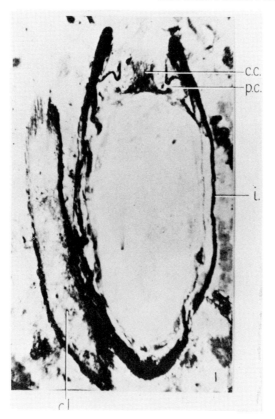

Figure 9.12. *Stamnostoma huttonense.* A longitudinal section of the seed in almost median section. CC = Central column; pc = pollen chamber; i = integument; sa = salpinx. × 26.5. (From Long, 1960a, Trans. Roy. Soc. Edinb. 64:210–15.)

female gametophyte is partially preserved and shows the remains of several archegonia. The seeds were partially enclosed in what seems to have been a rather primitive cupular organ; these were borne in pairs as shown in the restoration drawing (Fig. 9.11B) and although only one seed was actually found enclosed, scars suggest that there may have been four seeds per cupule.

Eurystoma angulare Long, 1960.

Although originally described in 1960, Long presented additional information on the cupular organ in 1965, and he reported (1969) a second species, *E. trigona*, which he interpreted as being borne on a frond of *Alcicornopteris.* Our summary here is taken from those three sources.

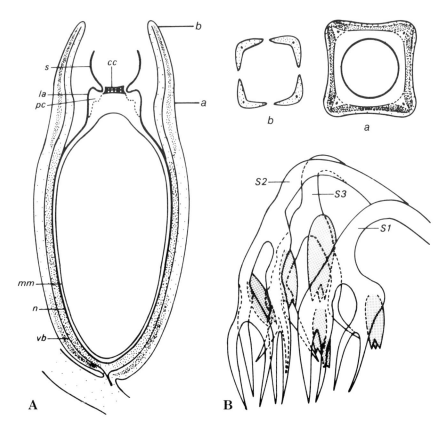

Figure 9.13. *Eurystoma angulare.* A. Diagrammatic restoration of median longitudinal section of a seed through diagonally opposite angles, and transverse sections at levels a and b. × 12.8. B. Restoration of a cupule showing three 'units', S1, S2, S3. Approximate position of ovules shown in stipple. × 3.5. (A. From Long, 1960, Trans. Roy. Soc. Edinb. 64:261–80. B. From Long, 1965, Trans. Roy. Soc. Edinb. 66:111–28.).

The *E. angulare* seed is 8 mm long and 2.5 mm in maximum diameter; it is radially symmetrical and quadrangular in cross section with a vascular strand in each angle of the integument (Fig. 9.13). The latter forms an enclosing sheath up to about the midregion of the pollen chamber and then divides into four lobes which curve inward slightly. The distal part of the lagenostome flares out distinctly to form a strongly developed salpinx (Fig. 9.13A).

The cupulate structures (Fig. 9.13B) were quite small, having a maximum width of 11 mm and a length of 10 to 15 mm; they seem to be little more than a terminal unit of slightly specialized branchlets partially enclosing the seeds. They were probably pendant and composed of as many as fifteen lobes bearing from two to ten seeds.

The seeds described as *Eurystoma trigona* (Fig. 9.14) were found in large quantities attached to foliage fragments identified by Long as *Alcicornopteris*. The individual seed or ovule has a maximum length of 4.5 mm and a width of 3.5 mm. In transverse section it is triangular with a vascular strand in each angle and a thin winglike flange extending outward. The integument is fused to the nucellus up to the base of the lagenostome and at a slightly higher level it divides into three separate lobes which do not come together to form a micropyle.

SOME PLATYSPERMIC SEEDS

Several platyspermic (bilaterally symmetrical) seeds are included in the assemblage that is now known and we include descriptions of representative ones.

Lyrasperma scotica (Calder) Long, 1960c.

This was originally described by Mary G. Calder in 1938 and more information has been obtained by Long from specimens collected from

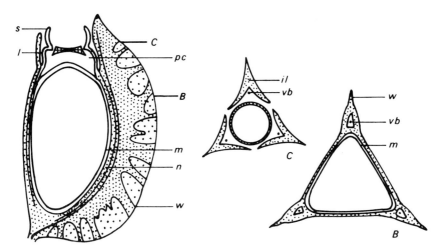

Figure 9.14. *Eurystoma trigona.* Diagrammatic restoration of longitudinal section through a 'wing,' and transverse sections at levels B and C. s = salpinx; l = lagenostome; pc = pollen chamber; m = megaspore; n = nucellus; w = wing of integument. × 14.4 (From Long, 1969, Trans. Roy. Soc. Edinb. 68:171–82.)

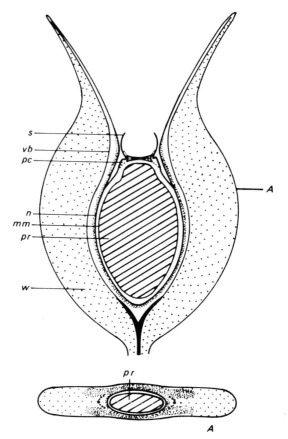

Figure 9.15. *Lyrasperma scotica.* Diagrammatic restoration of longitudinal section in major plane, and transverse section at level A. s = salpinx; vb = vascular strand; pc = pollen chamber; n = nucellus; mm = megaspore membrane; pr = prothallus; w = wing of integument. × 5.8. (From Long, 1960c, Trans. Roy. Soc. Edinb. 64:261–80.)

numerous Berwickshire localities. He notes:

> "The general impression gained by the wide occurrence of these seeds in Berwickshire is that the parent plant must have been an abundant, if not dominant, member of the Lower Carboniferous flora in this region." p. 262.

In its general external form it is likened to an elm (*Ulmus*) samara, (Fig. 9.15) the largest specimens being 12 mm wide and 18 mm long. They are curved and with an integument composed of two halves that are fused

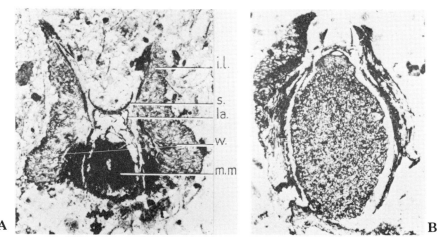

Figure 9.16. *Lyrasperma scotica.* A. Longitudinal section of upper half of seed in the major plane. × 9. B. Longitudinal section in major plane showing prothallus. × 9.6. (From Long, 1960c, Trans. Roy. Soc. Edinb. 64:261–80.)

to the nucellus up to the base of the lagenostome; beyond this they are separate and tend to flare outward, there being no micropyle. Much of the integument consists of a strongly developed sclerotesta composed of thick-walled cells. The single vascular bundle in the pedicel divides to form two branches which extend up into the tips of the two wings.

The distal part of the lagenostome forms a wide salpinx (Fig. 9.16) that is circular in cross section; thus, although the seed as a whole is platyspermic, the nucellus is radially symmetrical in its upper free region.

In some specimens the female gametophyte is quite well preserved (Fig. 9.16B) and displays parts of as many as three archegonia in some of which the jacket cells are recognizable. Of the numerous seeds that Long examined, pollen was found in the pollen chamber of ten, the largest number of grains in a single seed being 132. The pollen measure 56 μm in diameter.

Deltasperma fouldenense Long, 1961b.

This seed is known from some twenty-nine specimens found in two blocks of calciferous sandstone found near the River Whitadder in Berwickshire. They are 2.7 mm long with major and minor widths of 2.1 and 1.2 mm respectively. The platyspermy is slightly asymmetric, the degree of flattening being greater on one side than the other. The salpinx is not strongly developed being but a short extension of the lagenostome. The outer surface of the integument is characterized by conspicuous large epidermal cells with bulging walls that suggest mucilage cells similar to those of *Conostoma oblongum* (Oliver and Salisbury, 1911) and *Physostoma stellatum* (Holden, 1954).

Most of the seeds examined contained a female gametophyte, variously preserved, as well as pollen grains which seem to have a rather thin wall and measure 45 μm in diameter.

EVOLUTIONARY TRENDS IN EARLY SEEDS

It seems fitting at this point to bring together the most significant facts that we now have concerning the morphology and evolution of seeds up to early Carboniferous times, the significance of the distinctive structures encountered, and to speculate, at least cautiously, on the paths of evolution.

Taking the definition of an ovule (= seed in this discussion) as 'an integumented megasporangium' the most primitive seed that we know about is *Genomosperma kidstoni* in which the megasporangium is only 'loosely' enclosed by sterile integumentary filaments. Most botanists seem to agree that this kind of structure was arrived at through a chain of development beginning with plants such as the homosporous rhyniophytes with a terminally borne single sporangium passing through stages in heterospory in which the number of megaspores was ultimately reduced to one. We have no positive information on such a chain of development as far as *Genomosperma* itself is concerned.

Theories of seed evolution have centered around the origin of the integument. Albert Long (1966) has reviewed these and his account may be referred to for a more detailed description than we can include here. Briefly, drawing from a summary discussion by Worsdell (1904), Long notes the following theories as being in vogue at the beginning of the century: the *axial theory*, the *sui generis theory*, and the *foliar theory*. It would seem that these are now only of historical interest, having been based on comparative morphology of living plants and some imagination at a time when the fossil record was able to contribute very little.

Later theories that have achieved some favor include the *synangial theory*, the *nucellar modification theory*, the *double integument theory*, and the *telomic theory*. In reference to the lagenostomalean pteridosperm seeds (such as those described in this chapter) there is little doubt that the *telomic theory* is preferable. In Long's words this is briefly stated as follows: "that the ovule is derived from an apical megasporangium borne on a fertile telome surrounded by two or more sterile telomes which may or may not have cohered and become adnate to the megasporangium." (p. 358) If not actually proven, one can say that the evidence conforms to the concept very closely and it constitutes a triumph derived from the astute use of paleobotanical data. Numerous investigators have contributed to the development of the concept starting with Zimmerman and his telome theory. (For further discussion see Long, 1966; Pettitt, 1970; D. L. Smith, 1964; Taylor, 1981).

There are several other features of these early seeds and their evolution that are of exceptional interest and about which we have learned much in the last two decades.

It is evident that the gross form of a seed and particularly its symmetry in transverse section, that is, whether platyspermic or radio-spermic, in itself is not adequate for use as a criterion of classification. In this connection Long has demonstrated very nicely that the distal part of the nucellus is essentially identical in: *Lyrasperma scotica*, a bilaterally symmetrical seed with two lobes; *Eurystoma trigona*, a seed that is triangular in transverse section; and *Eurystoma angulare*, a typically radiospermic seed with four lobes in transverse section. In consequence he notes with apparent justification: ". . . that there seems little doubt that all three species must be related despite the difference in number of the integumental lobes." (1969, p. 176).

Modifications of the distal part of the integument and nucellus, which seem to be related to the problem of capturing pollen, present a most intriguing series of experiments on the part of these early seed plants. We may never have enough information on the exact stratigraphic occurrence of the various seeds to determine the precise sequence of events, but the specific "efforts" made by the plants seem to have been as follows: Starting with seeds such as *Genomosperma kidstoni* in which the integument consists of separate lobes that are strongly divergent at the apex, a fusion of the separate segments took place resulting in seeds such as *Eurystoma angulare* in which the lobes are nearly fused, to one such as *Stamnostoma huttonense* in which the lobes are completely fused, thus forming a large mouth into which the pollen could be blown.

Concurrently, at least in some plants, the lagenostome became extended as a distal funnel, the salpinx. And judging from the numerous pollen grains that are found in the pollen chamber of many of the seeds this functioned quite efficiently to direct the pollen to a position where they could effectively perform their function of producing male sex cells to fertilize the eggs in the gametophyte at the distal end of the prothallus.

With the more complete enclosure of the nucellar apex by the integument, resulting in a distinct micropyle, and apparently the evolution of a pollen-drop mechanism, the salpinx became, as Long so aptly puts it, 'redundant'. (1961). The pollen-drop mechanism is present in most of the groups of living gemnosperms and, briefly, involves a disintegration of some of the apical nucellar tissue to form a semi-fluid which exudes through the micropyle and serves to trap pollen that are blown into it. These tend to be sucked back into the seed when the pollen drop dries thus bringing them to a suitable place for germination and fertilization.

Figure 9.17. *Callospermarion undulatum.* A Middle Pennsylvanian age seed with the pollination drop at the apex. Ohio University Paleobotanical Herbarium No. 1027. × 30. (Courtesy of Gar W. Rothwell.)

Although some remarkable cases of fossilization have been reported, the pollen drop is hardly one that could be expected. It is thus of exceptional interest to note Rothwell's report (1977) of an apparent pollen drop (Fig. 9.17) extending through the micropyle of a *Callospermarion*-type ovule (a pteridosperm of the Callistophytaceae) from a Middle Pennsylvanian horizon. Aside from its general appearance which is quite convincing, it also contains several microspores.

That so distinctive a structure as the salpinx should have evolved when so urgently needed, served a highly useful purpose for a period of time, and then gone out of existence when a (presumably) more efficient mechanism (the pollen drop) was evolved, may cause one to wonder about the nature of the directive force that is responsible.

As an interesting corollary to this account, Niklas (1981) made models of some of the early seeds and simulated their pollination in a wind tunnel to test these hypotheses. The data obtained generally support them and offer new ideas about possible evolutionary 'trade-offs' between pollen trapping and ovule-size parameters, although the experiments suffer from the same constraints that descriptive studies have, namely our lack of knowledge of developmental sequences in early ovules and of the way in which ovules were borne on their parent plants.

There are several other aspects of these lagenostomalean seed plants that are especially interesting, or about which we need to obtain much more information. The small size of the seeds is notable and there is some evidence that they were borne in large numbers on an individual plant. And the small size of the pollen, which seems to be more pteridophytic than pollenlike in its gross morphology, is distinctive.

Finally, perhaps the greatest gap in our knowledge of these plants is in their habit as a whole. We try to fill this gap in a small way with the following discussion of *Pitus*.

GREAT FOREST TREES OF THE EARLY CARBONIFEROUS

Pitus Witham, 1833.
Pitus antiqua Witham, 1833.

Among the rather few fossil assemblages that give us an overall picture of an early seed plant (pteridosperm), the genus *Pitus* and some associated taxa are especially interesting and seemingly significant.

Not all of the parts that may compose one plant are known in organic connection. Briefly, the huge stems with portions of the fronds are known from especially well preserved fossils; the kinds of seeds that were borne on the fronds is reasonably certain; the general morphology of the sterile frond and the nature of the pollen organs is less definite but there are suggestions as to what they were. As a matter of record, and reference for those who may wish to obtain a more detailed account than we can give here, we include a chronological summary of the more important literature:

1830. First description of *Pitus* by Witham.

1833. First use of the generic name *Pitus* by Witham.

Summary descriptions by Seward, 1917, Scott, 1923.

1935. Detailed account of the stem anatomy of several species of *Pitus* by Gordon.

1960. Description of probable seed and cupule, *Stamnostoma huttonense*, by Long.

1962. Description of terminal portion or rachis (*Tristichia ovensi*) bearing the seed and cupule, by Long.

1963. Description of petiole-rachis, *Lyginorachis papilio*, by Long.

1979. Most recent summary account, by Long, including discussion of possible pollen organs.

Petrified stumps and trunks known under the name *Pitus* have attracted the attention of naturalists, and those who were simply curious, from the early part of the last century. Our information, as indicated above, comes from numerous sources but the recent summary (1979) by Albert Long is especially informative.

Our knowledge of the several species of *Pitus* is based on petrified remains found at various Lower Carboniferous localities in northern England and southern Scotland. Some of the earliest known specimens were discovered in quarrying operations and were known by the locality; thus the so-called Craigleith Tree was exposed in the Craigleith Quarry located within the city of Edinburgh. A specimen, found in the same quarry in 1830, had a maximum length (of the part of the trunk that was preserved) of 59 feet and a base diameter of 6 feet. Part of this has been recently (1977) erected as an open air exhibit at the Royal Botanic Garden in Edinburgh.

Many of the localities of the early years, where classic discoveries of fossil plants have been made, have long since been lost or obliterated. But in the case of *Pitus*, what is probably the greatest discovery was made a little more than a decade ago when J. B. W. Day found a deposit of spectacular specimens in the bed of the King Water (Fig. 9.18), a small stream located between Spadeadam and Gilsland in Cumbria (extreme northeast England). One of us (HNA) had the good fortune to visit the site on a cold January day in 1976 under the guidance of Albert Long.

Ten petrified stumps of *Pitus primaeva* have been found here in the position in which they grew and Long says of them:

"... it is evident that *Pitus* must have formed forests or groves at altitudes low enough to allow them to be buried in a marginal environment near the Lower Carboniferous sea ..." (1979, p. 113).

Being mostly in the bed of the stream (Fig. 9.19) it is not a place where the fossils will withstand erosion for long and it is hoped that something may be done to preserve one or more, the largest of which measures a little over 8 feet in diameter. It is, indeed, a fine display and for those who have an interest in other structures of the distant past (although not quite so far back in time) the Roman wall of Hadrian lies a short distance to the north.

There are four or five apparently distinct species of *Pitus* known from well-preserved stem structure and much of the following description is taken from W. T. Gordon's detailed study of 1935. The pith may be as much as 2 inches in diameter (Fig. 9.20) and is parenchymatous with the cells containing a variety of contents. A ring of primary mesarch vascular bundles, numbering as many as fifty, is located near the periphery, some

Figure 9.18. A view of the King Water, Cumbria, northeast England. (Courtesy of Albert Long.)

being in contact with the secondary wood and others slightly immersed in the pith parenchyma (Fig. 9.21).

The secondary xylem, which of course attains tremendous proportions in the larger specimens, consists of tracheids that are described as having the araucarioxylon type of pitting; that is, they are multiseriate, tending to be hexagonal due to the crowding. The ray structure is especially significant. In *Pitus dayi* the rays are four to six cells broad and up to thirty-six cells high: in *P. withami* they are mostly two to three cells broad, and in *P. rotunda* ten to fifteen cells broad. These differences are quite striking and Gordon notes: "It is the breadth and depth of the medullary rays, when viewed in longitudinal tangential section, which alone distinguishes the several species of *Pitys*; . . ."

The leaves or fronds that were borne on the *Pitus* branches are believed to have been of three kinds: the sterile and presumably photosynthetic ones: those bearing the seeds in cupules; and those bearing the pollen organs. It should perhaps be emphasized that our knowledge of these organs is based in part on anatomical comparisons of

Figure 9.19. Albert G. Long and a stump of *Pitus primaeva* in the bed of the King Water, Cumbria, northeast England. (Courtesy of Albert Long.)

branch fragments, and associations of compression and petrified remains. The tentative conclusions that can be drawn amount to much more than guesses but we admittedly do not have a complete picture based on organic connection of the various parts.

Thus, returning to a stem in transverse section, when a primary vascular strand passes out through the secondary wood and cortex to the base of the petiole, it will divide into three to five strands (Fig. 9.22 A). And at this point (i.e. the base of the petiole or rachis) the anatomy compares very closely with fossils previously described as *Lyginorachis papilio* (Long, 1963). But within a short distance of this proximal region the several bundles coalesce to form a U-shaped strand that is accompanied by secondary wood (Figs. 9.28B & C). Pinna traces are given off sparingly, and at a point about 20 cm from the stem the rachis dichotomizes.

The fertile (seed bearing) fronds believed by Long to belong to *Pitus* are characterized by having a central axis between the two branches of the dichotomy (Fig. 9.23). This central axis has a vascular strand in the form

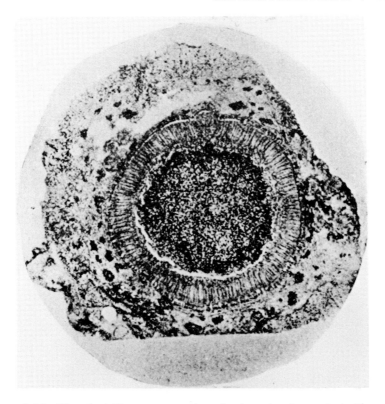

Figure 9.20. *Pitus dayi.* Transverse section of a stem showing central pith, wood and cortex. × 4. (From D. H. Scott, 1923, vol. II.)

of a three-angled protostele, and is believed to have borne clusters of cupulate organs. It seems most helpful to quote from Long's account (1962), and we have also reproduced here a photo of a portion of such a branch showing a terminal pair of cupules (Fig. 9.24). Of these he writes:

> "Cupules containing seeds are borne on bifurcating axes similar in size and form to the vegetative axes already described. . . . The cupules are borne on naked stalks about 1 mm wide. The length of the cupules is from 10 to 15 mm and the pedicels from 3 to 7 mm. . . . The cupules seem always to be borne in pairs with one cupule slightly overtopping the other and they are usually in such close contact that it is not always easy to distinguish them. . . . These, and others, show that the principal lobes of the cupules are separate from the base and sub-divided above into a number of fine elongated segments. Dr. Kidston was able to show that the cupules sometimes contain seeds and this fact I have

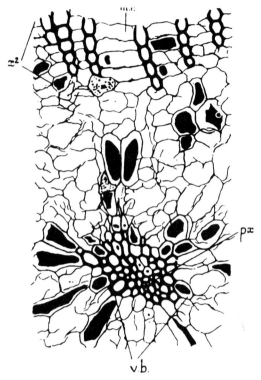

Figure 9.21. *Pitus dayi.* Transverse section of part of a stem showing the periphery of the pith with a primary xylem strand and the innermost part of the secondary xylem. × 89. (From D. H. Scott, 1923, vol. II.)

confirmed by the extraction of a total of twenty-four seeds. Of the cupules which yielded these seeds two gave two seeds and the rest one from each. It is probable that the cupules were multiovular." (p. 483)

Long's identification of the seeds as such is based largely on the megaspores and pollen that he was able to macerate out and he concludes ". . . there is a very close resemblance to the cupules of *S [tamnostoma] huttonense* which I described from petrified specimens before having seen the compressions." (p. 484). We have described the seeds of *Stamnostoma huttonense* on a previous page.

Long (1979) has found synangia associated with the *Tristichia* fossils which may have been the pollen organs of *Pitus*. These consist of eight partially united and somewhat elongate sporangia. It is worth noting that one of his sections of the microsporangia also contains a section of a *Stamnostoma* seed.

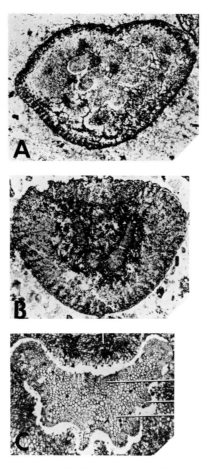

Figure 9.22. *Lyginorachis papilio.* Transverse sections of petiole showing stages in the fusion of vascular bundles and development of secondary xylem at different levels. A. × 3.8. B. × 4.4. C. × 11. (From Long, 1963, Trans. Roy. Soc. Edinb. 65:211–24.)

We conclude with a final note on the nomenclature. In his 1979 summary, Long reviews the more significant items in the older literature and some concept of the complexity may be gained from this statement:

"Throughout the remainder of the nineteenth century, the genus *Pitus* had a very confused history. At various times the species were referred to other genera of fossil stems. Thus in 1872 J. H. Balfour [in what was the first paleobotanical textbook in English] referred to them under five different generic names—*Pinites* L. & H. 1831, *Peuce* L. & H. 1832,

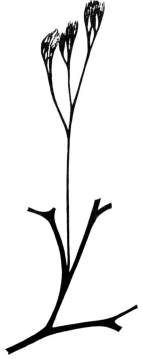

Figure 9.23. Long's concept of the way cupules were borne on *Pitus*. × 0.67. (From Long, 1963, Trans. Roy. Soc. Edinb. 65:211–24.)

> *Dadoxylon* Endlicher 1847, *Pissadendron* Endlicher 1847, and *Araucarioxylon* Kraus 1870 (see Balfour 1872 . . .). In 1881 Goeppert reverted to the name *Pitus* but changed its spelling to *Pitys*. This, however, was erroneous since the name *Pitys* had already been used by Endlicher for certain Pinaceous cones." (p. 112)

What emerges from all of this is some understanding of great forest trees, large for any age or plant group, which probably grew at low altitudes in what is now central Britain.

SUMMARY

One of the most important developments in paleobotany of the past two decades has been the discovery of a considerable assemblage of seeds and seed-bearing organs in horizons of late Devonian and Lower Carboniferous age. We have introduced a representative selection of the

Figure 9.24. *Stamnostoma huttonense.* Two of the terminal cupules. × 2.7. (From Long, 1962, Trans. Roy. Soc. Edinb. 64:477–89.)

Lower Carboniferous studies in order to present a significant unit of vascular plant evolution based on well-preserved compression or petrifaction fossils that reveal a considerable amount of cell structure.

An especially well preserved example, *Archaeosperma arnoldii*, comes from the Upper Devonian of Pennsylvania and is known from seeds, with a distally lobed integument, partially enclosed by telomic branch endings that are regarded as composing a primitive cupule. Fossils of a similar nature have also been found in the Upper Devonian of West Virginia, and isolated megaspores that probably represent a similar plant have been reported from Quebec. In all three areas *Archaeopteris* is also present; there is therefore a suggestion that some species of this genus may have been seed bearing.

We have reviewed rather briefly the more important aspects of the great series of studies of British Lower Carboniferous plants made by Albert G. Long in the two decades starting in 1960. A considerable number of these deal with seed-bearing organs of the early pteridosperms.

The information gleaned from his studies of large numbers of petrified seeds contributes significantly to our knowledge of their early evolution: the enclosure of the megasporangium (nucellus) by a ring of telomes or terminal branchlets to form the integument; the development (in several modifications) of the distal end of the nucellus into a *salpinx* which served as a pollen collection chamber prior to the evolution of the pollen-drop mechanism. In some cases the seeds were partially enclosed by an outer ring of branchlets representing early stages in the development of the cupule. Platyspermic as well as radially symmetrical seeds have been reported from both late Devonian and Lower Carboniferous deposits.

Much remains to be learned about the gross habit of the plants that bore these seed organs. However, a significant clue comes from numerous discoveries of petrified stumps and tree trunks, assigned to the genus *Pitus*, in Lower Carboniferous deposits of northern England and southern Scotland. The most interesting of these is a recently found assemblage of petrified stumps in the King Water in northeast England some of which have a trunk-base diameter of eight feet, thus representing apparently primitive gymnosperms of considerable magnitude.

10 / Heterospory in the Devonian

Heterospory is a phenomenon in which the spores in vascular plants are of two kinds: one is small and develops into a male gametophyte, while the other is significantly larger and develops into a female gametophyte. There is, however, considerable variation in different genera relative to spore size, the numbers produced, and sexual behavior in the gametophytes. Since the Devonian is rich in the origins of heterospory in several apparently unrelated groups we have considered it appropriate to discuss it in a separate chapter. It may also be noted that some of these Devonian plants, although very well preserved, cannot be satisfactorily classified at present in any of the established major groups. We take the opportunity here to describe them in some detail.

INTRODUCTORY COMMENTS

It may be helpful for those who are not familiar with heterospory to describe very briefly the nature of its occurrence in some living pteridophytic plants.

In *Selaginella*, a very widely distributed genus of some 700 species, the strobilus or cone may bear two kinds of sporangia: megasporangia which usually contain four large megaspores, and microsporangia which contain hundreds of small spores. The former develop into a female gametophyte and the latter into a male gametophyte. This contrasts with homosporous plants where spores are more uniform in size and germinate to produce bisexual gametophytes.

Selaginella has long been used in textbooks as a kind of classic example of heterospory. There is actually known to be considerable variation in the number of megaspores per sporangium both within a species and in different species. For further information reference may be made to Duerden's (1929) survey.

Numerous species of fossils have been described under the generic name *Selaginellites* from varous horizons extending back to the Carboniferous. Some of these appear to be closely related if not congeneric with the living *Selaginella* and suggest the existence of this lycopod line since the Carboniferous.

Heterospory occurs in the living species of *Isoetes* (the quillworts), a unique group whose relationships are somewhat controversial although most botanists ally it with the lycopods; they are mostly aquatic or found in moist places. The sporangia are exceptionally large, being up to 7 mm long. They are characteristically chambered and the megasporangia may produce 50 to 300 megaspores in different species, while the microsporangia may produce 150,000 to a million spores (Bold, 1973). The great size and spore productivity of the sporangia remind one of some of the early Devonian plants in this feature.

Several studies have been made of the spores and resultant gametophytes of the sphenopsid (Sphenophyla) genus *Equisetum* (the horsetail rushes). The overall results indicate strong uniformity in spore size, that is, they are homosporous, but the gametophytes show some tendency to be unisexual. Three recent studies may be cited that are especially informative. Duckett (1970a) studied the spores of ten species and found no evidence of heterospory, spore size being quite constant both within a species and in different species. In another study (1970b) dealing with the gametophytes of five species, he found that some spores develop into male gametophytes while others are bisexual; the former produce only antheridia while the latter initially produce archegonia followed by antheridia. And Hauke (1968), in a study of *E. bogotense* from Costa Rica, found that the gametophytes are strictly male or female, and also differ somewhat in their gross morphology.

The living polypodiaceous fern *Platyzoma microphyllum* from Queensland bears sporangia of somewhat different sizes; the larger ones contain spores in the range of 163 to 183 μm and the smaller ones contain spores in the range of 71–101 μm. The larger spores develop into spatulate gametophytes that bear archegonia while the smaller spores develop into smaller, filamentous gametophytes that bear antheridia (Tryon, 1964, 1967). It is also of interest to add that the 'megasporangia' here may contain some of the smaller spores and the 'microsporangia' may contain some of the larger ones. This mixture also may be present in a Devonian plant, *Chaleuria*, that is described on a later page.

Returning to the Devonian, we will first summarize the information that we have concerning the origins of heterospory in two groups of plants that are especially important in that the lines of evolution they represent continued on to produce seed plants.

In the lycopod group the earliest evidence that we have for heterospory comes from the arborescent (or semi-arborescent) *Cyclostigma* (see Chapter 5), known from the Upper Devonian of Ireland and Bear Island in the Arctic. Whether this was actually a progenitor of the great *Lepidodendron* and related forest trees of the Carboniferous is uncertain but it is clear that it was in the main stream of lycopod evolution that was headed in that direction. Taking as a definition for the seed (in paleobotany, seed often = ovule) 'an integumented megasporangium', the Carboniferous fossils known as *Lepidocarpon* qualify, although they are quite distinct from the seeds of the pteridosperms. The abundant data that is now at hand on the evolutionary stages which led to the lycopod seed, such as reduction in megaspore number per sporangium, increase in size of the remaining spores, and enclosure of the megasporangium by the sporophyll, is one of the most clearly defined chapters in plant evolution that the fossil record has revealed. Further details may be found in research reports of Abbot (1963), Ramanujam and Stewart (1969), and Phillips (1979).

In our treatment of the progymnosperms and the early seed plants (Chapter 8 and 9) we have described certain Upper Devonian and Lower Carboniferous plants that represent an especially important stream of evolution partially bridging the gap between pteridophytic and seed plants: the former, represented by *Archaeopteris*, are pteridophytic in their reproduction (being heterosporous), and with their cambium they produced stem structure that is gymnospermous. As to the latter, the earliest seed plants, such as *Archaeosperma, Genomosperma*, and numerous others that are cited, reveal significant stages in the development of an integument and the enclosing cupule, and thus offer clues to the origin of the angiosperm carpel.

An objection may be raised that, in interpreting apparent heterospory in fossil plants, we cannot germinate the fossil spores, and thus we cannot be certain that sexuality in the developing gametophytes was linked with spore size. This objection is largely eliminated by two factors: 1) by the information that we have from living heterosporous plants and from numerous instances of fossil megaspores containing well-preserved female gametophytes in the lycopod and progymnosperm/pteridosperm lines; and 2) ultrastructure of spore walls of living and fossil plants thus far demonstrates that megaspore walls are generally organized differently than are those of microspores (Pettitt, 1966; Taylor et al. 1980). It is not implied that there are not gaps in our knowledge, and in the descriptions of Devonian fossils that follow we have tried to distinguish clearly between observable facts and speculations concerning development in the plants when they were alive.

There are now several Devonian plants that are known from quite well preserved fossils and which display unique features, thus contributing significantly to the diversity of plants in that period, but which cannot be classified satisfactorily in any of the major groups. It is customary to establish new families and even higher taxa for such plants and we have no serious objection to such a procedure, but in the case of the plants described in the following pages this does not seem to advance our understanding of their affinities. We have, therefore, chosen to include them here simply because they are heterosporous and do contribute significantly to our understanding of that phenomenon.

SOME HETEROSPOROUS DEVONIAN PLANTS

Chaleuria **Andrews, Gensel, and Forbes 1974.**
Chaleuria cirrosa **Andrews, Gensel, Forbes, 1974.**

Our knowledge of this plant is based on abundant and well-preserved fossils and they present the oldest, or one of the oldest, instances of megafossils that provide significant information on gross morphology of the plant as well as good evidence of having been heterosporous. It was found in an outcrop, only a few feet above high-tide mark, near Dalhousie Junction, New Brunswick in the Campbellton Formation. This was thought to be of Middle Devonian age at the time, but more recent unpublished studies by D. C. McGregor of the dispersed spore assemblage of that area (discussed in an article by Gensel, 1982a) reveal the age as probably late Emsian (late Lower Devonian). The generic name is taken from the locality, which is at the head of Chaleur Bay.

A representative specimen is shown in Fig. 10.2A; the main axis is broken at both ends and is part of a plant that probably attained a height of at least one meter. The main axis as shown here is about 1 cm in diameter and bears primary branches in a close spiral. These branches are, in different specimens, either wholly sterile (and presumably photosynthetic) or wholly fertile. The sterile ones bear numerous secondary branchlets, about 8 mm long, that dichotomize once or twice. In the case of the fertile specimens the primary branches bear comparable branchlets, the ultimate divisions terminating in a pair of sporangia. (Fig. 10.1, 10.2B)

This is a good example of a Devonian plant with three-dimensional morphology in which the branches were apparently rather rigid and were preserved accordingly in the enclosing sediments. Thus only a small

Figure 10.1. *Chaleuria cirrosa.* Restoration of part of the plant showing a main axis and densely spiralled sterile and fertile branches. (From Andrews et al. 1974, Palaeontology.)

portion of the entire branch system is revealed when the rock is first split open.

Many of the sporangia were removed from several specimens, cleared for study, and the results as described below were the same in all cases. The sporangia are ovoid, frequently curved, and dehisced longitudinally; one sporangium from a pair is shown, after having been cleared, in Fig. 10.3A. This one contains only two or three spores but many of the sporangia were found to have been preserved prior to dehiscence. Some were found to contain numerous small spores in the range of 30 to 48 μm in diameter; others contain significantly larger ones ranging from 60 to 156 μm; and in a few cases a mixture of the two sizes was observed.

There are also morphological differences between the two spore sizes (Fig. 10.3B & C). The larger ones are circular to subcircular in outline and the exine outside the contact areas is sculptured with closely packed minute granules or rods; the smaller spores are much more strongly

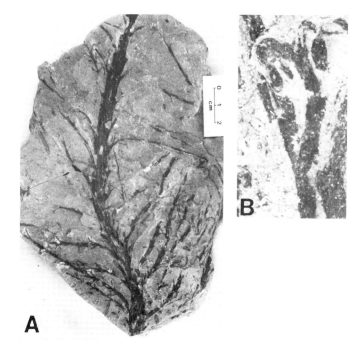

Figure 10.2. *Chaleuria cirrosa.* A. Main axis and densely spiralled branches. × 0.5. B. Detail of a fertile branchlet. × 4.3. (A. NCUPC Collections. B. From Andrews et al. 1974, Palaeontology.)

triangular in outline and the exine outside the contact areas is sculptured with baculi, coni, and, rarely, spinae.

The closest comparison that we can draw with any other fossil is with *Arctophyton gracile* as described by Schweitzer (1968) from the Middle Devonian of Spitsbergen. The two are similar in their general habit but the ultimate sterile appendages of *Arctophyton* are more profusely branched and the spores are not preserved.

In their original description of the plant the authors stated: "We are not able to fit *Chaleuria* satisfactorily into any existing scheme of classification." (p. 406) But on the basis of heterospory they suggested a relationship with the progymnosperms. We now feel that even that suggestion was misleading; it is an intriguing and unique plant that expands our knowledge of Devonian plants in general and of heterospory in particular but its affinities are not known.

Barinophyton White, 1905.

This genus was founded by David White on specimens from the Upper Devonian of Perry, Maine, the type species being *B. richardsoni.*

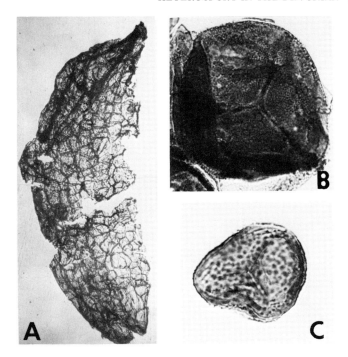

Figure 10.3. *Chaleuria cirrosa.* A. A single sporangium from which most of the spores have been lost. × 36. B. A presumed megaspore. × 398. C. A presumed microspore with distinct coni. × 763. (From Andrews et al. 1974, Palaeontology.)

Although many Devonian plants seem strange and are difficult to classify, the genus *Barinophyton* is especially complex and puzzling. Following White's brief introduction several other paleobotanists have dealt with it, among the more important being studies by Arnold (1939), Pettitt (1965), Stockmans (1948), and most recently by Brauer (1980, 1981).

Barinophyton citrulliforme, Arnold, 1939.

This species was established on specimens found near Rock City, Cattaraugus County, New York State. Brauer's account (1980) of it deals with specimens collected from a roadside locality near Burtville, Potter County, Pennsylvania, the horizon being the Catskill Formation of the Upper Devonian. His fossils were much better preserved than those available to any previous investigator; his treatment of them stands as a model of paleobotanical excellence; and the information that is given in his 1980 report greatly advanced our knowledge of the genus. A considerable portion of the data given here comes from that source.

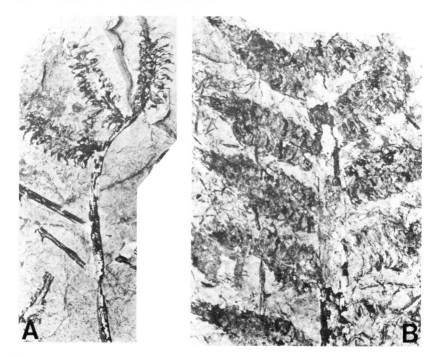

Figure 10.4. *Barinophyton citrulliforme.* A,B. Belgian specimens showing a main axis and laterally borne strobili. × 0.6 and × 0.8. (From Stockmans, 1948, Mem. Mus. Roy. Hist. Nat. Belg.)

Most of our knowledge of the plant pertains to the unique spore-bearing 'strobilar complex' which consists of a main axis bearing the strobili in a spiral arrangement. Fig. 10.5A shows a portion of the complex with four of the strobili attached to a central axis, the latter being 5.5 to 8.5 mm in diameter. It is appropriate to note that *B. citrulliforme* has also been reported from the Upper Devonian of Belgium by Stockmans (1948). His account is rather brief but includes photos of good specimens, two of which we have included here (Fig. 10.4); it shows parts of four strobili and indicates that the complex or group of strobili was borne terminally on a branch or main stem of the plant.

The individual strobilus consists of an axis that bears two rows of 'sporangiferous appendages' (see Fig. 10.6) which in turn bear the sporangia. The sporangiferous appendages are arranged in an alternate fashion, two-ranked along the axis, and they recurve abaxially around the axis. The appendages have a total length of 10 to 12 mm and each bears one large sporangium inside the curve. There are twenty-five in each row making a total of fifty per strobilus. Each sporangium measures about 5 mm wide, 8 mm long, and 2 to 3 mm deep.

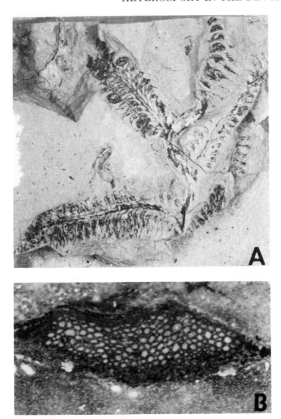

Figure 10.5. *Barinophyton citrulliforme.* A. Part of a terminal 'strobilar complex,' showing four attached strobili. From New York State. × 0.7. B. A section through the main axis between two strobili showing the xylem. × 30. (From Brauer, 1980, Amer. Journ. Bot.)

Brauer was able to remove numerous fragments of the individual sporangial contents and in each case these revealed several complete megaspores and hundreds of microspores. The microspores range from 33 to 48 μm and the megaspores are 700 to 900 μm in diameter. He estimates a total of twenty to thirty megaspores per sporangium and thousands of microspores. The size difference is quite striking: Fig. 10.7B shows two megaspores with several hundred microspores or microspore casts; Fig. 19.7A shows a single megaspore with numerous microspores clinging to it; and Fig. 10.7C is a photo of microspores at a higher magnification. The microspores are smooth or finely granulate, trilete, and curvaturate. They compare best with the dispersed spore genus *Retusotriletes.* The megaspores also are smooth walled, with a very small

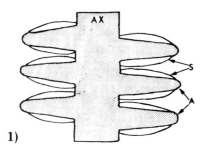

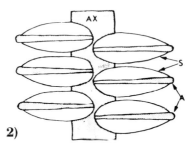

Figure 10.6. *Barinophyton citrulliforme.* Restorations of part of a strobilus showing 1) adaxial surface and 2) abaxial surface. × 2.7. (From Brauer, 1980, Amer. Journ. Bot.)

trilete mark, and abundant folds. A study by T.N. Taylor and D.F. Brauer (1983) showed the two types of spores differ ultrastructurally; the megaspores exhibit a spongy wall pattern similar to that seen in megaspores of other plants while the microspore walls are homogeneous.

There is no structural evidence of sporangial dehiscence and Brauer says: "Thus, it would seem possible that the sporangia of *B. citrulliforme* were released unopened, that the sporangia functioned to prevent dessication during seasonal dry periods and that the spores were later released after the breakdown of the sporangial wall." (1980, p. 1204)

An especially curious aspect of *Barinophyton*, and one that has concerned previous investigators, is the apparent presence of both micro- and megaspores in the same sporangium. Pettitt (1965) sums this up very well in his description of specimens of *B. richardsoni* from the Upper Devonian of Perry, Maine: "If each carbonaceous mass between successive appendages is the remains of one sporangium as is suggested in the descriptions of this genus by Arnold (1939) and by Kräusel and Weyland (1941), it is difficult to explain the presence of both microspores and megaspores in every sporangial fragment. The possibility that the sporangia are bisexual cannot be ruled out, but it would certainly be unusual." (p. 85)

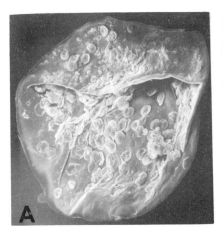
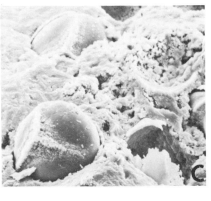
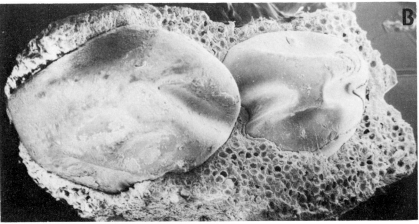

Figure 10.7. *Barinophyton citrulliforme.* A. A single megaspore with numerous microspores adhering to it. × 100. B. A sporangial fragment from a single sporangium showing two megaspores surrounded by microspores and microspore molds. × 61. C. An enlargement of part of B showing microspores. × 700. (From Brauer, 1980, Amer. Jour. Bot.)

Brauer's fossils are permineralized and they show clearly that the two kinds of spores were indeed produced in each sporangium. One may agree with Petitt that this is "unusual" but the fossil record reveals many things that are unusual in terms of what is known only from living plants.

The vascular tissue of the main axis (Fig. 10.5B) is a somewhat flattened exarch protostele of 1.5 to 2.0 mm in diameter and the tracheids are described as having a basically annular pattern with some connected annular thickenings.

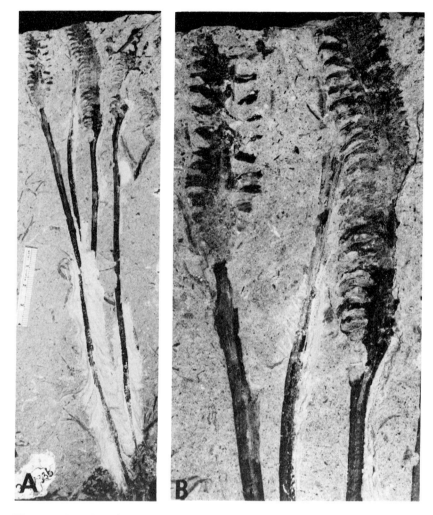

Figure 10.8. *Protobarinophyton pennsylvanicum.* A. Plant axes shown bearing terminal strobilus. × 0.4. B. Detail of strobili from A. × 1.5. (From Brauer, 1981, Rev. Palaeobot. Palynol.)

Although we still know rather little about the plants as a whole, Brauer has given a very significant lead in a second contribution (1981). He figures a specimen of the closely related genus *Protobarinophyton* in which there is an axis 29 cm long attached below the proximal strobilus (Fig. 10.8). It appears as a flattened stem without appendages, a little under 1 cm in diameter and shows a prominent midrib. He points out the close comparison of this axis to those classified under the generic name *Taeniocrada* and he also notes: "Vegetative specimens referable to

TABLE 10.1

Species, Reference	Microspores, μm	Megaspores, μm	Origin
B. citrulliforme			
(Arnold) Brauer, 1980	33.0–48.5	700–900	Burtville, Pa
			Upper Devonian
B. cf. obscurum (Dun)			
White.	47.0–65.0	480–600	Burtville, Pa
Brauer, 1981			Catskill Fm
			Upper Devonian
B. richardsoni (Dawson)	48.0–62.0	220–250	
White.			Perry, Maine
Pettitt, 1965			Frasnian

Taeniocrada have been described at almost every locality for *Barinophyton* and *Protobarinophyton.*" (1981, p. 359) under the generic name *Taeniocrada* and he also notes: "Vegetative specimens referable to *Taeniocrada* have been described at almost every locality for *Barinophyton* and *Protobarinophyton.*" (1981, p. 359)

This is an intriguing suggestion. The naked, sterile, flat axes with a narrow but conspicuous midrib that are usually referred to as *Taeniocrada* are known from many localities in the Devonian. The present writers have encountered them in our explorations in northern Maine and along the southeast coasts of Canada, but with no fertile parts to allow even a guess as to their relationships. Brauer's observations do not necessarily imply that all *Taeniocrada*-like axes represent the stems that bore the *Barinophyton* fructifications but we have strong evidence that some did.

In the several species of *Barinophyton* that have been described perhaps the most interesting variation is that of spore size, as shown in Table 10.1.

We do not have exact stratigraphic data on the relative ages of these three species but there is a strong suggestion of an evolutionary trend in megaspore size. Future investigations may be expected to shed more light on this aspect of *Barinophyton.*

Barinophyton is one of several problematical genera (including *Pectinophyton* Høeg, *Protobarinophyton* Ananiev, and *Barinostrobus* Kräusel and Weyland) that are placed in the family Barinophytaceae of the Order *incertae sedis* Barinophytales. A summary treatment of these fossils is given by Høeg in Vol. II of the *Traité de Paléobotanique.* In a discussion of the relationships of the family Brauer (1981) says: ". . . Barinophytaceae can be considered a natural grouping of plants that have in common sporangia borne on short sporangiferous appendages and aggregated into strobili." (p. 361). But: "The Barinophytaceae do not show any close affinities to any fossil or extant plants." (p. 359).

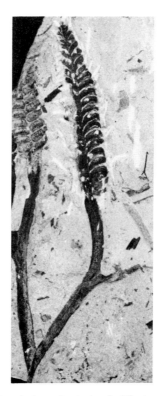

Figure 10.9. *Protobarinophyton obrutschevii.* Photo of a fertile specimen provided by D. F. Brauer from a USSR specimen in the collections of the British
Museum of Natural History. × 0.9.

Protobarinophyton Ananiev, 1955.
Protobarinophyton obrutschevii Ananiev, 1955.

This was established for specimens collected from a Lower Devonian
horizon near Torgashino, Siberia. The genus has been dealt with by other
authors, including a report on a new species, *P. pennsylvanicum*, by
Brauer in 1981. It is not as well known as *Barinophyton*, the most
distinctive difference being, in the present stage of our knowledge, that
the stems of *Protobarinophyton* bear a single terminal strobilus (Fig.
10.9).

Enigmophyton Høeg, 1942.
Enigmophyton superbum Høeg, 1942.

Among the more interesting fossil plants to have been found in
Spitsbergen is an assemblage consisting of branching axes bearing

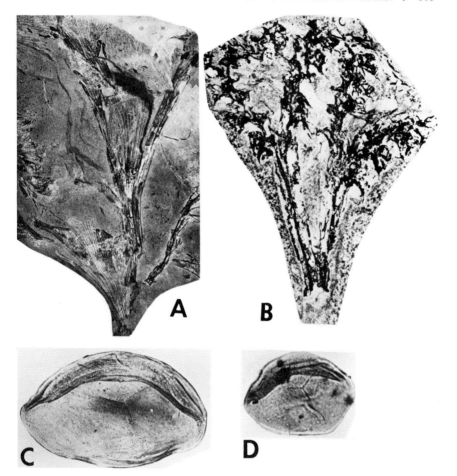

Figure 10.10. *Enigmophyton superbum.* A. Stem with a leaf attached. × 0.4. B. Bifurcating axis terminating in strobili. × 0.9. C,D. Presumed megaspore and microspore, respectively. × 166 and × 395. (A,B from Høeg, 1942, Norges Svalbard-og Ishavs-Unders. C,D from Vigran, 1964, Norsk Polarinst. Skr.)

flabelliform leaves that are associated with unique heterosporous fructifications (Figs. 10.10 and 10.11). The former were given the name *Enigmophyton superbum* and the latter were described as *"Enigmophyton* (?), fructification." We are using the terminology here but it should be understood that the two are not known in organic connection. They compose a dominant element of the flora of Planteryggen, Mimerdalen, and are of upper Middle Devonian (or lowermost Upper Devonian) age, and are among the many interesting plants described by Høeg in his 1942 volume.

Figure 10.11. *Enigmophyton superbum.* Restoration of a part of the plant as envisaged by Høeg. The fertile branch is not actually known to be attached. (From Høeg, 1942, Norges Svalbard-og Ishavs-Under.)

E. *superbum* consists of stem specimens 5 mm broad, the longest fragment that was found measuring 27 cm. Two kinds of branching have been observed; the most frequent one is a regular dichotomy that takes place about every 6 cm, with the branches forming an acute angle that results in a nearly erect habit (Fig. 10.11). In addition, smaller branches occasionally depart from the stems at the point of dichotomy; these are 1 to 2 mm broad and the longest ones are 2 cm. Their significance is not understood.

The stems bear distinctive fan-shaped leaves measuring at least 16 cm long and 12 cm broad (Fig. 10.10A). They have no petiole as such, the basal constricted part being attached to the stem at the points of dichotomy. Bifurcating veins are present which maintain a distance of 1.0 to 1.5 mm, and the distal margin of the leaf is variously split into lobes of different sizes.

The fructification ("*Enigmophyton* ?") consists of bifurcating axes (Fig. 10.10B) with a total length of about 2 cm. The axis bears sporophylls that were apparently arranged in a low spiral and they depart from the axis at right angles, being described as follows: "The base of the sporophyll seems to be thick, while the distal part is quite thin; most probably it forms a flat, thin lamina, expanded transversely . . ." (p. 117–118).

The sporophyll is about 2 mm long and apparently bore one sporangium on the upper side. Some of the sporangia contain spores that measure about 250 μm in diameter while the spores in others are 50 to 85 μm. Numerous spores were obtained from the sporangia and the difference in size seems to leave no doubt as to the heterosporous nature of the fructification.

Vigran (1964) described the spores (Fig. 10.10C, D) in more detail; the megaspores are smooth to punctate, with a short trilete mark and with the areas around the laesurae darker than the rest of the wall. She called them *Enigmophytospora*.

Several generic entities have been described from Devonian horizons which consist of more or less flabelliform leaves, under such names as *Psygmophyllum, Platyphyllum,* and *Ginkgophyllum.* This type of leaf may constitute one of the earliest occurrences of laminate leaves in the fossil record. Very little is known about the plants as a whole or their affinities and one can only say that they add to the known diversity of Devonian plant life. Høeg has reviewed the literature and his account may be referred to for such additional information as is available.

Kryshtofovichia Nikitin, 1934.
Kryshtofovichia africani Nikitin, 1934.

The fossils described under this name are distinctive but problematical mega- and microsporangia from an Upper Devonian horizon in the Veronezh district of Central Russia near Petino village. The megasporangia, which are round to ovoid, are about 3 by 2 mm and contained one hundred or more spores. The megaspores are about 0.5 mm in diameter and are distinguished by having conspicuous hooked spines and an apical three-valved apparatus (microspore chamber) that Nikitin calls an "androcamera" and that represents raised lips in the trilete region. The author suggests that these cameras "when under the water level were able to open while, when getting dry, they became closed." The microsporangia are ellipsoidal and measure about 2.0 × 1.7 mm and contain large numbers of microspores which measure about 25 × 45 μm. It is reported that the microsporangia are often found attached to each other in a linear group of four. The microspores are reported to be often found adhering in masses to the surface of the megaspores and are found in the apical camera. They are of particular interest in being monolete, comparing closely to the dispersed spore genus *Archaeoperisaccus* (McGregor, 1969). Taylor et al. (1980) described the ultrastructure of megaspores similar to those of *Kryshtofovichia*, the megaspores being called *Nikitinsporites*; again, they exhibit a spongy wall pattern and elaborate aperture.

There seems to be little doubt that the micro- and megasporangia are correctly referred to the same plant but the relationships of the latter are very much of a question. Nikitin notes: "The same cameras are typical according [to] the author's opinion to the spores of *Lepidodendron* (*Lagenicula* Kidst.)", which suggest lycopod affinities. Other aspects of the fossils, such as the apparent grouping of the sporangia do not support this.

EVIDENCE FROM DISPERSED SPORES

The evidence afforded by dispersed fossil spores indicates that there were more heterosporous (as well as homosporous) plants in the early Devonian than are presently known from the macrofossil record. Our brief summary is taken from Chaloner's account of 1967, "Spores and Land-Plant Evolution."

The data gathered together by Chaloner is based on the generally accepted but arbitrary demarcation line of 200 μm in diameter for the identification of a megaspore. On this basis two sets of data are of particular interest: There are no species as such known below the Emsian with a mean spore size above 200 μm. This suggests that there were no vascular plants below that time point that were consistently hetero-sporous. There are a few species, first appearing in the Siegenian, in which spore size ranges up to or larger than 200 μm, and Chaloner notes: "These probably represent plants with only weakly differentiated hetero-spory, or in which the size range of megaspores and microspores overlaps." (p. 88).

Although no spore-species above the 200 μm size appear in the fossil record below the Emsian, from that time on the number increases significantly through the Devonian. And as to the spore record in the Devonian generally: "It may also be noted that here, as elsewhere in the fossil record, the macrofossil remains probably greatly under-represent the diversity of heterosporous plants." (Chaloner, 1967, p. 90). This statement is supported by the rather numerous fragments of stems, and other organs that are encountered, that indicate different plants which can be significantly understood only when more complete specimens are found.

SUMMARY

We are concerned here primarily with the origins of heterospory in Devonian plants but a few notations may be included on its occurrence in other plant groups of later geologic periods.

It is evident that heterospory has originated independently in numerous groups of plants beginning, according to the present state of our knowledge, in the Lower Devonian. Within that time range it seems to have reached a clearly defined morphological state in the Emsian as indicated by such well-preserved megafossils as *Chaleuria cirrosa*, while the dispersed spore records suggest a somewhat faltering start in the Gedinnian. It therefore follows that this differentiation of spores into two distinct sizes and functions was initiated quite soon after the origin of homosporous vascular plants, which probably became established in the late middle Silurian.

In the Devonian, the appearance of heterosporous lycopods and progymnosperms seems especially important since these two rather distantly related groups were successful in carrying heterospory to its ultimate goal, the development of the seed. In the case of the lycopods the Upper Devonian *Cyclostigma*, an arborescent or near-arborescent plant, may represent the early stages in the evolution of the great Carboniferous forest trees such as the Lepidodendrons with their unique type of seed. The fossil herbaceous representatives (generally assigned to *Selaginellites*) continued on to the present merging into *Selaginella* with its numerous and widely distributed species. In the case of the progymnosperms, the Upper Devonian genus *Archaeopteris* represents that line of plants that probably led to the Carboniferous pteridosperms.

We have described three other Devonian macrofossils, *Chaleuria*, *Barinophyton*, and *Enigmophyton* that were heterosporous. There is perhaps a closer comparison to be drawn between *Barinophyton* and *Enigmophyton* than there is between either of these and *Chaleuria*. Thus they represent two and perhaps three distinct lines of vascular plants in which heterospory evolved and it is our opinion that, at the present time, there is little to be gained from speculations as to their relationships to other groups.

In the Coenopteridopsida, as this assemblage is treated by Taylor (1981), *Stauropteris burntislandica* from the Lower Carboniferous of Pettycur, Scotland, is a heterosporous member of this group. It is distinctive with its spherical microsporangia containing numerous microspores, and the megasporangium which is a slightly asymmetrical, ovoid organ with two large functional megaspores. (See Surange, 1952; *Traité de Paléobotanique*, Vol IV).

Also, in the Upper Carboniferous heterospory appears in the articulates in *Calamostachys*, a genus of numerous species that are reviewed by Boureau in Vol. III of the *Traité de Paléobotanique*. Some of these were definitely heterosporous while others were apparently homosporous. *Calamostachys casheana* was described from specimens found in Lancashire coal-ball petrifications many years ago by W.C. Williamson.

His well-known drawing of a longitudinal section through the cone, showing the microsporangia above and the megasporangia below, appears as a frontispiece in Scott's *Studies in Fossil Botany* (Vol. I, 1920), and elsewhere. *C. americana* Arnold (1958), from Illinois, U.S.A., is another heterosporous species. It is a large and very well preserved cone with microspores 85 to 114 μm and megaspores 150 to 260 μm in diameter.

Among living plants, heterospory is also found in the two pteridophytic orders Marsiliales and Salviniales which include several genera of aquatic plants commonly known as 'water ferns', but whose natural relationships to other groups of plants are problematical. They present many unique features and certainly are not closely related to any of the other known groups in which heterospory is found. Fossils referable to these groups first appear in Cretaceous age rocks, and some especially interesting petrified remains of *Azolla intertrappea* from the early Tertiary of India were described by Sahni in 1941.

Heterospory appears quite early in the Devonian and the available evidence indicates that it originated in several different groups in that period, and it originated in other groups from the Carboniferous on. As additional sources of information on aspects of heterospory that we are concerned with here, as well as others, two recent surveys may be referred to: "Heterospory and the Origin of the Seed Habit," by John Pettitt (1970), and "The Origin and Development of Heterospory in Vascular Plants," by Ian Sussex (1966). There is much that remains to be learned about heterospory and further studies of both living and fossil plants may be expected to contribute to that end.

11 / Palynological Considerations

Palynology is the study of microscopic, acid-resistant organic remains and as such encompasses spores, pollen grains, algal remains such as dinoflagellates and acritarchs, and other entities whose affinities are not well known. Its primary application is in stratigraphy—these small organic remains are produced in relatively large numbers and different morphologies and are deposited as sedimentary particles in a wide range of rock types independent of the deposition of megafossils. In fact, in some areas, dispersed spore assemblages are most abundant and best preserved in sediments lacking megafossils and relatively sparse and poorly preserved in sediments rich in megafossils (McGregor, 1979c). Spores and pollen grains are extracted from the rocks by chemical means, mounted on microscope slides and studied with light or electron microscopy. They can be compared to *in situ* spores or collectively characterized and compared to other dispersed spore assemblages. More precise stratigraphical correlation results when spore assemblages are correlated to faunal assemblages whose stratigraphy is well worked out.

An increasing number of paleobotanists consider spore evidence when discussing megafossils, and palynologists are increasingly concerned with the parent plant affinities or other biological implications of dispersed spore types. It is clear that studies of spores found dispersed in Devonian sediments are an important source of supplemental information on the nature of Devonian plant life and major events in their evolution. Such studies also provide a means for improved biostratigraphic correlation, and thus dating, of Devonian rocks. Characterization of spores from sporangia of Devonian plants has provided considerable insight on an important phase of their life history and offers data on reproductive processes.

Devonian sediments contain mostly spores (i.e. a single-celled reproductive disseminule) and acritarchs (affinities unknown, possibly resting cysts of algae). Many Early Devonian plants produced spores of one size (they were homosporous), but with the advent of heterosporous plants in the Lower and Middle Devonian, megaspores and microspores appear. Most of the spores are trilete, i.e. they exhibit a Y-shaped region of weakness representing the site of germination on the proximal face (side nearest the center of the tetrad). In extant plants, trilete spores are produced mainly by ferns and fern allies, and by a few bryophytes (especially liverworts). Rare monolete spores (with a single suture line) appear in dispersed spore assemblages from late Lower Devonian on and one type is briefly abundant in certain regions in the Upper Devonian. Early seed plants of Carboniferous age produced trilete pollen and it is assumed that the seed plants of the Late Devonian may have also. If found dispersed, they would be difficult to recognize as being different from pteridophyte spores. Most later gymnosperms produce pollen with a distal germination area or no obvious germination area. Thus the following discussion on Devonian spores and spore assemblages will include possible pre-pollen, microspores and megaspores of heterosporous plants in addition to spores of homosporous plants. Any spore under 200 μm in diameter, whose source is not known, is often referred to as a *miospore* or *isospore*.

GENERAL SURVEYS OF PALYNOLOGICAL DATA PERTAINING TO VASCULAR PLANT EVOLUTION

Application of the data obtained from assemblages of dispersed spores to our general understanding of vascular plant evolution is fairly recent. Chaloner (1967) in his paper "Spores and Land Plant Evolution" demonstrates that the dispersed spore record mostly confirms the evolutionary advances seen in the megafossil record through the Devonian, although at times a major evolutionary event may be reflected slightly earlier in time by the spore record (advent of heterospory, diversity of land plants in the Late Silurian). The dispersed spore record shows that spores increase in number, complexity and diversity from the Late Silurian to the Carboniferous. Spores increase in size, reaching 200 μm by the end of the Siegenian, and exceed it in subsequent stages. In Late

Devonian and Early Carboniferous times spores fall into two size classes. These data suggest the advent and establishment of heterospory; as noted in Ch. 10, the earliest possibly heterosporous plants known from megafossils are Emsian in age (*Chaleuria cirrosa, Barinophyton*-like plants). Chaloner also discussed possible microfossil evidence for the existence of seed plants and noted the occurrence of the seed megaspore *Cystosporites devonicus* in the Upper Devonian. We now know that at least three types of ovules existed in Late Devonian times, and that one of them, *Archaeosperma*, possesses a seed megaspore identical to *C. devonicus*. The pollen produced by the plants bearing these ovules is unknown and may be identical to the trilete spores as mentioned previously.

Several studies have shown that spores are more diverse in the Late Silurian than megafossil remains are (Richardson and Lister, 1969; Richardson and Ioannides, 1973). Gray, et al. (1982) have advocated that dispersed spores may indicate an earlier time for the establishment of plants on the land (see Ch. 3) based on studies of spores from the Ordovician and Silurian of the eastern USA, Sweden, and Libya. Prior to the lower Silurian, tetrads of spores occur, with isolated spores becoming prominent in Llandovery time. Similar spores have been recorded from Silurian continental sediments by several workers, including Pratt, et al. (1978), Strother and Traverse (1979) and McGregor and Narbonne (1978).

Richardson (1969) presented a comprehensive survey of dispersed spores found in Devonian sediments and discussed spore-parent plant relationships. He noted, as have subsequent workers, that many dispersed spore types of stratigraphic importance (they are distinctive, widely occurring, and of relatively short duration) have not been found in sporangia (e.g. *Emphanisporites, Ancyrospora*, some pseudosaccate spore types, some with proximal papillae between the trilete rays). Gensel (1980) and Allen (1980) survey the occurrence of *in situ* spores which further address spore-parent plant relationships.

Richardson and, later, McGregor (1979a) note that some spore types are cosmopolitan while others are more restricted geographically. One example of the latter is the occurrence of the monolete spore *Archaeoperisaccus* and the megaspore *Nikitinsporites* in lower Upper Devonian (Frasnian) rocks in regions restricted to 30° or more north paleolatitude (see map in McGregor, 1979). These types of spores have been found in the presumed lycophyte cone *Kryshtofovichia* Nikitin (McGregor, 1969); at present, nothing is known about the plant which bore this cone.

Some spores appear to be restricted to certain types of sedimentary facies, reflecting specific environments—in Frasnian and Famennian deposits, spores referable to the dispersed spore genera *Ancyrospora* and

Rhabdosporites (the latter is produced by the progymnosperms *Tetra-xylopteris* and *Rellimia*) are most abundant in sediments indicating deposition in freshwater lakes. Spores of comparable age and referable to *Geminospora* (affinities uncertain, perhaps microspores of *Archaeopteris* in part), are most abundant in nearshore marine environments. This correlation of spore type and sedimentary environment has been noted in Russia, New York State, and Scotland.

BIOSTRATIGRAPHY

McGregor (1979a) points out that great progress has been made in the past decade on improved biostratigraphic correlations using Devonian spores, but more work is needed on vertically long sequences, and ones with accompanying lithological and faunal studies. The most extensive advances have been made in delimiting the Lower-Middle Devonian boundary and the Upper Devonian-Lower Carboniferous boundary. This is partly a result of the activities of working groups devoted to strati-graphic or morphologic problems. Most of the work being done centers on sediments that form the perhiphery of the Old Red Sandstone landmass, and much remains to be done in other regions. The same is true of megafossil studies.

In regard to spore distribution in time and space, some near-cosmopolitan forms are also long ranging. However, we know that a given spore type may be produced by different types of plants and result in apparent longevity. Other spore genera are quite short-lived. For example, the genera *Retispora* and *Vallatisporites* existed only for latest Devonian times (Fa2d to Tn1b) but were nearly worldwide. It would be of great interest to know what plants produced these distinctive spores. They die out in the Early Carboniferous and are replaced by other distinctive and sometimes short-lived forms. Perhaps this signals an extinction event and subsequent replacement by other types of plants.

McGregor noted also that some spore assemblages occur in regions where the paleopole has been placed and that the spores present are not entirely different from ones occurring in presumed equatorial zones. Similar discrepancies exist in regard to megafossils also. Several inter-pretations can be suggested: some Devonian plants may have exhibited broad climatic tolerances, similar spores located at the poles and at the equator may have been produced by different plants, environmental differences between pole and equator were less extreme in the Devonian, or further refinement of paleocontinental positions is needed.

Although much more information is needed, it seems possible, as suggested for *Retispora* and *Vallatisporites*, that some changes in spore assemblages might indicate major events in vascular plant evolution—at least in terms of extinction and appearance of new types, both locally and worldwide. Further examples include: Spore assemblage composition changes in the upper part of the Battery Point Formation, Gaspé, Canada, (McGregor, 1973, 1977), and comparable changes have been recorded in Belgium, Germany, and possibly in the USSR, Algeria, Scotland, and northern Spain. This is close to the Lower- Middle Devonian boundary. Stratigraphic resolution of megafossil assemblages has not been precise enough to document similar obvious changes although in general, Middle Devonian floras contain many new types of plants with some carryover of Lower Devonian ones (Banks, 1980b).

The great abundance of pseudosaccate and zonate spores and ones with large bifurcate spines in the Middle and Upper Devonian signals the advent of some distinctive plant groups, some of the former produced by the progymnosperms (the pseudosaccate spore *Rhabdosporites* is produced by the aneurophytes *Tetraxylopteris* and *Rellimia*). The types of spores produced by many Middle Devonian plant types are unknown, perhaps because many megafossil taxa are based on permineralized vegetative axes.

IN SITU SPORE STUDIES

Since the work of Kidston and Lang in 1917, many Devonian paleobotanists have been interested in determining if spores were present in the structures they interpreted as sporangia, both to prove they represented sporangia and to assess the nature of their reproductive disseminules. Despite this interest, the number of plants from which *in situ* spores (spores from sporangia) have been obtained and thoroughly described is comparatively low—reaching today about 50% of all megafossil genera. This partly results from the fact that some plant genera are based on entirely vegetative remains while in other genera preservation has not allowed for extraction of spores. A number of Devonian plants have yielded *in situ* spores which are either immature or too poorly preserved to describe. *In situ* spores are most extensively known from Early Devonian plants, especially those of Siegenian and Emsian age, and least known from Middle and Late Devonian plants. Despite these drawbacks, *in situ* spore studies have provided a significant source of supplemental data, especially regarding aspects of reproduction in early land plants.

Numerous papers have listed Devonian plants for which *in situ* spores are known, generally including comparison to dispersed spore genera or characterizing them in terms of morphological complexity. The most

TABLE 11.1

Taxon	Dispersed Spore Type	Reference
RHYNIOPHYTES		
Cooksonia pertoni	illustrated, not compared	Lang (1937)
Horneophyton lignieri	cf. *Apiculiretusispora* and ?*Emphanisporites decoratus*	Bhutta (1973b), see also Eggert (1974)
Rhynia gwynne-vaughanii	indeterminate; *Ambitisporites*	Allen (1980)
Rhynia major	a) *Geminospora* or	a) Bhutta (1973a)
	b) *Perotrilites microbaculatus*	b) Edwards and Richardson (1974)
Cooksonia crassiparietalis	a) illustrated, not compared;	a) Iurina (1964), b) McGregor (1973)
	b) *Apiculiretusispora* cf. *plicata*	
Hsüa robusta	spores illustrated, not compared; small, round, smooth, 18–36 μm	Li, Cheng-sen (1982)
Nothia aphylla	a) spores in section view illustrated, not compared, walls corroded;	a) El-Saadawy and Lacey (1979a), b) Bhutta, (1969)
	b) *Apiculiretusispora*	
Renalia hueberi	*Retusotriletes*	Gensel (1976)
ZOSTEROPHYLLOPHYTES		
Gosslingia breconensis	illustrated, not compared	Edwards (1970b)
Oricilla bilinearis	*Calamospora*, *Retusotriletes*	Gensel (1982a)
Rebuchia ovata	a) illustrated, described, not compared, poorly preserved or immature?; b) ?*Retusotriletes* sp.	a) Hueber (1972), b) Allen (1980)
Sawdonia acanthotheca	*Calamospora atava*, *C. pannucea*, *Retusotriletes rotundus*	Gensel et al. (1975), McGregor and Camfield (1976)
Sawdonia ornata	a) described, not illustrated or compared;	a) Hueber (1971), b) McGregor (1973)
	b) *Calamospora*, *Retusotriletes*	
Serrulacaulis furcatus	illustrated, described, not compared; *Calamospora* or *Retusotriletes*	Hueber and Banks (1979)

320

Species	Spores	References
Zosterophyllum divaricatum	*Retusotriletes*	Gensel (1982b)
Zosterophyllum cf. *fertile*	*Retusotriletes* cf. *R. dubius*	Edwards (1969b)
Zosterophyllum llanoveranum	a) illustrated, described, not referable to existing sporae dispersae taxon; b) ?*Leiotriletes* sp.	a) Edwards (1969a), b) Allen (1980)

LYCOPHYTES

Species	Spores	References
Leclerqia complexa	*Aneurospora (Acanthotriletes)* cf. *heterodonta*	Banks et al. (1972), Streel (1972)
Barsostrobus famennensis	not compared; trilete, verrucate megaspore	Fairon-Demaret (1977)
Cyclostigma kiltorkense	*Lagenicula*-type megaspore	Chaloner (1968)

*TRIMEROPHYTES

Species	Spores	References
Psilophyton charientos	*Apiculiretusispora brandtii*	Gensel (1979)
Psilophyton crenulatum	*Apiculiretusispora* cf. *A. plicata*	Doran (1980)
Psilophyton dawsonii	*Retusotriletes* cf. *R. triangulatus, Apiculiretusispora*	Banks et al. (1975), Streel (1967), Gensel and White (1983)
Psilophyton forbesii	*Apiculiretusispora brandtii*	Gensel (1979)
Psilophyton princeps	*Retusotriletes* sp., *Calamospora atava, Apiculiretusispora*	Hueber (1968), Gensel and White (1983)
Pertica varia	*Apiculiretusispora plicata/arenorugosa*	Granoff et al. (1976)
Pertica dalhousii	*A. brandtii/A. plicata*	Doran et al. (1978)
Trimerophyton robustius	a) illustrated, described, not compared b) *Calamospora atava/pannucea; Apiculiretusispora* sp., *Retusotriletes*	a) Hopping (1956), b) McGregor (1973), Allen (1980)

*A feature of the spores of all of these plants is that the outer sculptured layer of exine tends to slough off to varying degrees (least so in *Psilophyton forbesii*, most in *P. princeps* and *P. dawsonii*).

TABLE 11.1 (continued)

Taxon	Dispersed Spore Type	Reference
PROGYMNOSPERMS		
Archaeopteris halliana *Archaeopteris* cf. *jacksoni* *Archaeopteris latifolia* *Archaeopteris macilenta* *Svalbardia polymorpha*	Megaspores: *Biharisporites, Contagisporites* Microspores: *Cyclogranisporites/Geminospora* type; *Apiculatasporites; Geminospora?* a) *Lycospora svalbardeae*; b) *Geminospora* *svalbardeae*	Pettitt (1965), Allen (1980) Phillips et al. (1972), McGregor (1979b), Allen (1980) a) Vigran (1964), b) Allen (1980)
Tetraxylopteris schmidtii *Rellimia thomsoni* *Aneurophyton germanicum*	*Rhabdosporites langi* *Rhabdosporites langi* *Aneurospora goensis*	Bonamo and Banks (1966) Leclercq and Bonamo (1971) Streel (1964)
INCERTAE SEDIS OR PLANTS OF VARIOUS AFFINITIES		
Calamophyton bicephalum (Cladoxylopsida)	*Dibolisporites*	Bonamo and Banks (1966)
Cathaiopteridium (Protopteridium) *minutum* Halle (Obhrel)	*Retusotriletes simplex*	McGregor (1973)
Chaleuria cirrosa	large: *Apiculiretusispora gaspiensis* and *Apiculatisporis microconus* small: *Camarozonotriletes sextantus*	Andrews et al. (1974), also McGregor and Camfield (1976)

322

Eviostachya hoegii (Sphenopsida)	*Acanthotriletes*	Leclercq (1957)
Oocampsa catheta	*Samarisporites praetervisus* or *Grandispora macrotuberculatus*; *?Perotrilites eximius*	Andrews et al. (1975), McGregor (1979b), Allen (1980)
Rhacophyton ceratangium	*Perotrilites* cf. *perinatus*	Andrews and Phillips (1968)

BARINOPHYTACEAE:

Barinophyton-like plant of Hueber (illustr. in Pettit, 1970)	a) like microspores of *B. richardsonii*; b) *Calamospora pannucea*	a) Pettitt (1970), b) McGregor (1973)
Barinophyton richardsonii	cf. *Calamospora* (large ones); small ones not referable to any dispersed spore taxon, without wrinkled layer are like *Calamospora*; *Retusotriletes*	Pettitt (1965, 1970)
Enigmophyton superbum	mega: *Enigmophytospora simplex*; micro: *Retusotriletes* (*Phyllothecotriletes microgranulatus*)	Vigran (1964)
Krithodeophyton croftii	cf. *Apiculiretusispora brandtii*	Edwards (1968)
Protobarinophyton obrutschevii	*Calamospora atava*	McGregor (1973)
Protobarinophyton timanicum	*Calamospora* cf. *pannucea*	Allen (1980)

current of these are Gensel (1980), Allen (1980, 1981) and McGregor (1979b) and should be referred to for those details which cannot be included here.

Spores have been described from sporangia of about 40 species and 30 genera of the 120 genera of named Devonian plants and are referable to about 30 of the approximately 100 Devonian dispersed spore genera. Table 11.1 lists plants with *in situ* spores and comparable dispersed spore taxa; some plants with undetermined spores are included where they provide information pertinent to this account. The basic characteristics of spores of the major groups of Devonian plants are summarized next, based on our current knowledge, with mention of problematical aspects or areas of particular interest.

Spores of the rhyniophytes are relatively small (22–77 μm in diameter), smooth or finely ornamented, and often curvaturate (= small ridges which outline the area of contact of spores while in the tetrad). Their walls may be thicker distally, of equal thickness, or exhibit a slight thickening equatorially (a crassitude). Spores of *Cooksonia* seem generally similar to the dispersed spore genus *Ambitisporites* whose stratigraphic range extends from lower Silurian (and perhaps upper Ordovician) through the Lower Devonian. Spores from the Rhynie Chert plants are rather poorly preserved, so not all external details can be determined, especially for *Rhynia, Nothia,* and *Asteroxylon.* The studies of Bhutta (1973a, b) suggest they are constructed similarly to those of *Cooksonia,* perhaps lacking any differential wall thickening. *Horneophyton* spores are clearly ornamented with apiculi or coni. They occur in both tetrahedral and decussate tetrads.

Zosterophyllophyte spores are larger (30–77 μm diameter) than the preceeding and are either smooth or finely ornamented. Some are curvaturate, but otherwise their walls appear to be evenly thickened. Commonly the spores are covered with abundant globules of tapetal residue which usually can be distinguished from ornamentation. While many zosterophyllophyte sporangia have yielded spores, they have not always been thoroughly characterized and more work is needed.

Spores obtained from trimerophytes are quite similar to one another, being composed of a thin outer ornamented layer which covers all but the contact areas, and a thicker inner layer which is homogeneous and smooth. The outer layer commonly detaches to some degree and may even be entirely lost. Examination of spores from four species of *Psilophyton* suggests, however, that the ornamentation initially is present; degree of maturity or type of preservation may cause it to detach.

Some trimerophyte spores exhibit a triangular dark area at the juncture of the trilete rays. This is variable within genera and even within sporangia. They are referable to three major genera of dispersed spores (Table 11.1 and Gensel, 1980).

Spores of a very few lycophytes are known. The Middle Devonian genus *Leclercqia* exhibits a spore with fairly robust spines or cones, some of which are biform, and thus differs significantly from spores of the previous plant groups. Both *in situ* and dispersed spores show that by Late Devonian times, heterosporous lycophytes were well established. Some, such as *Cyclostigma*, produced lageniculate megaspores, a type commonly encountered among Carboniferous lycophytes. As mentioned previously, the presumed lycophyte cone *Kryshtofovichia* produces trilete megaspores with long spines, and monolete microspores of a distinctive morphology.

Not only are very few representatives of sphenopsids found in the Devonian, but of them *in situ* spores are known from one genus, *Eviostachya*. They are small, trilete, and spiny, differing greatly from the types of spores found in Carboniferous sphenophytes (*Calamospora*, elater-bearing spores, and perinate spores). Spores are known for the presumed cladoxylalean *Calamophyton*, and are trilete and finely apiculate. This plant has been considered by some to be a protoarticulate. Spore and megafossil evidence is too scanty at present to be of help in assessing the early history of this intriguing group of plants.

Rhacophyton produces a spore with a thin outer wall layer which may represent a perispore as is found in many modern pteridophytes.

Those progymnosperms for which spores are known possess spores which are two-layered but differ from the spores of *Rhacophyton*. *Tetraxylopteris* and *Rellimia* produce nearly identical spores (*Rhabdosporites langii*). The spores of *Svalbardia polymorpha* (=*Geminospora*) are rather similar to the microspores of *Archaeopteris*. Several species of *Archaeopteris* are heterosporous, including *A. latifolia, A. macilenta,* and *A. halliana.* It would be of interest to learn if the others are as well; if not, this might support the suggestion that the genus encompasses a wide variety of plants (Bonamo, 1975). Other plants which are heterosporous are discussed in the preceeding chapter.

Spores of *Oocampsa*, which is interpreted as being a possible intermediate between trimerophytes and progymnosperms (Andrews et al. 1975), exhibits a very distinctive, zonate spore with pronounced distal sculpture. It has been suggested that spores of *Barrandeina dusliana* are similarly constructed (Allen, 1980) but the two plants do not appear at all related.

Other plants of uncertain affinities exhibiting *in situ* spores of interest include *Chaleuria, Barinophyton, and Enigmophyton*; these are discussed in some detail in the preceding chapter.

OTHER CONSIDERATIONS

As presently known, some plants of very differing affinities produce very similar appearing spores (*Psilophyton, Krithodeophyton, Cooksonia crassiparietalis*, and some species of *Zosterophyllum* all produce spores of an *Apiculiretusispora* type). Similarly, some genera included within a single order or class exhibit very similar spores (*Archaeopteris, Svalbardia; Psilophyton, Pertica*). Conversely, spores from a single sporangium may be referable to several dispersed spore genera. The differences between some of these genera usually represent small morphologic changes, such as presence/absence of ornament, height of sculptural elements, or presence/absence of curvaturae.

ULTRASTRUCTURE

Ultrastructural studies of the pollen grains and spores of modern and fossil plants has provided an additional important source of information concerning their relationships and/or reproductive biology. Megaspores in extant plants and many extinct ones for example, possess a spongy wall which may be one to several layers thick. Microspores or miospores (of homosporous plants) differ ultrastructurally; while their walls may be more than one layer thick, they generally are homogeneous or lamellate. Pteridophyte spores also exhibit an extra wall layer called a perispore which differs from the remainder of the wall in organization and chemical composition. It is deposited after the rest of the wall is formed and may be very thin and tightly investing or thick, loose and ornate (Tryon and Tryon, 1982). We are only beginning to understand the nature and variation of this layer and its possible functional significance. Gymnosperm and angiosperm pollen grains exhibit walls which are variously organized, and usually distinguishable from one another (Doyle et al. 1975).

Only a few Devonian spores have been examined ultrastructurally. Allen (1965) illustrated sections of some dispersed spore taxa using light microscopy; these are useful in interpreting differences in wall thickness, separation of wall layers, nature of apertures, etc. Pettitt (1966) illustrated TEM sections of the mega- and microspores of *Archaeopteris* and of several younger fossil and extant plants. While both megaspores and microspores of *Archaeopteris* exhibit a two-layered wall, the layers are not separated from one another, and their detailed ultrastructure differs. Taylor et al. (1980) described the ultrastructure of the dispersed spore *Nikitinsporites*, which is similar to the megaspores of the presumed lycophyte cone *Kryshtofovichia*. It also has a spongy wall and distinctive long spine-like processes. Taylor and Brauer (1983) show that the megaspores of *Barinophyton* differ ultrastructurally (are spongy) from the microspores. Gensel and White (1983) describe the ultrastructure of *Psilophyton forbesii* spores, showing them to be homogeneous. Their apertures apparently are simple, or not well preserved, as no special organization in that region could be found.

Studies of other *in situ* and dispersed spores would provide useful information that would aid in interpreting wall organization and might also help in elucidating changes associated with their reproductive potential—such as diversification of homosporous vascular plants, advent of heterospory and the seed habit and development of pollen grains.

12 / Devonian Floras and Concluding Remarks*

We have dedicated our book in part to E. A. Newell Arber who prepared a similar account a little over sixty years ago. A comparison of his small volume with the present one reveals the vast increase in our knowledge of the origins and development of land plants, much of which has been reported in the past two or three decades. The contrast, however, is even greater than this comparison reveals in that we have necessarily given here only a representative selection of the more significant contributions. Our bibliography will lead the inquisitive reader to a more comprehensive understanding of Devonian plant life. And in addition to what has been said in the preceding chapters, there are certain topics included here that we have touched on but lightly or that may be brought together and summarized hopefully to produce a more complete and unified picture of Devonian vegetation.

BIOGEOGRAPHY

We deem it especially useful to begin with some consideration of the disposition of the land masses in the Devonian. Just where, on the surface of the Earth, were all these plants living? The answer is far from being entirely satisfactory but some information is available from studies of the rocks, including their sedimentary and tectonic history, paleomagnetism

---

*With the thought that it will make this chapter more readable we have listed at the end the more pertinent sources of information, rather than insert them through the text.

and other factors. For this account, we have chosen to use a paleo-continental reconstruction of Scotese et al. (1979 and pers. comm.) from the many that exist; this was derived from a combination of the above types of data plus considerations of biogeography and climatic inferences (including wind and ocean current patterns). For the Paleozoic they recognize six major paleocontinents, namely Baltica (northern Europe, Scandinavia), Siberia, China (China, Japan, southeastern Asia and Malaysia), Gondwana (South America, Florida, Africa, Arabia, India, Southern Europe, Australia and Antarctica), Laurentia (central and north America, Greenland, Iceland, Spitsbergen, parts of Ireland and Scotland) and Kazachstania (mostly central Russia). During the Devonian (Fig. 12.1), Baltica, Laurentia, and small landmasses of England and Avalonia (northeastern North America) are believed to have collided, forming extensive mountain chains at the site of the collision. This resulted in the formation of a larger continent, Laurussia (or Laurasia). Subsequent erosion of the mountains produced the extensive fluvial sediments referred to as Old Red Sandstone or "Old Red Land." Laurussia, Kazachstania, China, and the Gondwana paleocontinent were located at or near the equator from the late Silurian to the Carboniferous, with Gondwana extending from the equator to the polar regions. The South Pole was located in present-day South Africa. Siberia was located between 30° and 60° N. latitude and rotated 180° from its present position. These positions are approximate, and other workers show them at different latitudes, in different orientations or with the Asian continents joined. A major problem with Devonian paleogeographic reconstructions is that few paleomagnetic data points are available for North America.

If we consider some of the major plant-bearing localities mentioned in the preceding chapters, we find for example that southern Britain, the Gaspé Peninsula and northern Maine, Belgium, and Germany are all located at or near the equator, as is the southern tip of Kazachstania and parts of the paleocontinent of China. Africa and parts of southern Europe and Australia were about 30°–40° S. latitude with part of the latter extending towards the equator. What we infer from these data is that for much of the Devonian, at least, the land masses on which plants are known to occur were in tropical or subtropical regions. Sedimentary data suggest their climate ranges from moist to dry. The northerly position of Siberia, in relation to the plants found there, is puzzling.

Undoubtedly the distribution of organisms, including plants, was influenced by the changes in continental positioning, and especially the collision of Baltica and Laurentia, that took place during the Devonian. It is hoped that future studies with an emphasis on tying together geological and biological information, will show these interactions in some detail.

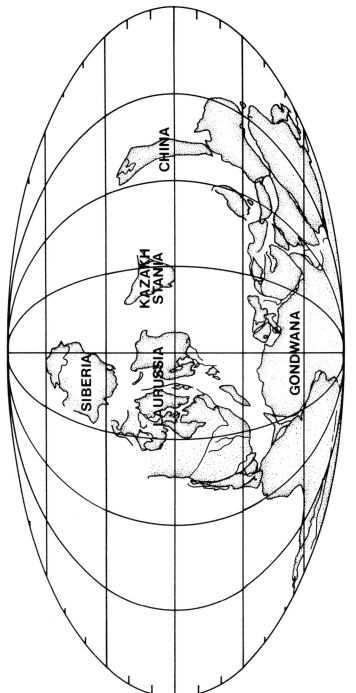

Figure 12.1. Proposed locations for the major continental masses during the early Devonian. (Courtesy of C.R. Scotese.)

Although we have described a considerable number of genera and species of Devonian plants and some concept may have been gained from our previous discussions concerning the composition of plant communities from late Silurian to early Carboniferous times, we will now try to bring some of this information together to portray some representative landscapes. Which plants grew together? In what types of environments did they grow? What factors controlled their distribution? These are some of the questions that arise. We may be at a stage to give some partial answers but for these and other related questions the answers for the most part remain unclear. The reasons are many. Not all plant-bearing localities are accurately dated. The extent of sediment and the time represented at a 'locality' is often now known or is variable. How close to the fossil source the plants originally grew is generally uncertain and probably variable. Correlation or comparison of compression-impression remains with petrifaction taxa is not easy. It will also have been noted that large areas of the Earth are sparsely represented in the record. Rather little has been reported from land areas presently in the southern hemisphere and from some regions in the northern one such as southeastern Asia and China: however some very encouraging recent developments in the latter country promise much for the future.

It is quite beyond the scope of our account to give a detailed discussion of the Devonian landscapes over a period of fifty million years, which at best could only be fragmentary. However, we will try to briefly describe a selected and presumably representative flora from the three major units of Devonian time. These are drawn from many sources as well as our own experience. Aside from the many references in our literature section, we would mention especially the accounts cited at the end of this chapter by Arber, Stockmans, Kräusel, Edwards, Banks, Li and Cai, A. Scott, and Andrews et al. as sources for this type of information.

AN EARLY DEVONIAN LANDSCAPE

Upper Silurian and Lowermost Devonian floras presently are composed of a mixture of non-vascular and vascular plants, including *Prototaxites, Pachytheca, Cooksonia,* and, in the Lower Devonian, *Zosterophyllum.* While non-vascular plants occur throughout the Devonian, our emphasis will be on vascular ones. Greatest diversification of vascular plant groups seems to have occurred in the Siegenian-Emsian (Knoll et al. in press) and we will concentrate therefore on the Emsian floras of eastern North America, an area which has yielded a considerable array of Early Devonian vascular plants.

TABLE 12.1. Occurrence of plant genera in the Early Devonian of eastern North America.

Genus	Affinities	Stratigraphic Range	Trout Valley Fm., Maine	Gaspé Bay Area, Quebec	North Shore, New Brunswick	Elsewhere in World
Psilophyton	T	Siegenian-Eifelian	+	+	+	+
Pertica	T	Emsian-Givetian	+	+	+	+
Sawdonia	Z	Siegenian-Frasnian	+	+	+	+
Drepanophycus	L?	Siegenian-Frasnian	+?	+	+	+
Taeniocrada	?	Late Silurian-Frasnian	+	+	+	+
Prototaxites	Th	Late Silurian-Frasnian	–	+	+	+
Pachytheca	A	Late Silurian-Emsian	+	–	+	+
Kaulangiophyton	L?	Emsian-	+	–	+	–
Leclercqia	L	Emsian-Givetian	+	–	+	+
Thursophyton	?	?Gedinnian-Frasnian	+	–	–	+?
Renalia	R	Emsian	–	+	–	+?
Trimerophyton	T	Emsian	–	+	–	+
Sciadophyton	?	Siegenian-Emsian	+?	+	–	–
Eogaspesiea	R	Emsian	–	+	–	–
Crenaticaulis	Z	Emsian	–	+	–	–
Chaleuria	?	Emsian	–	+	+	–
Oocampsa	?	Emsian	–	–	+	–
Zosterophyllum	Z	Gedinnian-Emsian	–	–	+	+
Loganophyton	?	Emsian	–	–	+	–
Oricilla	Z	Emsian	–	–	+	–
Total # genera			10	12	14	12

*Stratigraphic ranges from Banks 1980 and Chaloner & Sheerin 1979. L = lycophyte, Th = thallophyte incertae sedis, A = alga, T = trimerophyte, R = rhyniophyte, Z = zosterophyllophyte. Designations represent authors' opinion. (Modified from Gensel, 1983, 3rd NAPC Proceedings).

Table 12.1 lists plant genera which have been described from northern Maine, northern New Brunswick, and the Gaspé Peninsula, Quebec. These are approximately contemporaneous, being Emsian in age, and all come from localities which may represent deposition and preservation in coastal plain rivers, lakes, or swamps. Several genera are present in all three areas, especially *Psilophyton, Pertica, Drepanophycus,* and *Leclercqia.* Different genera of zosterophyllophytes occur in each area, with perhaps *Sawdonia* being present in all.

Andrews et al. (1977) described a landscape typical for northern Maine as consisting of large stands of *Psilophyton* with smaller clumps of other plant taxa interspersed, such as *Thursophyton, Psilophyton dapsile,* axes referred to *Taeniocrada,* and *Kaulangiophyton.* The general aspect would be similar to a present day coastal marshland with mostly low-lying vegetation. This type of landscape seems appropriate for the other areas as well. For example, the several outcrops in northern New Brunswick, which range from earliest to latest Emsian in age (based on dispersed spore data provided by D. C. McGregor) vary individually in their floral components but generally one taxon predominates and others occur in smaller amounts. Table 12.2 shows these localities, and several from Gaspé, with the plants known thus far to occur in each. Locality A in New Brunswick contains at one level *Sawdonia acanthotheca,* overlain by axes of *Zosterophyllum divaricatum.* This might indicate that these plants co-existed in different areas and were preserved at slightly different times. Alternatively, they may have grown in the same general area at different times within the time span represented by the sedimentary unit. We are inclined to prefer the former interpretation, as the *Sawdonia* remains appear to have been deposited near their site of growth, and the *Zosterophyllum* ones to have been transported for at least a short distance.

A nearly contemporaneous outcrop in Gaspé contains abundant remains of *Psilophyton forbesii, Crenaticaulis verruculosus, Drepanophycus* and vegetative axes referable to *Hostinella.* These are only rarely intermixed and similar interpretations to the ones above may apply.

Another locality in New Brunswick (B), of latest Emsian age, contains *Psilophyton charientos, Kaulangiophyton, Leclercqia,* and the 'endemic' *Oocampsa,* in roughly equal abundance.

In terms of relative size, these plants were generally a few cm to a meter tall, with the exception of *Pertica* and *Trimerophyton,* which attained a height of about two meters. *Drepanophycus, Kaulangiophyton,* and many of the zosterophyllophytes had rhizomes and aerial shoots, forming dense, low-lying mats. Thus, some stratification of plants into two layers may have existed.

A MIDDLE DEVONIAN LANDSCAPE

Very few floras of Eifelian (earliest Middle Devonian) age are known but many Givetian ones are. The well-known floras of Belgium and Germany may be considered representative and include the following genera:

Drepanophycus
Leclercqia (includes some former species of *Protolepidodendron*)
Estinnophyton (includes some former species of *Protolepidodendron*)
Cladoxylon scoparium
Hyenia
Calamophyton
Pseudosporochnus
Rellimia
Aneurophyton
flabellate leaves

Some smooth axes with a dark central strand have been referred to *Taeniocrada*, and certain smooth and spiny axes to species of *Psilophyton*; since these are vegetative, their affinities remain unclear.

Many of the plants listed above, plus additional ones, have been found in Middle Devonian localities in New York State. In addition, several aneurophytes, lycopods, and the plants referred to the iridopterids occur there. The flora of Gilboa, New York is particularly rich in lycopods (Grierson and Banks, 1963; Matten, 1974). Some apparently distinctive genera also occur in the USSR (Petrosian, 1968).

Many Middle Devonian plants attained the size of shrubs or small trees, including some aneurophytes, cladoxyls and lycopods. Others are too incompletely preserved to assess size. How these plants occurred in relation to one another also is not clear.

AN UPPER DEVONIAN LANDSCAPE

We suspect there was some differentiation into distinct plant communities during the Upper Devonian, and perhaps earlier (Matten, 1974), and that more than one major flora existed during that time. Frasnian sediments often contain abundant *Archaeopteris* and some other plants, while mid-late Famennian floras have seed-bearing plants also. We will compare the composition of floras from New York State and

TABLE 12.2. Stratigraphic Relationships of Plant-Bearing Localities in New Brunswick and Gaspé

Series	Stage	Provisional Spore Assemblage Zones and Subzones (McGregor, 1973, 1977)	Plant Genera from Outcrops Along Restigouche R., no. New Brunswick[1] (each outcrop designated by a letter)	Plant Genera from Outcrops Along North and South Shore of Gaspé Bay, P.Q.[2] (each outcrop designated by a letter)
Middle Devonian	Eifelian	*Grandispora*	[B] *Oocampsa catheta, Psilophyton charientos, (Leclercqia),[3] (Kaulangiophyton)*	
	Late Emsian	*lindlarensis*	H,I [I] *Sawdonia ornata, (Psilophyton* sp.); [H] (New genus—zosterophyll), *(Psilophyton* sp.)	
			A,E [A] *Chaleuria cirrosa, Pertica dalhousii, Loganophyton, (Psilophyton* sp.); [E] *Oricilla bilinearis, (Psilophyton* sp.), (various lycophyte fragments)	

[X] *Psilophyton dawsonii*

V-W [V] *Renalia hueberi*;
[W] *Prototaxites*

S-T [S] *Sawdonia ornata, Psilophyton princeps, Trimerophyton robustius;*
[T] *Pertica varia,* cf. *Pertica* sp.

U,Y [U] *Psilophyton forbesii, Crenaticaulis verruculosus, Drepanophycus spinaeformis,* (*Hostinella*); [Y] *Eogaspiesia gracilis*

[F] *Drepanophycus gaspianus*

[K] *Psilophyton princeps,*
Drepanophycus spinaeformis

[Q] *Psilophyton crenulatum*

[M] *Sawdonia acanthotheca, Zostero-phyllum divaricatum*

sextantii

annulatus-

emsiensis
caperatus

Lower Devonian		
	Early Emsian	
		Siegenian

[1]Number of Genera = 13+ (approximately 10 more identified but not described)
[2]Number of Genera = 10 (a few more identified but not described)
[3]Plants in parentheses have not been published in detail

——— = zones
------ = subzones

(Dispersed spore correlations courtesy of D. C. McGregor.)

Figure 12.2. A Devonian 'botanical garden', showing major types of Lower (left), Middle (center) and Upper (right) Devonian plants. (From Chaloner and Macdonald, 1980. With permission Controller H.B.M. Stationery Office.)

Pennsylvania, West Virginia, and Belgium to those of Kerry Head, Ireland and Bear Island to illustrate this.

The Frasnian-Famennian flora of the Catskill Delta of New York and Pennsylvania (summarized by Banks, 1966) is dominated by the progymnosperm *Archaeopteris*, which may have been the main component of the extensive forests of that time (Beck, 1964a). Aneurophytes such as *Tetraxylopteris* and *Aneurophyton* occur in Frasnian sequences as do several lycopods, including *Colpodexylon, Drepanophycus,* and *Archaeosigillaria*. Zosterophyllophytes, namely *Serrulacaulis* and *Sawdonia ornata*, several cladoxyls and iridopterids also were present. In the Famennian, *Archaeopteris, Barinophyton,* and some poorly understood lycopods which may have been quite large (*Protolepidodendropsis, Sublepidodendron*) occur. Many other taxa, especially zosterophylls, apparently are no longer present. Famennian sediments in West Virginia have yielded *Archaeopteris* and Rhacophyton (Phillips et al. 1972; Andrews and Phillips, 1968), lagenostomalean type seeds, foliage referable to *Spenopteris*, some lycopod remains, *Barinophyton*, and other less

well-preserved remains (Gillespie et al. 1981). The Upper Devonian flora of Belgium is similar (Stockmans, 1948).

The late Famennian flora of Bear Island contains *Archaeopteris, Rhacophyton,* foliage referable to *Sphenopteridium, Sphenophyllum, Pseudobornia,* and *Sublepidodendron.* A slightly different flora has been described from Kerry Head, Ireland, where the seed *Hydrasperma tenuis* occurs as well as rachides referable to *Lyginorachis papilio,* another type of presumed pteridosperm, and the enigmatic *Buteoxylon*, thus suggesting an increase in presumed seed plants. It is worth noting that this flora is based on petrifactions while others are predominatly compressions and it is not clear to what extent preservation type biases one's assessment of taxonomic affinities.

The pictorial representation of Devonian plants that was published by Chaloner and Macdonald (1980) that is included here (Figure 12.2) provides a general view of each of the plants typical of Lower, Middle, and Upper Devonian and supplements our description. A major difference, however, is that one should consider that each of the plant genera probably grew in greater numbers in a given area, and that we still have much to learn about which ones grew together and also what environmental preferences each type exhibited.

EVOLUTIONARY CONSIDERATIONS

Of the various texts, summary, and encyclopedic accounts that have appeared since Arber's *Devonian Floras* of 1921, we wish to give special attention to one more—Chaloner and Sheerin's "Devonian Macrofloras" of 1979. This brings together a tremendous amount of morphological and anatomical data revealing evolutionary trends and proposed phylogenetic relationships. They record the first appearance of various anatomical or morphological features of Devonian plants and while these undoubtedly will require modification as new discoveries are made, they provide a concise summary of the changes we have detailed in the previous chapters. We have therefore included several of their charts as Figures 12.3–12.6.

It is also useful to consider these various plants collectively in terms of their overall phylogenetic relationships—to what groups, if any, did they give rise? Again, Chaloner and Sheerin's account is useful and we include their phylogenetic chart as Figure 12.6. It will be noted that many lineages are uncertain and some of the early vascular plants probably represent 'experiments' which eventually died out, while others seem central to the origins of extant groups.

FUTURE POSSIBILITIES

Several genera of Devonian plants of considerable interest have not been treated in detail in our account, especially where they are not well understood. We mention as examples the genus *Taeniocrada*, which has become a catch-all for broad smooth axes with a dark central strand. As evidence accumulates, it is clear that perhaps more than one type of plant has been included in this taxon—see for example our discussion of *Barinophyton*. The Early Devonian genus *Sciadophyton* also is intriguing—recent workers (Remy, et al., 1980; Schweitzer, 1980) have proposed that it might represent a gametophyte. This is an aspect of the life history of early vascular plants about which we. know little; if their gametophytes (haploid plant bodies) resemble modern ones, there is less chance of their being preserved, and also some difficulty, in the absence of excellent preservation, in ascertaining their nature. Again, as mentioned in Chapter 4, further evidence is needed to assess the nature of the plants that have been considered as gametophytes.

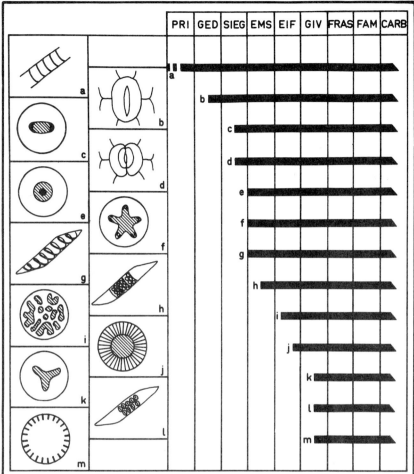

	PRI	GED	SIEG	EMS	EIF	GIV	FRAS	FAM	CARB

Diagrammatic record of the time of first appearance of various anatomical features of Devonian fossil plants: (a) tracheids with annular thickenings (e.g. in axes of Lang 1937; see also Edwards and Davies 1976); (b) stoma with apparently interconnected guard cells (e.g. in *Zosterophyllum*, Lele and Walton 1961); (c) stele with xylem of elliptical cross-section with exarch protoxylem (e.g. in *Gosslingia*, Edwards 1970b); (d) stoma with paracytic subsidiary cells (e.g. in *Drepanophycus*, Stubblefield and Banks 1978); (e) stele with xylem cylindrical and endarch (e.g. in *Rhynia*, Kidston and Lang 1917); (f) stele with xylem of stellate cross-section (e.g. in *Asteroxylon*, Kidston and Lang 1920); (g) tracheids with spiral thickenings (e.g. in *Asteroxylon*, Kidston and Lang 1920); (h) tracheids with mesh-like wall, simulating pits, between scalariform bars (e.g. in *Psilophyton*, Banks et al. 1975); (i) polystele (e.g. in *Calamophyton*, Schweitzer 1973); (j) secondary xylem (e.g. in *Rellimia*, Mustafa 1975); (k) trilobed stele related to branch arrangement (e.g. in *Triloboxylon*, Matten 1974); (l) tracheids with bordered pits (e.g. in *Leclercqia*, Grierson 1976); (m) secondary cortex or periderm (e.g. in *Triloboxylon*, Scheckler and Banks 1974).

Figure 12.3. Charts showing the first appearance of various anatomical and morphological features of Devonian plants. (From Chaloner and Sheerin, 1979, *Special Papers in Palaeontology*.)

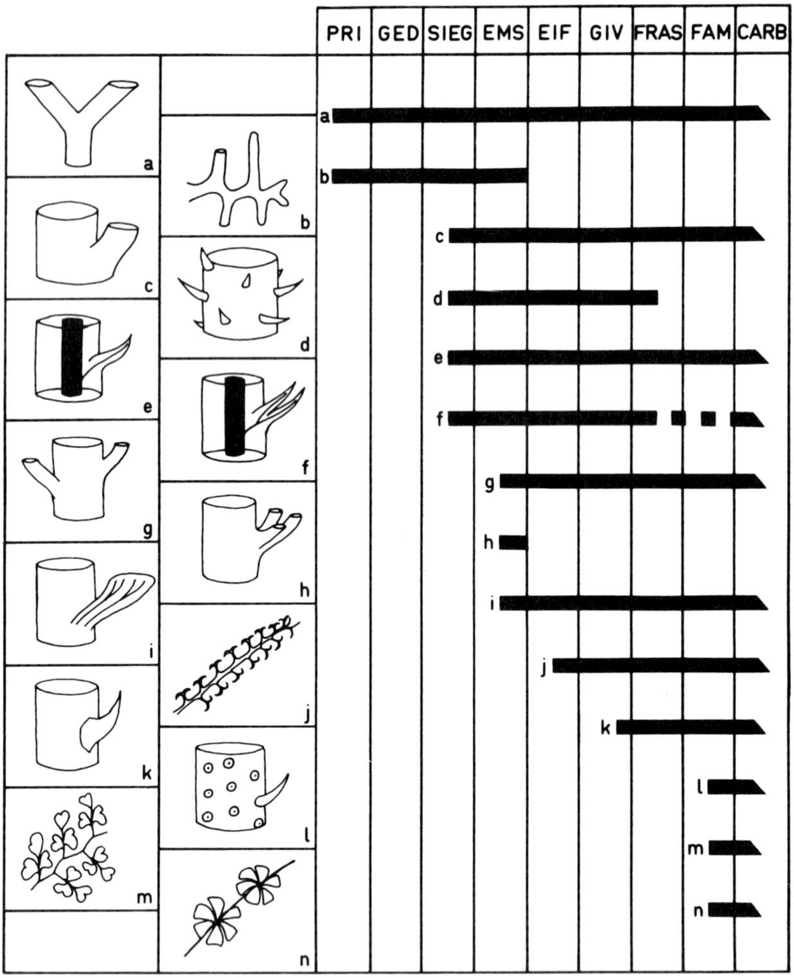

Diagrammatic record of the time of first appearance of various features of the vegetative morphology of Devonian fossil plants: (*a*) equal dichotomy (e.g. in *Cooksonia*, Obrhel 1962); (*b*) H-branching (e.g. in *Zosterophyllum*, Høeg 1942); (*c*) overtopping: ratio of main axis diameter to that of overtopped lateral 2:1 (e.g. in 'Psilophyton princeps', Strathmore Group, of Lang 1932, but see also earlier records of the genus *Psilophyton*, e.g. Banks 1979, where age and/or branch status is less secure); (*d*) enations (e.g. in *Sawdonia*, Hueber 1971); (*e*) vascularized microphylls (e.g. in *Drepanophycus*, Kräusel and Weyland 1935*b*); (*f*) forked microphylls (e.g. in *Sugambrophyton*, Schmidt 1954); (*g*) overtopping: ratio of main axis to overtopped lateral *c.* 4:1 (e.g. in *Pertica*, Kasper and Andrews 1972); (*h*) 'trichotomy' of overtopped laterals (e.g. in *Trimerophyton*, Hopping 1956); (*i*) vascularized wedge-shaped megaphylls (e.g. in *Platyphyllum*, Høeg 1942); (*j*) planated, ?pinnate megaphyll (e.g. in *Aneurophyton*, Leclercq 1940); (*k*) expanded leaf base, i.e. leaf cushion or leaf cushion precursor? (e.g. in *Archaeosigillaria*, Grierson and Banks 1963); (*l*) leaf abscission (e.g. in *Cyclostigma*, Chaloner 1968); (*m*) sphenopterid megaphylls (e.g. in *Sphenopteris*, Baily 1861); (*n*) whorled leaves (e.g. in *Sphenophyllum*, Nathorst 1902).

Figure 12.4. Charts showing the first appearance of various anatomical and morphological features of Devonian plants. (From Chaloner and Sheerin, 1979, *Special Papers in Palaeontology.*)

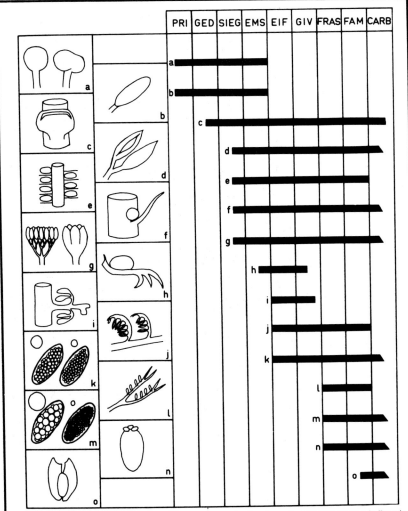

		PRI	GED	SIEG	EMS	EIF	GIV	FRAS	FAM	CARB

Diagrammatic record of the time of first appearance of various types of reproductive structure in Devonian fossil plants: (a) terminal rounded sporangia (e.g. in *Cooksonia*. Obrhel 1962); (b) terminal ellipsoid sporangia (e.g. in *Eorhynia*. Ishchenko 1975); (c) lateral reniform sporangia with transverse dehiscence (e.g. in *Zosterophyllum*. Edwards 1975); (d) terminal fusiform sporangia with longitudinal dehiscence (e.g. in *Dawsonites*. Croft and Lang 1942); (e) small lateral sporangia associated with appendages in biseriate arrangement (e.g. in *Krithodeophyton*. Edwards 1968); (f) sporangia in axils of microphyllous leaves (e.g. in *Baragwanathia*. Lang and Cookson 1935); (g) corymbose aggregation of fusiform sporangia (*Hedeia*) and synangial aggregation of sporangia (*Yarravia*) (Cookson 1935; Lang and Cookson 1935); (h) sporangia borne singly on adaxial surface of sporophylls (e.g. in *Leclercqia*. Banks et al. 1972); (i) sporangia borne on recurved bifurcating side-branches of a dichotomous sporangiophore (e.g. in *Calamophyton*. Leclercq and Andrews 1960); (j) fusiform sporangia borne in clusters on pinnate fertile organs (*Rellimia*. Leclercq and Bonamo 1971); (k) sporangia with morphologically distinct spores of two sizes (e.g. in *Chaleuria*. Andrews et al. 1974); (l) sporangia on megaphylls (e.g. in *Archaeopteris*. Phillips et al. 1972); (m) sporangia with mega- and microspores (e.g. in *Archaeopteris*. Pettitt 1965); (n) megaspore tetrad: one large, presumed fertile, three small, presumed abortive (e.g. in *Cystosporites*. Chaloner and Pettitt 1964); (o) radiospermic (*Archaeosperma*) and platyspermic (*Spermolithus*) seeds (Pettitt and Beck 1968; Chaloner et al. 1977).

Figure 12.5. Charts showing the first appearance of various anatomical and morphological features of Devonian plants. (From Chaloner and Sheerin, 1979, *Special Papers in Palaeontology.*)

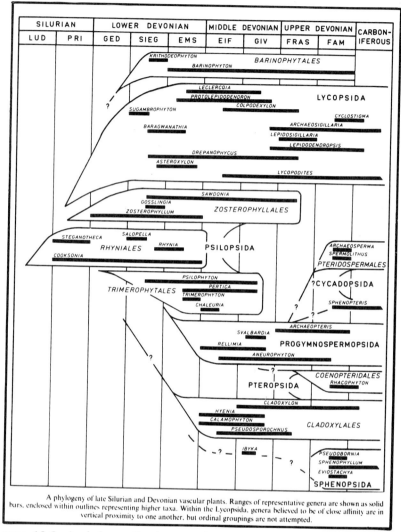

Figure 12.6. A phylogenetic chart emphasizing relationships of late Silurian and Devonian plants. (From Chaloner and Sheerin, 1979, *Special Papers in Palaeontology*.)

The middle Devonian taxa *Barrandeina* and *Duisbergia* are also interesting but problematical forms: the former bears some resemblance to a poorly organized or preserved archaeopteridalean progymnosperm, while the latter is intriguing in being of considerable size and in the nature of its leaves. Lastly, the anatomy of the many lycopods known from the

Devonian would be a most useful addition to our understanding of early evolution in that distinctive group of plants.

In addition to the above considerations, the serious reader will wonder what the future may bring. Will our knowledge in this area of botanical-geological investigation continue to expand? There is no doubt as to a positive answer. New fossil localities will be discovered and presently known areas will continue to yield both new plants and expand our understanding of those that are now known only from fragments. Let us take as an example the areas with which we are particularly well acquainted, northern Maine and southeastern Canada. Of the several square miles encompassed by the Trout Brook Formation only a very small portion has been exposed and explored. The many miles of coastal exposures in northern New Brunswick and the Gaspé Peninsula erode very rapidly every year to reveal previously unknown deposits. It is perhaps possible that a locality may become "worked out', but for the most part we are inclined to concur with Professor Tom Harris in his appraisal of collecting along the Yorkshire coast, concerning which he wrote ". . . I am convinced that it is the collector rather than the locality which is exhausted."

Also, the development of new study techniques since Newell Arber's time has been responsible for much of our progress. Methods dealing with pyritic petrifactions, maceration techniques, and degaging have yielded a great wealth of information. The scanning and transmission electron microscopes have enabled us to see structural details previously withheld to us. New techniques will almost certainly appear from younger workers with fresh ideas.

The information that has been garnered thus far about the Devonian vegetation has been largely descriptive, supplying us with botanical and stratigraphical information. We need a great deal more of this 'raw data' but it may also be expected that there will be a growing trend to use this information for the solution of broader questions and problems about the development of life on the land. One of our colleagues has given us a rather long list of questions that she would like to see dealt with in a book such as this. For the most, only partial answers, if any, can be given but they may be answered in the future. We mention a few for those who come after us to deal with: How did the distinct differentiation between root and stem take place? This would seem to have an important bearing on the establishment of plants on dry land. When and how did the deciduous habit evolve? What kind of ecological niches existed in the Devonian and how did they develop and expand during the first 50 million years? Actually we have given answers already to some very basic matters, including the origins of heterospory, the evolution of the cambium, the arborescent habit, and structural adaptations that enabled aquatics to advance to a life on the land. Much has been learned in the past sixty years and much more will be forthcoming in the next sixty.

PERTINENT SOURCES OF INFORMATION FOR CHAPTER 12

Devonian paleogeography
Bambach et al, 1980
Heckel and Witzke, 1979
Johnson, 1980
Scotese et al, 1979
Smith et al, 1981

Floristic Accounts
Arber, 1921
Andrews et al, 1977
Banks, 1966
Edwards, 1973, 1980
Kräusel, 1936
Kräusel and Weyland, 1948
Li and Cai, 1978
Petrosian, 1968
Stockmans, 1940, 1948, 1968
Scott, A. C. 1980

Stratigraphic accounts
Banks 1980
Stockmans, 1940, 1948, 1968

Evolutionary trends (plant structures)
Chaloner and Sheerin, 1979

Literature Cited

Abbott, M. L. 1963. Lycopod fructifications from the Upper Freeport (No. 7) coal in southeastern Ohio. *Palaeontographica* 112B: 93–118.

Aderca, B. 1932. Contribution à la connaissance de la flore dévonienne belge. *Ann. Soc. Géol. Belgique* 55: M1–16.

Allen, K. C. 1965. Lower and Middle Devonian spores of north and central Vestspitsbergen. *Palaeontology* 8: 687–748.

_____. 1980. A review of *in situ* late Silurian and Devonian spores. *Rev. Palaeobot. Palynol.* 29: 253–70.

_____. 1981. A comparison of the structure and sculpture of *in situ* and dispersed Silurian and Early Devonian spores. *Rev. Palaeobot. Palynol.* 34: 1–9.

Ananiev, A. R. 1955. Rasteniia. Atlas rukovod. form iskopaemykh fauny i flory zapadnoi Sibiri. 1: 279–96.

_____. 1957. New plant fossils from the Lower Devonian of Torgachino in the southeastern part of Western Siberia. *Bot. Zh. SSSR* 42 (5): 691–702.

_____. 1964. Recent studies on the Devonian floras of Siberia. *Tenth Intern. Bot. Congress, Edinburgh,* p. 17–18 (Abstr).

Ananiev, A. R. and Stepanov, S. A. 1968. Finds of sporiferous organs in *Psilophyton princeps* Dawson emend. Halle in the Lower Devonian of the southern Minusinsk Trough, Western Siberia. In: New materials on the stratigraphy and paleontology of the Lower and Middle Paleozoic of Western Siberia. *Tomsk State Univ. Trudy* 202: 30–46.

_____. 1969. The first find of *Psilophyton* flora from the Lower Devonian of the Salair Mountain range. *Treatises of the Order of the Red Banner of Labor, Tomsk State Univ. Trudy* 203: 13–28.

Andrews, H. N. 1947. *Ancient plants and the world they lived in.* Cornell Univ. Press, Ithaca, 279 p.

_____. 1960. Notes on Belgian specimens of *Sporogonites. The Palaeobotanist* 7: 85–9.

_____. 1961. *Studies in paleobotany.* John Wiley & Sons, New York, 487 p.

_____. 1963. Early seed plants. *Science* 142: 925–31.

_____. 1974. Palaeobotany, 1947–72. In: M. Crosby et al. (Ed.), *25 Years of Botany. Annals Missouri Bot. Garden.* 61: 179–202.

_____1980. *The fossil hunters.* Cornell Univ. Press, Ithaca, 421 p.

Andrews, H. N. and Mahabale, T. S. 1961. The extra-stelar tissues of certain American calamite axes. *Professor S. P. Agharkar Commemoration Volume,* 89–96. Bombay, India.

———, and Phillips, T. L. 1968. *Rhacophyton* from the Upper Devonian of West Virginia. *Bot. Journ. Linnean Soc. London* 61: 37–64.

Andrews, H. N., Phillips, T. L., and Radforth, N. W. 1965. Paleobotanical studies in Arctic Canada. I. *Archaeopteris* from Ellesmere Island. *Canadian Journ. Bot.* 43: 545–56.

———, Kasper, A., and Mencher, E. 1968. *Psilophyton forbesii*, a new Devonian plant from northern Maine. *Torrey Bot. Club Bull.* 95: 1–11.

———, Read, C. B., and Mamay, S. H. 1971. A Devonian lycopod stem with well-preserved cortical tissues. *Palaeontology* 14: 1–9.

———, Gensel, P. G., and Forbes, W. H. 1974. An apparently heterosporous plant from the Middle Devonian of New Brunswick. *Palaeontology* 17: 387–408.

———, Gensel, P. G., and Kasper, A. E. 1975. A new fossil plant of probable intermediate affinities (Trimerophyte-Progymnosperm). *Can. J. Bot.* 53: 1719–28.

Andrews, H. N., Kasper, A., Forbes, W. H., Gensel, P. G., and Chaloner, W. G. 1977. Early Devonian flora of the Trout Valley Formation of northern Maine. *Rev. Palaeobot. Palynol.* 23: 255–85.

Arber, E. A. N. 1921. *Devonian floras.* Cambridge Univ. Press, 100 p.

———, and Goode, R. H. 1915. On some fossil plants from the Devonian rocks of North Devon. *Cambridge Phil. Soc. Proceed.* 18: 89–104.

Arnold, C. A. 1930a. The genus *Callixylon* from the Upper Devonian of central and western New York. *Papers Mich. Acad. Sci. Arts Letters* 11: 1–50.

———. 1930b. Bark structure of *Callixylon. Bot. Gaz.* 90: 427–31.

———. 1931. On *Callixylon newberryi* (Dawson) Elkins et Wieland. *Contrib. Mus. Paleont. Univ. Mich.* 3: 207–32.

———. 1933. A lycopodiaceous strobilus from the Pocono sandstone of Pennsylvania. *Amer. Journ. Bot.* 20: 114–17.

———. 1934. *Callixylon whiteanum* sp. nov., from the Woodford chert of Oklahoma. *Bot. Gaz.* 96: 180–85.

———. 1935a. Some new forms and new occurrences of fossil plants from the Middle and Upper Devonian of New York State. *Buffalo Soc. Nat. Sci.* 17: 1–12.

———. 1935b. On seedlike structures associated with *Archaeopteris* from the Upper Devonian of northern Pennsylvania. *Contrib. Mus. Paleont. Univ. Mich.* 4: 283–86.

———. 1936a. Observations on fossil plants from the Devonian of eastern North America. I. Plant remains from Scaumenac Bay, Quebec. *Contrib. Mus. Paleont. Univ. Mich.* 5: 37–48.

———. 1936b. Observations, etc. II. *Archaeopteris macilenta* and *A. sphenophyllifolia* of Lesquereux. *Contrib. Mus. Paleont. Univ. Mich.* 5: 49–56.

———. 1937. Observations, etc. III. *Gilboaphyton goldringiae*, gen. et sp. nov. from the Hamilton of eastern New York. *Contrib. Mus. Paleont. Univ. Mich.* 5: 75–8.

———. 1939. Observations, etc. IV. Plant remains from the Catskill Delta deposits of northern Pennsylvania and southern New York. *Contrib. Mus. Paleont. Univ. Mich.* 5: 271–313.

_____. 1940. Structure and relationships of some Middle Devonian plants from western New York. *Amer. Journ. Bot.* 27: 57–63.

_____. 1941. Observations, etc. V. *Hyenia banksii* sp. nov. *Contrib. Mus. Paleont. Univ. Mich.* 6: 53–7.

_____. 1947. *An introduction to paleobotany*. McGraw-Hill Book Co., New York, 433 p.

_____. 1952a. Observations, etc. VI. *Xenocladia medullosina* Arnold. *Contrib. Mus. Paleont. Univ. Mich.* 9: 297–309.

_____. 1952b. A specimen of *Prototaxites* from the Kettle Point Black shale of Ontario. *Palaeontographica* 93B: 45–56.

_____. 1958. Petrified cones of the genus *Calamostachys* from the Carboniferous of Illinois. *Contrib. Mus. Paleont. Univ. Mich.* 14: 149–65.

_____. 1960. A lepidodendrid stem from Kansas and its bearing on the problem of cambium and phloem in Paleozoic lycopods. *Contrib. Mus. Paleont. Univ. Mich.* 15: 249–67.

Baily, W. H. 1861. Explanations to accompany sheets 147 and 157 of the maps of the Geological Survey of Ireland. 12–18.

Balfour, J. H. 1872. *Introduction to the study of palaeontological botany*. Adam and Charles Black, Edinburgh.

Bambach, R. K., Scotese, C. R., and Ziegler, A. M. 1980. Before Pangea: The Geographies of the Paleozoic World. *Amer. Scientist* 68: 26–37.

Banks, H. P. 1944. A new Devonian lycopod genus from southeastern New York. *Amer. Journ. Bot.* 31: 649–59.

_____. 1960. Notes on Devonian lycopods. *Senck. Leth.* 41: 59–88.

_____. 1964. Putative Devonian ferns. *Torrey Bot. Club. Mem.* 21(5): 10–25.

_____. 1966. Devonian flora of New York State. *The Empire State Geogram* 4(3): 10–24.

_____. 1968a. The early history of land plants, in: *Evolution and Environment*, Ellen T. Drake (Ed.), Yale Univ. Press, p. 73–107.

_____. 1968b. Anatomy and affinities of a Devonian *Hostinella. Phytomorphology* 17: 321–30.

_____. 1972. The stratigraphic occurrence of early land plants. *Palaeontology* 15: 365–77.

_____. 1973. Occurrence of *Cooksonia*, the oldest vascular land plant macrofossil, in the Upper Silurian of New York State. *Journ. Indian Bot. Soc. Golden Jubilee Vol.* 50A: 227–35.

_____. 1975a. Paleogeographic implications of some Silurian-Early Devonian floras, in: *Gondwana Geology*, K. S. W. Campbell (Ed.), Australian National Univ. Press, Canberra, p. 75–97.

_____. 1975b. The oldest vascular land plants: a note of caution. *Rev. Palaeobot. Palynol.* 20: 13–25.

_____. 1975c. Reclassification of Psilophyta. *Taxon* 24: 401–13.

_____. 1978. The oldest stomata (paracytic) with paired guard cells. *The Palaeobotanist* 25: 27–31.

_____. 1980a. The role of *Psilophyton* in the evolution of vascular plants. *Rev. Palaeobot. Palynol.* 29: 165–76.

_____. 1980b. Floral assemblages in the Siluro-Devonian. in Dilcher, D. L. and T. N. Taylor (Eds.) *Biostratigraphy of Fossil Plants*, pp. 1–24. Dowden, Hutchinson and Ross, Stroudsburg, Pa.

_____. 1981. Peridermal activity (wound repair) in an early Devonian (Emsian) trimerophyte from the Gaspé Peninsula, Canada. *The Palaeobotanist.* 28–29: 20–25.

Banks, H. P., and Davis, M. R. 1969. *Crenaticaulis*, a new genus of Devonian plants allied to *Zosterophyllum*, and its bearing on the classification of early land plants. *Amer. Journ. Bot.* 56: 436–49.

Banks, H. P., Bonamo, P. M., and Grierson, J. D. 1972. *Leclercqia complexa* gen. et sp. nov., a new lycopod from the late Middle Devonian of eastern New York. *Rev. Palaeobot. Palynol.* 14: 19–40.

_____, Leclercq, S., and Hueber, F. M. 1975. Anatomy and morphology of *Psilophyton dawsoni*, sp. n. from the late Lower Devonian of Quebec (Gaspé), and Ontario, Canada. *Palaeontographica Americana* 8 (48), p. 77–127.

Barber, C. A. 1889. The structure of *Pachytheca*. *Ann. Bot.* 3: 141–48.

_____. 1891. The structure of *Pachytheca*. II. *Ann. Bot.* 5: 145–62.

Barghoorn, E. S., and Darrah, W. C 1938. *Horneophyton*, a necessary change of name for *Hornea*. *Bot. Mus. Leaflets Harvard Univ.* 6: 142–44.

Barnard, P. D. W., and Long, A. G. 1975. *Triradioxylon*—a new genus of Lower Carboniferous petrified stems and petioles together with a review of the classification of early Pterophytina. *Trans. Roy. Soc. Edinburgh* 69: 231–50.

Beck, C. B. 1957. *Tetraxylopteris schmidtii* gen. et sp. nov., a probable pteridosperm precursor from the Devonian of New York. *Amer. Journ. Bot.* 44: 350–67.

_____. 1960a. Connection between *Archaeopteris* and *Callixylon*. *Science* 131: 1524–25.

_____. 1960b. The identity of *Archaeopteris* and *Callixylon*. *Brittonia* 12: 351–68.

_____. 1962a. Plants of the New Albany shale II. *Callixylon arnoldii* sp. nov. *Brittonia* 14: 322–27.

_____. 1962b. Reconstructions of *Archaeopteris* and further consideration of its phylogenetic position. *Amer. Journ. Bot.* 49: 373–82.

_____. 1963. *Ginkgophyton* (*Psygmophyllum*) with a stem of gymnospermic structure. *Science* 141: 431–33.

_____. 1964a. Predominance of *Archaeopteris* in Upper Devonian flora of western Catskills and adjacent Pennsylvania. *Bot. Gaz.* 125: 126–28.

_____. 1964b. The woody, fern-like trees of the Devonian. *Torrey Bot. Club. Mem.* 21: 26–37.

_____. 1967. *Eddya sullivanensis*, gen. et sp. nov., a plant of gymnospermic morphology from the Upper Devonian of New York. *Palaeontographica* 121B: 1–22.

_____. 1970. The appearance of gymnospermous structure. *Biol. Rev.* 45: 379–400.

————. 1971. On the anatomy and morphology of lateral branch systems of *Archaeopteris. Amer. Journ. Bot.* 58: 758–84.

————. 1976. Current status of the Progymnospermopsida. *Rev. Palaeobot. Palynol.* 21: 5–23.

————. 1981. *Archaeopteris* and its role in vascular plant evolution. In: *Paleobotany, Paleoecology, and Evolution.* K. J. Niklas (Ed.). Vol. 1, p. 193–230. Praeger Special Studies, New York.

Berry, E. W. 1927. Devonian floras. *Amer. Journ. Sci.* 14: 109–20.

Bertrand, P. (1913) 1914. Étude du stipe de l'*Asteropteris noveboracensis.* 12th *Congr. Geol. Canada. Compt. Rend.,* 909–24.

————. 1935. Contribution à l'étude des Cladoxylées de Saalfeld. *Palaeontographica* 80B: 101–70.

Bhutta, A. A. 1972. Observations on the sporangia of *Horneophyton lignieri* (Kidston and Lang) Barghoorn and Darrah, 1938. *Pakistan Journ. Bot.* 4: 27–34.

————. 1973a. On the spores (including germinating spores) of *Rhynia major* Kidston and Lang. *Biologia* 19: 47–55.

————. 1973b. On the spores (including germinating spores) of *Horneophyton* (*Hornea*) *lignieri* (Kidston and Lang) Barghoorn and Darrah (1938). *Pakistan Journ. Bot.* 5: 45–55.

Bold, H. C. 1973. *Morphology of plants.* 3rd ed. Harper and Row, New York, 668 p.

————, Alexopoulos, C. J., and Delevoryas, T. 1980. *Morphology of plants and fungi.* 4th ed. Harper and Row, New York, 819 p.

Bonamo, P. M. 1975. The Progymnospermopsida: building a concept. *Taxon* 24: 569–79.

————. 1977. *Rellimia thomsonii* (Progymnospermopsida) from the Middle Devonian of New York State. *Amer. Journ. Bot.* 64: 1272–85.

————. 1983. *Rellimia thomsonii* (Dawson) Leclercq and Bonamo (1973): the only correct name for the aneurophytalean progymnosperm. *Taxon* 32: 449–72.

Bonamo, P. M. and Banks, H. P. 1966. *Calamophyton* in the Middle Devonian of New York State. *Amer. Journ. Bot.* 53: 778–91.

————. 1967. *Tetraxylopteris schmidtii*: its fertile parts and its relationships within the Aneurophytales. *Amer. Joun. Bot.* 54: 755–68.

————, and Grierson, J. D. 1981. Leaf variation in *Leclercqia complexa* and its possible significance. *Bot. Soc. Amer. Misc. Ser. Publ.* 160: 42 (abstract).

Boureau, E. (ed.) *Traité de Paléobotanique.* The following sections are especially pertinent to the present work: Vol. II, 1967. Psilophyta by Høeg, O. A., pp. 191–433. Lycophyta by Chaloner, W. G. in collaboration with Boureau, E. pp. 435–862. Vol. III. 1964. Sphenophyta by E. Boureau. pp. 1–477. Vol. IV. Fasc. 1, 1975. Filicophyta by Andrews, H. N., Arnold, C. A., Boureau, E., Doubinger, J., and Leclercq, S. pp. 1–491.

Boureau, E., Lejal-Nicol, A., and Massa, D. 1978. À propos du Silurien et du Dévonian en Libye. *C. R. Hébdomadaire des Séances de l'Acad. des Sci.* 186D: 1567–71.

Brauer, D. F. 1980. *Barinophyton citrulliforme* (Barinophytales *incertae sedis,*

Barinophytaceae) from Upper Devonian of Pennsylvania. *Amer. Journ. Bot.* 67: 1186–1206.

———. 1981. Heterosporous, barinophytacean plants from the Upper Devonian of North America and a discussion of the possible affinities of the Barinophytaceae. *Rev. Palaeobot. Palynol.* 33: 347–62.

Bridge, J. S., Van Veen, P. M., and Matten, L. C. 1980. Aspects of the sedimentology, palynology and palaeobotany of the Upper Devonian of southern Kerry Head, Co. Kerry, Ireland. *Geol. Journ.* 15: 143–70.

Calder, M. G. 1938. On some undescribed species from the Lower Carboniferous flora of Berwickshire; together with a note on the genus *Stenomyelon* Kidston. *Trans. Roy. Soc. Edinb.* 59: 309–31.

Carluccio, L. M., Hueber, F. M., and Banks, H. P. 1966. *Archaeopteris macilenta,* anatomy and morphology of its frond. *Amer. Journ. Bot.* 53: 719–30.

Carruthers, W. 1872a. Notes on some fossil plants. *Geol. Mag.* 9: 46–59.

———. 1872b. On the history, histological structure and affinities of *Nematophycus logani* Carr. (*Prototaxites logani* Dawson), an alga of Devonian age. *Month. Micr. Journ.* 8: 160–72.

Chaloner, W. G. 1967. Spores and land plant evolution. *Rev. Palaeobot. Palynol.* 1: 83–93.

———. 1968. The cone of *Cyclostigma kiltorkense* Haughton, from the Upper Devonian of Ireland. *Journ. Linnean Soc. London* (Botany) 61: 25–36.

———. 1970. The rise of the first land plants. *Biol. Rev.* 45: 353–77.

———. 1972. Devonian plants from Fair Isle, Scotland. *Rev. Palaeobot. Palynol.* 14: 49–61.

Chaloner, W. G., and Pettitt, J. M. 1964. A seed megaspore from the Devonian of Canada. *Palaeontology* 7: 29–36.

———, and Sheerin, A. 1979. Devonian macrofloras. In: *The Devonian System,* House, M. R., Scrutton, C. T. and Bassett, M. G. (Eds), *Special Papers in Palaeontology* 23: 145–61.

———, and Macdonald, P. 1980. *Plants invade the land.* H. M. Stationery Office, The Royal Scottish Museum, Edinburgh, 17 p.

Chaloner, W. G., Hill, A. J., and Lacey, W. S. 1977. First Devonian platyspermic seed and its implications in gymnosperm evolution. *Nature* 265: 233–35.

———, Hill, A., and Rogerson, E. C. W. 1978. Early Devonian plant fossils from a southern England borehole. *Palaeontology* 21: 693–707.

Cookson, I. C. 1926. On the occurrence of the Devonian genus *Arthrostigma* in Victoria. *Proc. Roy. Soc. Victoria* (n.s.) 38: 65–8.

———. 1935. On plant-remains from the Silurian of Victoria, Australia, that extend and connect floras hitherto described. *Phil. Trans. Roy. Soc. London* 225B: 127–48.

———. 1949. Yeringian (Lower Devonian) plant remains from Lilydale, Victoria, with notes on a collection from a new locality in the Siluro-Devonian sequence. *Mem. Nat. Mus. Melbourne* 16: 117–30.

Cornet, B., Phillips, T. L., and Andrews, H. N. 1976. The morphology and variation in *Rhacophyton ceratangium* from the Upper Devonian and its bearing on frond evolution. *Palaeontographica* 158B: 105–29.

Corsin, P. 1945. Les algues de l'Eodevonien de Vimy (P. de C.). *Soc. Sci. Agr. et Arts de Lille, Mem.* ser 5, No. 9. pp. 1–86.

Crépin, F. 1875. Observations sur quelques plants fossiles des dépôts dévoniens. *Soc. Roy. Bot. Belgique Bull.* 14: 214–30.

Croft, W. N., and Lang, W. H. 1942. The Lower Devonian flora of the Senni Beds of Monmouthshire and Breconshire. *Phil. Trans. Roy. Soc. London* 231B: 131–63.

Daber, R. 1960. *Eogaspesiea gracilis* n.g. n. sp. *Geologie* 9: 418–25.

Dawson, J. W. 1859. On fossil plants from the Devonian rocks of Canada. *Quart. Journ. Geol. Soc. London* 15: 477–88.

_____. 1870. The primitive vegetation of the earth. *Nature* 2: 85–8.

_____. 1871. The fossil plants of the Devonian and Upper Silurian Formations of Canada. *Geol. Survey Canada,* p. 1–92., Montreal.

_____. 1881. Notes on new Erian (Devonian plants. *Canadian Naturalist and Quart Journ. Sci.* n. ser. 9: 475–76.

_____. 1892. *The Geological History of Plants,* 294 pp. Appleton and Co., New York.

Dittrich, H. S., Matten, L. C. and Phillips, T. L. 1983. Anatomy of *Rhacophyton ceratangium* from the Upper Devonian (Famennian) of West Virginia. *Rev. Palaeobot. Palynol.* 40: 127–147.

Don, A. W. R. and Hickling, G. 1917. On *Parka decipiens. Quart. Journ. Geol. Soc. London* 71: 648–66.

Doran, J. B. 1980. A new species of *Psilophyton* from the Lower Devonian of northern New Brunswick, Canada. *Canadian Journ. Bot.* 58: 2241–62.

Doran, J. B., Gensel, P. G., and Andrews, H. N. 1978. New occurrences of trimerophytes from the Devonian of eastern Canada. *Canadian Journ. Bot.* 56: 3052–68.

Dorf, E. 1933. A new occurrence of the oldest known terrestrial vegetation, from Beartooth Butte, Wyoming. *Bot. Gaz.* 95: 240–57.

Dorf, E., and Rankin, D. W. 1962. Early Devonian plants from the Traveler Mountain area, Maine. *Journ. Paleont.* 36: 999–1004.

Doyle, J. A., Van Campo, M., and Lugardon, B. 1975. Observations on exine structure of *Eucommidites* and Lower Cretaceous angiosperm pollen. *Pollen et Spores* 17: 429–86.

Duckett, J. G. 1970a. Spore size in the genus *Equisetum. New Phytol.* 69: 333–46.

_____. 1970b. Sexual behavior in the genus *Equisetum,* subgenus *Equisetum. Journ. Linnean Soc. London Botany,* 63: 327–52.

Duerden, H. 1929. Variations in megaspore number in *Selaginella. Ann. Bot.* 43: 451–57.

Edwards, David S. 1980. Evidence for the sporophyte status of the Lower Devonian plant *Rhynia gwynne-vaughanii* Kidston and Lang. *Rev. Palaeobot. Palynol.* 29: 177–88.

Edwards, Dianne. 1968. A new plant from the Lower Old Red Sandstone of South Wales. *Palaeontology* 11: 683–90.

_____. 1969a. Further observations on *Zosterophyllum llanoveranum* from the Lower Devonian of South Wales. *Amer. Journ. Bot.* 56: 201–10.

_____. 1969b. *Zosterophyllum* from the Lower Old Red Sandstone of South Wales. *New Phytologist* 68: 923–31.

————. 1970a. Fertile Rhyniophytina from the Lower Devonian of Britain. *Palaeontology* 13: 451–61.

————. 1970b. Further observations on the Lower Devonian plant *Gosslingia breconensis* Heard. *Phil. Trans. Roy. Soc. London* 258B: 225–43.

————. 1973. Devonian floras. In: Hallam, A., (ed.) Atlas of Palaeobiogeography. p. 105–15, Amsterdam: Elsevier.

————. 1975. Some observations on the fertile parts of *Zosterophyllum myretonianum* Penhallow from the Lower Old Red Sandstone of Scotland. *Trans. Roy. Soc. Edinburgh* 69: 251–65.

————. 1976. The systematic position of *Hicklingia edwardii* Kidston and Lang. *New Phytol.* 76: 173–81.

————. 1979. A late Silurian flora from the Lower Old Red Sandstone of South-west Dyfed. *Palaeontology* 22: 23–52.

————. 1980a. Early Land Floras. In: Panchen, A. L., (ed.) *The terrestrial environment and the origin of land vertebrates.* Systematics Assoc. Special Vol. No. 15, p. 55–85, London, Academic Press.

————. 1980b. Studies on Lower Devonian petrifactions from Britain. I. Pyritized axes of *Hostinella* from the Brecon Beacons quarry, Powys, South Wales. *Rev. Palaeobot. Palynol.* 29: 189–200.

Edwards, D., and Richardson, J. B. 1974. Lower Devonian (Dittonian) plants from the Welsh Borderland. *Palaeontology* 17: 311–24.

Edwards, D. and Davies, E. C. W. 1976. Oldest recorded *in situ* tracheids. *Nature, Lond.* 263: 494–95.

————. and Rogerson, E. C. W. 1979. New records of fertile Rhyniophytina from the late Silurian of Wales. *Geol. Mag.* 116: 93–8.

Edwards, D., Bassett, M. G., and Rogerson, E. C. W. 1979. The earliest vascular land plants: continuing the search for proof. *Lethaia* 12: 313–24.

————, Edwards, D. S., and Raynor, R. 1983a. The cuticle of early vascular plants and its evolutionary significance. In: Cutler, D. F., Alvin, K. L., and Price, C. E., (eds.) *The Plant Cuticle.* Linnean Soc. Symposium, Ser. 10, New York: Academic Press. pp. 341–61.

————, Feehan, J., and Smith, D. G. 1983b. A late Wenlock flora from Co. Tipperary, Ireland. *Bot. Journ. Linn. Soc.* 86: 19–36.

Edwards, W. N. 1924. On the cuticular structure of the Devonian plant *Psilophyton. Journ. Linnean Soc. London (Botany)* 46: 377–85.

Eggert, D. A. 1974. The sporangium of *Horneophyton lignieri* (Rhyniophytina). *Amer. Journ. Bot.* 61: 405–13.

Eggert, D. A. and Taylor, T. N. 1971. *Telangiopsis* gen. nov., an upper Mississippian pollen organ from Arkansas. *Bot. Gaz.* 132: 30–37.

———— and Kanemoto, N. Y. 1977. Stem phloem of a Middle Pennsylvania *Lepidodendron. Botanical Gazette* 138(1): 102–11.

El-Saadawy, W. and Lacey, W. S. 1979a. Observations on *Nothia aphylla* Lyon ex Høeg. *Rev. Palaeobot. Palynol.* 27: 119–47.

————. 1979b. The sporangia of *Horneophyton lignieri* (Kidston and Lang) Barghoorn and Darrah. *Rev. Palaeobot. Palynol.* 28: 137–44.

Fairon-Demaret, M. 1971. Quelques charactères du *Drepanophycus spinaeformis* Göppert. *C. R. Acad. Sci. Paris* Ser. D, 273: 933–35.

_____. 1972. Les feuilles fertiles de *Protolepidodendron wahnbachense* Kräusel, R. et Weyland, H., 1932 du Dévonien Inférieur de Belgique. *Bull. Inst. R. Sci. Nat. Belg.* 48: 1–9.

_____. 1974. Nouveaux specimens du genre *Leclercqia* Banks, H. P., Bonamo, P. M., et Grierson, J. D., 1972, du Givetien (?) du Queensland (Australie). *Bull. Inst. R. Sci. Nat. Belg.* 50: 1–4.

_____. 1977. A new lycophyte cone from the Upper Devonian of Belgium. *Palaeontographica* 162B: 51–63.

_____. 1978. *Estinnophyton gracile* gen. et sp. nov., a new name for specimens previously determined *Protolepidodendron wahnbachense* Kräusel and Weyland from the Siegenian of Belgium. *Bull. Acad. Roy. Belg.* ser. 5, 64: 597–609.

_____. 1979. *Estinnophyton wahnbachense* (Kräusel et Weyland) comb. nov., une plant remarkable du Siegenien d'Allemagne. *Rev. Palaeobot. Palynol.* 28: 145–60.

_____. 1980. A propos des spécimens déterminés *Protolepidodendron scharianum* par Kräusel et Weyland, 1932. *Rev. Palaeobot. Palynol.* 29: 201–20.

_____. 1981. Le genre *Leclercqia* Banks, H. P., Bonamo, P. M. et Grierson, J. D. 1972 dans le Devonien Moyen de Belgique. *Bull. Inst. R. Sci. Nat. Belg.* 31-XII p. 1–10.

Fairon-Demaret, M. and Banks, H. P. 1978. Leaves of *Archaeosigillaria vanuxemii*, a Devonian lycopod from New York. *Amer. Journ. Bot.* 65: 246–49.

Fleming, J. 1831. On the occurrence of scales of vertebrated animals in the Old Red Sandstone of Fifeshire. *Edinburgh Journ. Nat. Geog. Sci.* 3: 81–6.

Foster, A. and Gifford, E. 1974. *Comparative Morphology of Vascular Plants*. 2nd ed. San Francisco: W. H. Freeman and Co., 751 p.

Galtier, J. 1981. Structures foliares de fougères et pteridospermales du Carbonifère inférieur et leur signification évolutive. *Palaeontographica* 180B: 1–38.

Galtier, J., and Hébant, C. 1973. Sur le phloème et le cambium d'une Calamopityacée, Pteridospermale probable du Carbonifère inférieur francais. *C. R. Acad. Sci. Paris* 276: 2257–59.

Garratt, M. J. 1978. New evidence for a Silurian (Ludlow) age for the earliest *Baragwanathia* flora. *Alcheringa* 2: 217–24.

Gensel, P. G. 1976. *Renalia hueberi*, a new plant from the Lower Devonian of Gaspé. *Rev. Palaeobot. Palynol.* 22: 19–37.

_____. 1977. Morphologic and taxonomic relationships of the Psilotaceae relative to evolutionary lines in early land vascular plants. *Brittonia* 29: 14–29.

_____. 1979. Two *Psilophyton* species from the Lower Devonian of eastern Canada with a discussion of morphological variation within the genus. *Palaeontographica* 168B: 81–99.

_____. 1980. Devonian *in situ* spores: a survey and discussion. *Rev. Palaeobot. Palynol.* 30: 101–32.

_____. 1982a. A new species of *Zosterophyllum* from the early Devonian of New Brunswick. *Amer. Journ. Bot.* 69: 651–69.

_____. 1982b. On the contributions of Sir J. W. Dawson to the study of early land plants (Devonian) and current ideas concerning their nature, diversity and evolutionary relationships. *Proceedings, 3rd North Amer. Paleontol. Conv.* vol. 1: 199–204.

_____. 1982c. *Oricilla*, a new genus referable to the Zosterophyllophytes from the late Early Devonian of northern New Brunswick. *Rev. Palaeobot. Palynol.* 37: 345–59.

Gensel, P. G. and White, A. R. 1983. The morphology and ultrastructure of spores of the Early Devonian trimerophyte *Psilophyton* (Dawson) Hueber and Banks. *Palynology* 7: 221–33.

Gensel, P. G., Kasper, A. E., and Andrews, H. N. 1969. *Kaulangiophyton*, a new genus of plants from the Devonian of Maine. *Torrey Bot. Club. Bull.* 96: 265–76.

_____, Andrews, H. N., and Forbes, W. H. 1975. A new species of *Sawdonia* with notes on the origin of microphylls and lateral sporangia. *Bot. Gaz.* 136: 50–62.

Gillespie, W. H., Rothwell, G. W., and Scheckler, S. E. 1981. The earliest seeds. *Nature* 293: 462–64.

Goeppert, H. R. 1850. Monographie der fossilen Coniferen. *Hollandsche Maat. Wetensch. Matuurk,* 6: 1–286.

_____. 1852. Fossile flora des Übergangsgebirges. *Nova Acta Leopoldina* 22: 1–299.

Goldring, W. 1924. The Upper Devonian forest of seed ferns. *New York State Mus. Bull.* 251: 50–72.

_____. 1927. The oldest known petrified forest. *Sci. Monthly* 24: 514–29.

Gordon, W. T. 1935. The genus *Pitys*, Witham emend. *Trans. Roy. Soc. Edinb.* 58: 279–311.

Gothan, W. 1927. Struksturzeigende Pflanzen aus dem Oberdevon von Wildenfels. *Sachsischen geol. Landesamts Abh.* 3: 1–12.

Granoff, J. A., Gensel. P. G., and Andrews, H. N. 1976. A new species of *Pertica* from the Devonian of eastern Canada. *Palaeontographica* 155B: 119–28.

Gray, J. and Boucot, A. J. 1971. Early Silurian spore tetrads from New York: earliest New World evidence for vascular plants? *Science* 173: 918–21.

_____. 1977. Early vascular land plants: proof and conjecture. *Lethaia* 10: 145–74.

Gray, J., Massa, D., and Boucot, A. J. 1982. Caradocian land plant microfossils from Libya. *Geology* 10: 197–201.

Grierson, J. D. 1976. *Leclercqia complexa* (Lycopsida, Middle Devonian): its anatomy, and the interpretation of pyrite petrifactions. *Amer. Journ. Bot.* 63: 1184–1202.

Grierson, J. D., and Banks, H. P. 1963. Lycopods of the Devonian of New York State. *Palaeontographica Americana* IV, no. 31: 219–95.

_____, and Hueber, F. M. 1968. Devonian lycopods from northern New Brunswick. In: *International Symposium on the Devonian System,* Oswald, D. H., (Ed.) 2: 823–36. Alberta Soc. Petroleum Geol., Calgary.

_____, and Bonamo, P. M. 1979. *Leclercqia complexa*: earliest ligulate lycopod (Middle Devonian). *Amer. Journ. Bot.* 66: 474–76.

Halle, T. G. 1916a. Lower Devonian plants from Röragen in Norway. *Kungl. Svenska Vetenskapsakad. Handl. Bd.* 57, no. 1: 1–46.

_____. 1916b. A fossil sporogonium from the Lower Devonian of Norway. *Bot. Notiser* pp. 79–81.

_____. 1936a. Notes on the Devonian genus *Sporogonites*. *Svensk Bot. Tidsk.* 30: 613–23.

_____. 1936b. On *Drepanophycus, Protolepidodendron* and *Protopteridium*, with notes on the Palaeozoic flora of Yunnan. *Palaeontologia Sinica*, ser. A. I. fasc. 4, 28 p. Nanking.

Hartman, C. M. 1981. The effect of pyrite on the tracheid structure of *Drepanophycus spinaeformis*, a long-ranging Devonian lycopod. *Rev. Palaeobot. Palynol.* 32: 239–55.

Hartman, C., and Banks, H. P. 1980. Pitting in *Psilophyton dawsoni*, an early Devonian trimerophyte. *Amer. Journ. Bot.* 67: 400–12.

Haughton, S. 1860. On *Cyclostigma*, a new genus of fossil plants from the Old Red Sandstone of Kiltorcan, Co. Kilkenny: and on the general law of phyllotaxis in the natural orders, Lycopodiaceae, Equisetaceae, Filices, etc. *Journ. Roy. Dublin Soc.*, Vol. ii. 407–20. Also published, without illustrations in *Nat. Hist. Rev.* (proceedings) 7: 209–22.

Hauke, R. L. 1968. Gametangia of *Equisetum bogotense*. *Torrey Bot. Club. Bull.* 95: 341–45.

Heard, A. 1925. *Psilophyton breconensis*. *Report Brit. Assoc. (Southampton)*, pp. 311–12.

_____. 1927. On Old Red Sandstone plants showing structure from Brecon (South Wales). *Quart. Journ. Geol. Soc. London* 83: 195–209.

Heckel, P. H., and Witzke, B. J. 1979. Devonian world paleogeography determined from distribution of carbonates and related lithic paleoclimatic indicators. In: House, M. R., Scrutton, C. T. and Bassett, M. G. (Eds.). The Devonian System. *Spec. Papers in Palaeontol.* 23: 99–123.

Heer, O. 1871. Fossile Flora der Bären Insel. *Kungl. Svenska Vetenskaps. Akad. Handl.* 9(5): 1–51.

Hirmer, M. 1927. *Handbuch der Palaobotanik*. Vol. I. Munich and Berlin· R. Oldenbourg.

Høeg, O. A. 1930. A psilophyte in South Africa. *Kgl. Norske Vid. Selsk. Forh.* 3: 92–4.

_____. 1931. Notes on the Devonian flora of western Norway. *Kgl. Norske Vid. Selsk. Skr.* 6: 1–18.

_____. 1935. Further contributions to the Middle Devonian flora of western Norway. *Norsk Geol. Tidsskr.* 15: 1–18.

_____. 1937. The Devonian floras and their bearing upon the origin of vascular plants. *Bot. Rev.* 3: 563–92.

_____. 1942. The Downtonian and Devonian flora of Spitsbergen. *Norges Svalbard-og Ishavs- Unserskelser* nr. 83, 229 p.

_____. 1945. Contributions to the Devonian flora of western Norway. III. *Norsk. Geol. Tidskrift* 25: 183–92.

———. 1952. *Psilophytites*, a new form genus of Devonian plants. *Paleobotanist I* (Birbal Sahni Mem. Vol.): 212–14.

———. 1967. Psilophyta. In *Traité de Paléobotanique*, vol. II: 191–433.

———. 1973. *Pertica* sp. in the Devonian of Mimerdalen, Spitsbergen. *Norsk Geol. Tidskrift* 53: 85–6.

Holden, H. S. 1954. The morphology of a new species of pteridosperm seed from the Yorkshire Coal Measures. *Ann. Bot. N.S.* 18: 407–15.

Holland, C. H. and Richardson, J. B. 1977. The British Isles. In: Martinsson, A. (ed.) *The Silurian-Devonian Boundary.* IUGS, Ser. A. No. 5, pp. 35–44. Stuttgart.

Hooker, J. D. 1861. In Salter, 1861. Note on the fossils found in the Worcester and Hereford Railway cuttings. *Quart. Journ. Geol. Soc. London* 17: 161–62.

———. 1889. Pachytheca. *Ann. Bot.* 3: 135–40.

Hopping, C. A. 1956. On a specimen of *"Psilophyton robustius"* Dawson, from the Lower Devonian of Canada. *Proceed. Roy. Soc. Edinb.* 66: 1–28.

House, M. R., Richardson, J. B., Chaloner, W. G., Allen, J. R. L., Holland, C. H., and Westoll, T. S. 1977. A correlation of Devonian rocks of the British Isles. *Geol. Soc. London Special Rept.* No. 7, 110 pp.

Hueber, F. M. 1961a. Psilophytes in the Upper Devonian of New York. *Amer. Journ. Bot* (abstracts), July, p. 541.

———. 1961b. *Hepaticites devonicus,* a new fossil liverwort from the Devonian of New York. *Ann. Mo. Bot. Gdn.* 48: 125–32.

———. 1964. The Psilophytes and their relationships to the origin of the ferns. *Torrey Bot. Club Mem.* 21: 5–9.

———. 1967. *Psilophyton*: the genus and the concept. In *International Symposium on the Devonian System.* ed. D. S. Oswald, Alberta Soc. Petrol. Geol. Calgary, Vol. 2: 815–22.

———. 1970. *Rebuchia*: a new name for *Bucheria* Dorf. *Taxon* 19: 822.

———. 1971. *Sawdonia ornata*: a new name for *Psilophyton princeps* var. *ornatum. Taxon* 20: 641–42.

———. 1972a. *Rebuchia ovata*, its vegetative morphology and classification with the Zosterophyllophytina. *Rev. Palaeobot. Palynol.* 14: 113–27.

———. 1972b. Early Devonian land plants from Bathurst Island, District of Franklin. *Geol. Surv. Canada.* Paper 71-28, pp. 1–17.

———. 1983. A new species of *Baragwanathia* from the Sextant Formation (Emsian) Northern Ontario, Canada. *Bot. J. Linn. Soc.* 86: 57–79.

Hueber, F. M., and Grierson, J. D. 1961. On the occurrence of *Psilophyton princeps* in the early Upper Devonian of New York. *Amer. Journ. Bot.* 48: 473–79.

———, and Banks, H. P. 1967. *Psilophyton princeps:* the search for organic connection. *Taxon* 16: 81–5.

———. 1979. *Serrulacaulis furcatus* gen. et sp. nov., a new zosterophyll from the lower Upper Devonian of New York State. *Rev. Palaeobot. Palynol.* 28: 169–89.

Itschenko, T. A. 1969. The Cooksonian paleoflora in the Skalskii horizon of Podolia and its stratigraphical significance. *Geol. Journ. Kiev* 29: 101–9.

_____. 1975. The late Silurian flora of *Podolia* (Ukraine), *Akad. Nauk. Ukrainskoi.* USSR, Kiev, 80 pp.

Iurina, A. L. 1964. New Devonian species of the genus *Cooksonia. Paleont. Journ.* Special Reprint No. 1, Akad. Nauk. USSR, pp. 107–13.

Jackson, B. D. 1965. *A Glossary of Botanic Terms.* 4th Ed. New York: Hafner Publishing Co.

Johnson, G. A. L. 1980. Carboniferous geography and terrestrial migration route. In The terrestrial environment and the origin of land vertebrates, ed. A. L. Panchen. Systematics Assoc. Spec. vol. no. 15, pp. 39–54. Academic Press.

Johnson, T. 1913. On *Bothrodendron (Cyclostigma) kiltorkense*, Haughton, sp. *Roy. Soc. Dublin, Sci. Proceed.* 13 (n.s.): 500–528.

_____. 1917. *Spermolithus devonicus*, gen. et sp. nov., and other pteridosperms from the Upper Devonian beds of Kiltorcan, Co. Kilkenny. *Roy. Soc. Dublin Sci. Proceed.* 15 (n.s.): 245–54.

Jonker, F. P. 1979. *Prototaxites* in the Lower Devonian. *Palaeontographica* 171B: 39–56.

Kasper, A. E. 1977. A new species of the Devonian lycopod genus *Leclercqia* from New Brunswick, Canada. *Amer. Journ. Bot.*, Pub. 154. p. 39 (abstract).

Kasper, A. E., and Andrews, H. N. 1972. *Pertica*, a new genus of Devonian plants from northern Maine. *Amer. Journ. Bot.* 59: 897–911.

_____, and Forbes, W. H. 1979. The Devonian lycopod *Leclercqia* from the Trout Valley Formation in Maine. *Maine Geology Bull.* (Geol. Soc. Maine) 1:49–59.

Kasper, A. E., Andrews, H. N., and Forbes, W. H. 1974. New fertile species of *Psilophyton* from the Devonian of Maine. *Amer. Journ. Bot.* 61: 339–59.

Kevan, P. G., Chaloner, W. G., and Savile, D. B. O. 1975. Interrelationships of early terrestrial arthropods and plants. *Palaeontology* 18: 391–417.

Kidston, R. 1894. On the occurrence of *Arthrostigma gracile* Dawson, in the Lower Old Red Sandstone of Perthshire. *Proc. Roy. Phys. Soc. Edin.* 12: 109.

_____. 1901. Carboniferous lycopods and sphenophylls. *Nat. Hist. Soc. Glasgow Trans.* 6 (n.s.): 25–140.

Kidston, R. and Lang, W. H. 1917–1921. On Old Red Sandstone plants showing structure, from the Rhynie chert bed, Aberdeenshire. *Trans. Roy. Soc. Edinburgh.* Part I. *Rhynia gwynne-vaughani*, K. and L. 1917. 51: 761–84. Part II. Additional notes on *Rhynia gwynne-vaughani*, K. and L.; with descriptions of *Rhynia major*, n. sp. and *Hornea lignieri*, n.g., n. sp. 1920. 52: 603–27. Part III. *Asteroxylon mackiei*, K. and L. 1920, 52: 643–80. Part IV. Restorations of the vascular cryptogams, and discussion of their bearing on the general morphology of the pteridophyta and the origin of the organization of land-plants. 1921. 52: 831–54. Part V. The Thallophyta occurring in the peat-bed; the succession of the plants throughout a vertical section of the bed, and the conditions of accumulation and preservation of the deposit. 1921. 52: 855–902.

_____. 1923. Notes on fossil plants from the Old Red Sandstone of Scotland. I.

Hicklingia edwardi K. and L. *Trans. Roy. Soc. Edinburgh* 53: 405–7.

———. 1924. Notes on fossil plants from the Old Red Sandstone of Scotland. II. *Nematophyton forfarense* Kidston sp. and III. On two species of *Pachytheca* (*P. media* and *P. fasciculata*) based on the characters of the algal filaments. *Trans. Roy. Soc. Edinburgh* 53: 603–14.

Klitzsch, E., Lejal-Nicol, A., and Massa, D. 1973. Le Siluro-Devonien à Psilophytes et Lycophytes du Bassin de Mourzouk (Libye). *C. R. Acad. Sci. Paris.* 227D: 2465–67.

Knoll, A. H., Niklas, K. J., Gensel, P. G., and Tiffney, B. In press. Character diversification and patterns of evolution in early vascular plants. *Paleobiology.*

Kräusel, R. 1936. Neue Untersuchungen zur paläozoischen Flora: Rheinische Devonfloren. *Sonderab. Bericht. der Deutsch. Bot. Gesell.* 54(5): 307–28.

———. 1937. Die Verbreitung des Devonflora. In *2nd Congr. Intern. Stratigr. Geol. Carb. Compte Rendu* (Heerlen 1935), pp. 527–37.

Kräusel, R. and Weyland, H. 1923. Beiträge zur Kenntnis der Devonflora. I. *Senckenbergiana* 5: 154–84.

———. 1926. Beiträge, etc. II. *Abh. Senckenb. Naturf. Ges.* 40: 115–45.

———. 1929. Beiträge, etc. III. *Abh. Senckenb. Naturf. Ges.* 41: 315–60.

———. 1930. Die Flora des deutschen Unterdevons. *Abh. Preuss. Geol. Landesan. Neue Folge.* H. 131: 5–90.

———. 1932. Pflanzenreste aus dem Devon. IV. *Protolepidodendron* Krejči. *Senckenbergiana* 14: 391–403.

———. 1934. Algen im deutschen Devon. *Palaeontographica* 79B: 131–42.

———. 1935. Neue Pflanzenfunde im rheinischen Unterdevon. *Palaeontographica* 80B: 171–90.

———. 1937. Pflanzenreste aus dem Devon. X. Zwei Pflanzenfunde im Oberdevon der Eifel. *Senckenbergiana* 19: 338–55.

———. 1938. Neue Pflanzefunde im Metteldevon von Elberfeld. *Palaeontogr.* 83B: 172–95.

———. 1940. Pflanzenreste aus dem Devon. XII. Die Gattung *Protolepidodendron* Krejči. *Senckenbergiana* 20: 6–16.

———. 1941. Pflanzenreste aus dem Devon von Nord Amerika. *Palaeontographica* 86B: 1–78.

———. 1948. Pflanzenreste aus dem Devon XIII. Die Devon floren Belgiens und des Rheinlandes, nebst Bemerkungen zu Einigenihrer Arten. *Senckenbergiana* 29: 77–99.

———. 1961. Über *Psilophyton robustius. Palaeontographica* 108B: 11–21.

Krejči, Jan. 1880. Notiz über die Reste von Landpflanzen in der böhmischen Silurformation. *Kgl. böhmischen Gesell. Wiss. Sitzunsber.* (1879), pp. 201–4.

Lang, W. H. 1925. Contributions to the study of the Old Red Sandstone flora of Scotland. I and II. *Trans. Roy. Soc. Edinburgh* 54: 253–79.

———. 1926. Contributions, etc. III, IV, V. *Trans. Roy. Soc. Edinburgh* 54: 785–99.

———. 1927. Contributions, etc. VI. On *Zosterophyllum myretonianum*, etc. *Trans. Roy. Soc. Edinburgh.* 55: 443–55.

_____. 1931. On the spines, sporangia, and spores of *Psilophyton princeps*, Dawson, shown in specimens from Gaspé. *Phil. Trans. Roy. Soc. London* 219B: 421–42.

_____. 1932. Contributions, etc. VIII. On *Arthrostigma, Psilophyton* etc. *Trans. Roy. Soc. Edinburgh* 57: 491–521.

_____. 1937. On the plant-remains from the Downtonian of England and Wales. *Phil. Trans. Roy. Soc. London* 227B: 245–91.

_____. 1945. *Pachytheca* and some anomalous early plants (*Prototaxites, Nemathothallus, Parka, Foerstia, Orvillea* n. gen.). *Journ. Linnean Soc. London* (Botany) 52: 535–52.

Lang, W. H., and Cookson, I. C. 1927. On some early Palaeozoic plants from Victoria, Australia. *Mem. Proc. Manchester Lit. Phil. Soc.* 71: 45–51.

_____. 1930. Some fossil plants of early Devonian type from the Walhalla series, Victoria, Australia. *Phil. Trans. Roy. Soc. London* 219B: 133–63.

_____. 1935. On a flora, including vascular land plants, associated with *Monograptus*, in rocks of Silurian age, from Victoria, Australia. *Phil. Trans. Roy. Soc. London.* 224B: 421–49.

Leclercq, S. 1940. Contribution a l'étude de la flore du Dévonien de Belgique. *Acad. Roy. Belg.* 12: 1–65.

_____. 1942. Quelques plants fossiles recueillies dans le Dévonien inférieur des environs de Nonceveux. *Ann. Soc. Géol. Belgique* 65: B193–211.

_____. 1951. Étude morphologique et anatomique d'une Fougère du Dévonien Superieur. *Ann. Soc. Geol. Belgique Mem.* 9: 1–62.

_____. 1954. Are the Psilophytales a starting or a resulting point? *Svensk Bot. Tidskrift* 48: 301–15.

_____. 1957. Étude d'une fructification de Sphenopside à structure conservée du dévonien supérieur. *Acad. Roy. Belg. Mem.* Vol. XIV (3): 39 pp.

_____. 1961. Strobilar complexity in Devonian sphenopsids. In *Recent studies in botany*. XII. Internat. Bot. Congress, Montreal, 1959. 2: 968–71. Toronto.

_____. 1964. Recent studies on Devonian sphenopsids. In *Tenth Internat. Bot. Congress*, Edinburgh, 1965. Abstracts of papers, p. 18.

Leclercq, S., and Bélliere, M. 1928. *Psygmophyllum gilkineti*, sp. n. du Dévonien moyen a facies Old Red Sandstone de Malonne environs de Namur, Belgique. *Linnean Soc. London Journ.* 48: 1–14.

_____, and Andrews, H. N. 1960. *Calamophyton bicephalum*, a new species from the Middle Devonian of Belgium. *Ann. Missouri Bot. Gard.* 47: 1–23.

_____, and Banks, H. P. 1962. *Pseudosporochnus nodosus* sp. nov., a Middle Devonian plant with cladoxylalean affinities. *Palaeontographica* 110B: 1–34.

_____, and Schweitzer, H. J. 1965. *Calamophyton* is not a Sphenopsid. *Acad. Roy. Belgique Bull. Classe des Sci.* ser. 5, 51: 1395–1403.

_____, and Lele, K. M. 1968. Further investigation on the vascular system of *Pseudosporochnus nodosus* Leclercq et Banks. *Palaeontographica.* 123B: 97–112.

_____, and Bonamo, P. M. 1971. A study of the fructification of *Milleria* (*Protopteridium*) *thomsonii* Lang from the Middle Devonian of Belgium. *Palaeontographica* 136B: 83–114.

———. 1973. *Rellimia thomsonii*, a new name for *Milleria* (*Protopteridium*) *thomsonii* Lang 1926 emend Leclercq and Bonamo 1971. *Taxon* 22: 435–37.

Lee, H., and Tsai C. 1978. III. Devonian floras of China. In *Papers for the International Symposium on the Devonian System*. Nanking Institute of Geology and Palaeontology, Academia Sinica, pp. 1–8.

Lele, K. M. and Walton, J. 1961. Contributions to the knowledge of "*Zosterophyllum myretonianum*" Penhallow from the Lower Old Red Sandstone of Angus. *Trans. Roy. Soc. Edinburgh* 64: 469–75.

Lemoigne, Y. 1968. Observation d'archégones portés par des axes du type *Rhynia gwynne-vaughanii* Kidston et Lang. Existence de gametophytes vascularisés au Devonien. *C. R. Acad. Sci. Paris* 266: 1655–57.

Li, Cheng -sen. 1982. *Hsüa robusta*, a new land plant from the Lower Devonian of Yunnan, China. *Acta Phytotaxonomica Sinica* 20(3): 331–42.

Li, X. and Cai, C. 1977. Early Devonian Zosterophyllum remains from southwest China. *Acta Palaeontologica Sinica* 16: 12–36.

Long, A. G. 1960a. *Stamnostoma huttonense* gen. et sp. nov.—a pteridosperm seed and cupule from the Calciferous Sandstone Series of Berwickshire. *Trans. Roy. Soc. Edinburgh* 64: 201–15.

———. 1960b. On the structure of *Calymmaththeca kidstoni* Calder (emended) and *Genomosperma latens* gen. et sp. nov. from the Calciferous Sandstone Series of Berwickshire. *Trans. Roy. Soc. Edinburgh* 64: 29–44.

———. 1960c. On the structure of "*Samaropsis scotica*" Calder (emended) and "*Eurystoma angulare*" gen. et sp. nov., petrified seeds from the Calciferous Sandstone Series of Berwickshire. *Trans. Roy. Soc. Edinburgh.* 64: 261–80.

———. 1961a. Some pteridosperm seeds from the Calciferous Sandstone Series of Berwickshire. *Trans. Roy. Soc. Edinburgh* 64: 401–19.

———. 1961b. On the structure of "*Deltasperma fouldenense*" gen. et sp. nov., and "*Camptosperma berniciense*" gen. et sp. nov., petrified seeds from the Calciferous Sandstone Series of Berwickshire. *Trans. Roy. Soc. Edinburgh* 64: 281–95.

———. 1962. "*Tristichia ovensi*" gen. et sp. nov., a protostelic Lower Carboniferous pteridosperm from Berwickshire and East Lothian, with an account of some associated seeds and cupules. *Trans. Roy. Soc. Edinburgh* 64: 477–89.

———. 1963. Some specimens of "*Lyginorachis papilio*" Kidston associated with stems of "*Pitys*". *Trans. Roy. Soc. Edinburgh* 65: 211–24.

———. 1965. On the cupule structure of *Eurystoma angulare*. *Trans. Roy. Soc. Edinburgh* 66: 111–28.

———. 1966. Some Lower Carboniferous fructifications from Berwickshire, together with a theoretical account of the evolution of ovules, cupules, and carpels. *Trans. Roy. Soc. Edinburgh* 66: 345–75.

———. 1969. *Eurystoma trigona* sp. nov., a pteridosperm ovule borne on a frond of *Alcicornopteris* Kidston. *Trans. Roy. Soc. Edinburgh* 68: 171–82.

———. 1979. Observations on the Lower Carboniferous genus *Pitus* Witham. *Trans. Roy. Soc. Edinburgh* 70: 111–27.

Lyon, A. G. 1964. The probable fertile region of *Asteroxylon mackiei*. *Nature* 203: 1082–83.

Mackie, W. 1914. The rock series of Craigbeg and Ord Hill, Rhynie, Aberdeenshire. *Trans. Edinburgh Geol. Soc.* 10: 205–36.

Mamay, S. H. 1962. Occurrence of *Pseudobornia* Nathorst in Alaska. *The Palaeobotanist* 11: 19–22.

Matten, L. C. 1968. *Actinoxylon banksii* gen. et sp. nov., a progymnosperm from the Middle Devonian of New York. *Amer. Journ. Bot.* 55: 773–82.

_____. 1973. The Cairo flora (Givetian) from eastern New York. I. *Reimannia*, terete axes, and *Cairoa lamenekii* gen. et sp. nov. *Amer. Journ. Bot.* 60: 619–30.

_____. 1974. The Givetian flora from Cairo, New York: *Rhacophyton*, *Triloboxylon* and *Cladoxylon*. *Journ. Linnean Soc. London* (Botany) 68: 303–18.

_____. 1981. *Svalbardia banksii* sp. nov. from the Upper Devonian (Frasnian) of New York State. *Amer. Journ. Bot.* 68: 1383–91.

Matten, L. C., and Banks, H. P. 1966. *Triloboxylon ashlandicum* gen. et sp. nov. from the Upper Devonian of New York. *Amer. Journ. Bot.* 53: 1020–28.

_____, and Schweitzer, H. J. 1982. On the correct name for *Protopteridium* (*Rellimia*) *thomsonii* (fossil). *Taxon* 31: 322–26.

Matten, L. C., Lacey, W. S., and Lucas, R. C. 1980. Studies on the cupulate seed genus *Hydrasperma* Long from Berwickshire and East Lothian in Scotland and County Kerry in Ireland. *Linnean Soc. London Biol. Journ.* 81: 249–73.

McGregor, D. C. 1969. Devonian plant fossils of the genera *Kryshtofovichia*, *Nikitinsporites* and *Archaeoperisaccus*. In: Norford et al., *Contrib. Canadian Paleont. Geol. Surv. Bull.* 182: 91–106.

_____. 1973. Lower and Middle Devonian spores of eastern Gaspé, Canada. I. Systematics. *Palaeontographica* 142B: 1–77.

_____. 1977. Lower and Middle Devonian spores of eastern Gaspé, Canada. II. Biostratigraphy. *Palaeontographica* 163B: 111–42.

_____. 1979a. Spores in Devonian stratigraphical correlation. In: House, M. R., Scrutton, C. T., and Bassett, M. G. (Eds), The Devonian System. *Spec. Papers in Palaeontology* 23: 163–84.

_____. 1979b. Devonian miospores of North America. *Palynology* 3: 31–52.

_____. 1979c. Devonian spores from the Barrandian region of Czechoslovakia and their significance for interfacies correlation. In *Current Research, Part B., Geological Survey of Canada.* Paper 79-1B, pp. 189–97.

McGregor, D. C., and Camfield, M. 1976. Upper Silurian? to Middle Devonian spores of the Moose River Basin, Ontario. *Geol. Surv. Canada Bull.* 263: 1–47.

_____, and Narbonne, G. M. 1978. Upper Silurian trilete spores and other microfossils from the Read Bay Formation, Cornwallis Island, Canadian Arctic. *Can. J. Earth Sci.* 15: 1292–1303.

Merker, H. 1958. Zum fehlenden Gleide der Rhynienflora. *Bot. Not.* 111: 608–18.

_____. 1959. Analyse der Rhynien-Basis und Nachweis des Gametophyten. *Bot. Not.* 112: 441–52.

Metcalf, C. R. and Chalk, L. 1950. *Anatomy of the dicotyledons.* Oxford: Clarendon Press, 2 vols.

Murchison, R. I. 1854. *Siluria—The history of the oldest fossiliferous rocks and their foundations, with a brief sketch of the distribution of gold over the earth.*

London: John Murray. 523 p.

Müstafa, H. 1975. Beiträge zur Devonflora, I. *Argumenta Palaeobot.* 4: 101–33.

Nathorst, A. G. 1884. Redogörelse för den tillsammans med G. de Geer år geologiska expeditionen till Spetsbergen. *Kgl. Svensk. Vet. Acad. Handl. Bihang.* 9(2): 1–78. Stockholm.

———. 1894. Zur paläozoischen Flora der arktischen zone. *Kgl. Svenska vetenskapsakad. Handlingar* 26: 1–80.

———. 1902. Zur oberdevonischen Flora der Baren-insel. *Kgl. Svenska Vetenskapsakad. Handlingar* 36: 1–60.

———. 1904. Die Oberdevonische Flora des Ellesmere Landes. *Rept. Second Norwegian Arctic Expedition in the Fram* (1898–1902), No. 1, pp. 1–22.

———. 1915. Zur Devonflora des westlichen Nowegens. *Bergens Mus. Arb.*, No. 9, pp. 1–34.

Nikitin, P. A. 1934. Fossil plants of the Petino horizon of the Devonian of the Voronezh Region. I. *Kryshtofovichia africani* nov. gen. et. sp. *Izvestia Acad. Nauk USSR*, No. 7A, Pt. 2: p. 1079–92.

Niklas, K. J. 1976. Morphological and ontogenetic reconstruction of *Parka decipiens* Fleming and *Pachytheca* Hooker from the Lower Old Red Sandstone, Scotland. *Trans. Roy. Soc. Edinburgh* 69: 483–99.

———. 1981. Airflow patterns around some early seed plant ovules and cupules: implications concerning efficiency in wind pollination. *Amer. J. Bot.* 68(5): 635–50.

Obrhel, J. V. 1961. Die Flora der Srbsko-Schichten (Givet) des mittelbohmischen Devon. *Sbornik Ústredniho Ustavu Geol.* 26: 7–46.

———. 1962. Die Flora der Pridoli-Schichten (Budnany Stufe) des Mittelbohmischen Silurs. *Geologie* 11: 83–97.

———. 1966. *Protopteridium hostinense* Krejči und Bemurkungen zu den übrigen Arten der Gattung *Protopteridium*. *Casopis pro mineralogii e Geologii* 4: 441–43.

Oliver, F. W. and Salisbury, E. J. 1911. On the structure and affinities of the Palaeozoic seeds of the *Conostoma* group. *Ann. Bot.* 25: 1–50.

Pant, D. D. 1962. The gametophyte of the Psilophytales. In *Proceedings of the Summer School of Botany, Darjeeling, 1960*, Maheshwari, P., ed., pp. 276–301. New Delhi.

Penhallow, D. P. 1889. (Introductory notes by Sir J. W. Dawson). On *Nematophyton* and allied forms from the Devonian (Erian) of Gaspé and Bay des Chaleurs. *Trans. Roy. Soc. Canada*, Sect. IV, 6: 27–47.

———. 1892. Additional notes on Devonian plants from Scotland. *Canadian Record of Sci.*, Jan. 1892, 5: 1–13.

———. 1893. Notes on Erian (Devonian) plants from New York and Pennsylvania. II. Notes on *Nematophyton crassum*. *Proc. U.S. Nat. Museum* 16: 105–14.

Petrosian, N. M. 1968. Stratigraphic importance of the Devonian Flora of the USSR. In *Internat. Symposium on the Devonian system*, Oswald, D. H. ed., vol. 1: 579–86. Calgary: Albta. Soc. Petrol. Geol.

Pettitt, J. M. 1965. Two heterosporous plants from the Upper Devonian of North America. *Bull. Brit. Mus. (Nat. Hist.) Geology* 10: 83–92.

———. 1966. Exine structure in some fossil and recent spores and pollen as revealed by light and electron microscopy. *Brit. Mus. Bull. Geol.* 13: 223–57.

_____. 1970. Heterospory and the origin of the seed habit. *Biol. Reviews* 45: 401–15.

Pettitt, J. M., and Beck, C. B. 1968. *Archaeosperma arnoldii*—a cupulate seed from the Upper Devonian of North America. *Contrib. Mus. Paleont. Univ. Mich.* 22: 139–54.

_____ and Lacey, W. S. 1972. A Lower Carboniferous seed compression from North Wales. *Rev. Palaeobot. Palynol.* 14: 159–69.

Phillips, T. L. 1979. Reproduction of heterosporous arborescent lycopods in the Mississippian-Pennsylvanian of Euramerica. *Rev. Palaeobot. Palynol.* 27: 239–89.

Phillips, T. L., Andrews, H. N., and Gensel, P. G. 1972. Two heterosporous species of *Archaeopteris* from the Upper Devonian of West Virginia. *Palaeontographica* 139B: 47–71.

Pichi-Sermolli, R. 1959. Pteridophyta. In: *Vistas in Botany*, 421.

Potonié, H. and Bernard, C. 1904. Flore dévonienne de l'étage H de Barrande. Supplement of *Système Silurien du centre de la Bohème*. Barrande, J. Prague. 68 p.

Pratt, L. M., Phillips, T. L., and Dennison, J. M. 1978. Evidence of non-vascular land plants from the Early Silurian (Llandoverian) of Virginia, U.S.A. *Rev. Palaeobot. Palynol.* 25: 121–49.

Ramanujam, C. G. K. and Stewart, W. N. 1969. A *Lepidocarpon* cone tip from the Pennsylvanian of Illinois. *Palaeontographica* 127B: 159–67.

Read, C. B. 1935. An occurrence of the genus *Cladoxylon* Unger, in North America. *Journ. Washington Acad. Sci.* 25: 493–97.

_____. 1936. A Devonian flora from Kentucky. *Journ. Paleont.* 10: 215–27.

_____. 1937. The flora of the New Albany Shale. Part 2. The Calamopityeae and their relationships. *U.S. Geol. Surv. Prof. Paper* 186E: 81–104.

_____. 1938. Some Psilophytales from the Hamilton group in western New York. *Torrey Bot. Club Bull.* 65: 599–606.

_____, and Campbell, G. 1939. Preliminary account of the New Albany shale flora. *American Mid. Naturalist* 21: 435–53.

Remy, W. 1982. Lower Devonian Gametophytes: Relation to the phylogeny of land plants. *Science* 215: 1625–27.

Remy, W. and Remy, R. 1980. Devonian gametophytes with anatomically preserved gametangia. *Science* 208: 295–96.

Remy, W., Remy, R., Hass, H., Schultka, St., and Franzmeyer, F. 1980. *Sciadophyton* Steinmann—a Gametophyt aus dem Siegen. *Argumenta Palaeobotanica* 6: 73–94.

Richardson, J. B. 1969. Devonian spores. In *Aspects of Palynology.* Tschudy, R. H., and Scott, R. A., eds. New York: Wiley-Interscience. pp. 194–222.

_____. 1974. The stratigraphic utilization of some Silurian and Devonian miospore species in the northern hemisphere: an attempt at a synthesis. In *Int. Sympos. Belgian Micropaleont. Limits, Publ. No. 9* (Namur 1974), pp. 1–13.

Richardson, J. B. and Lister, T. R. 1969. Upper Silurian and Lower Devonian spore assemblages from the Welsh Borderland and South Wales. *Palaeontology* 12: 201–52.

————, and Ioannides, N. 1973. Silurian palynomorphs from the Tannezuft and Acacus Formations, Tripolitania, North Africa. *Micropaleontology* 19: 257–307.

Rothwell, G. W. 1971. Additional observations on *Conostoma anglo-germanicum* and *C. oblongum* from the Lower Pennsylvanian of North America. *Palaeontographica* 131B: 167–78.

————. 1977. Evidence for a pollination-drop mechanism in Palaeozoic pteridosperms. *Science* 198: 1251–52.

Rudwick, M. J. S. 1979. The Devonian: a system born from conflict. *Special Papers in Palaeontology*, No. 23, pp. 9–21.

Sahni, B. 1941. Indian silicified plants. I. *Azolla intertrappea* Sahni and H. S. Rao *Proceed. Indian Acad. Sci.* 14: 489–501.

Scheckler, S. E. 1974. Systematic characters of Devonian ferns. *Ann. Missouri Bot. Gard.* 61: 462–73.

————. 1975a. *Rhymokalon*, a new plant with cladoxylalean anatomy from the Upper Devonian of New York State. *Canadian Journ. Bot.* 53: 25–38.

————. 1975b. A fertile axis of *Triloboxylon ashlandicum*, a progymnosperm from the Upper Devonian of New York. *Amer. Journ. Bot.* 62: 923–34.

Scheckler, S. E., and Banks, H. P. 1971a. Anatomy and relationships of some Devonian progymnosperms from New York. *Amer. Journ. Bot.* 58: 737–51.

————. 1971b. *Proteokalon*, a new genus of progymnosperms from the Devonian of New York State and its bearing on phylogenetic trends in the group. *Amer. Journ. Bot.* 58: 874–84.

————. 1972. Periderm in some Devonian plants. In *Advances in plant morphology.* pp. 58–64. Prof. V. Puri Commemoration Vol. Murty, Y. S., Johri, B. M., Mohan Ram, H. Y., and Varghese, T. M., eds. Meerut City, U. P., India.

Schmalhausen, J. 1894. Uber devonische Pflanzen aus dem Donetz-Becken. *Comité geol. Russie Mem.* 8: 1–36.

Schmid, R. 1976. Septal pores in *Prototaxites*, an enigmatic Devonian plant. *Science* 191: 287–88.

Schmidt, W. 1954. Pflanzen-reste aus der Tonschiefer-Gruppe (Unteres Siegen) des Siegerlandes. I. *Sugambrophyton pilgeri* n. gen. n. sp., eine Protolepidodendraceae aus den Hamburg-schichten. *Palaeontographica* 97B: 1–22.

Schultka, S. 1978. Beiträge zur anatomie von *Rhacophyton condrusorum* Crépin. *Argumenta Palaeobot.* 5: 11–22.

Schuster, R. 1966. *The Hepaticae and Anthocerotae of North America East of the Hundredth Meridian.* New York: Columbia Univ. Press.

Schweitzer, H.-J. 1965. Über *Bergeria mimerensis* und *Protolepidodendropsis pulchra* aus dem Devon Westspitzbergens. *Palaeontographica* 115B: 117–38.

————. 1966. Die Mitteldevon-Flora von Lindlar (Rheinland). 1. Lycopodiinae. *Palaeontographica* 118B: 93–112.

————. 1967. Die Oberdevon-Flora der Bäreninsel. I. *Pseudobornia ursina* Nathorst. *Palaeontographica* 120B: 116–37.

_____. 1968. Pflanzenreste aus dem Devon Nord-Westspitzbergens. *Palaeontographica* 123B: 43–75.

_____. 1972. Die Mitteldevon-Flora von Lindlar (Rheinland). 3. Filicinae— *Hyenia elegans* Kräusel & Weyland. *Palaeontographica* 137B: 154–75.

_____. 1979. Die Zosterophyllaceae des rheinischen Unterdevons. *Bonner Paläobot. Mitteil.*, 32 p.

Schweitzer, H. J. 1980a. Die Gattungen *Taeniocrada* White und *Sciodophyton* Steinmann im Unterdevon des Rheinlandes. *Bonner Palaeobot. Mitteil*, No. 5. 38 pp.

_____. 1980b. Über *Drepanophycus spinaeformis* Goeppert. *Bonner Paläeobot. Mitteil.* no. 7, 29 p.

_____. 1981. Der generationwechsel rheinischer Psilophyten. *Bonner Palaebot. Mitteil.* 8: 19pg.

Schweitzer, H. J., and Geisen, P. 1980. Über *Taeniophyton inopinatum, Protolycopodites devonicus* und *Cladoxylon scoparium* aus dem Mitteldevon von Wuppertal. *Palaeontographica* 173B: 1–25.

_____, and Matten, L. C. 1982. *Aneurophyton germanicum* and *Protopteridium thomsonii* from the Middle Devonian of Germany. *Palaeontographica* 184B: 65–106.

Scotese, C. R., Bambach, R. K., Barton, C., Vander Voo, R., and Ziegler, A. M. 1979. Paleozoic Base Maps. *Journ Geol.* 87(3): 217–77.

Scott, A. C. 1980. The ecology of some Upper Paleozoic floras. In *The terrestrial environment and the origin of land vertebrates.* Panchen, A. L. (Ed.), pp. 87–115. Syst. Assoc. Spec. Pap. No. 15. London and New York: Academic Press.

Scott, D. H. 1897. On *Cheirostrobus*, a new type of fossil cone from the Lower Carboniferous strata (Calciferous Sandstone Series). *Phil. Trans. Roy. Soc. Lond.* 189: 1–34.

_____. 1923. *Studies in Fossil Botany.* 2 vols., 3rd ed. London: A. and C. Black. 434 p., 446 p.

Serlin, B. S. and Banks, H. P. 1978. Morphology and anatomy of *Aneurophyton*, a progymnosperm from the late Devonian of New York. *Palaeontographica Americana* 8: 343–59.

Seward, A. C. 1910. *Fossil Plants,* II. Cambridge Univ. Press, 624 p.

_____. 1917. *Fossil Plants,* III. Cambridge Univ. Press, 656 p.

Skog, J. E. and Banks, H. P. 1973. *Ibyka amphikoma*, gen. et sp. n., a new protoarticulate precursor from the late Middle Devonian of New York State. *Amer. Journ. Bot.* 60: 366–80.

Smith, A. G., Hurley, A. M., and Briden, J. C. 1981. Phanerozoic paleocontinental world maps. *Cambridge Earth Science Series.* Cambridge Univ. Press.

Smith, D. L. 1962. Three fructifications from the Scottish Lower Carboniferous. *Palaeontology* 5: 225–37.

_____. 1964. The evolution of the ovule. *Biol. Reviews* 39: 137–59.

Smoot, E. L., Taylor, T. N., and Serlin, B. S. 1982. Archaeocalamites from the Upper Mississippian of Arkansas. *Rev. Palaeobot. Palynol.* 36: 325–34.

Solms-Laubach, H. 1895. Ueber devonische Pflanzenreste aus der Lenneschiefern der Gegend von Grafrath im Niederhein. *Kgl. Preuss. Geol. Landesanst.*, Jarhb. fur 1894, 15: 67–99.

Stein, W. E. Jr. 1981. Reinvestigation of *Arachnoxylon kopfi* from the Middle Devonian of New York State, U.S.A. *Palaeontographica* 177B: 90–117.

_____. 1982a. *Iridopteris eriensis* from the Middle Devonian of North America, with systematics of apparently related taxa. *Bot. Gaz.* 143: 401–16.

_____. 1982b. The Devonian plant *Reimannia*, with a discussion of the Class Progymnospermopsida. *Palaeontology* 25: 605–22.

Stein, W. E., Wight, D. C., and Beck, C. B. 1983. *Arachnoxylon* from the Middle Devonian of southwestern Virginia. *Can. Journ. Bot.* 61(4): 1283–99.

Steinmann, G. 1929. Die organischen Einschlusse. In: Neue bemerkenswerte Funde im ältesten Unterdevon des Wahnbachtales bei Siegburg, by G. Steinmann and W. Elberskirch. *Sitzber. Niederrhein. Geol. Ver.*, 1927–28C: 1–74.

Stockmans, F. 1940. Végétaux Éodevoniens de la Belgique. *Mém. Mus. Roy. Hist. Nat. de Belg.* 93: 1–90.

_____. 1946.Tour d'horizon paleobotanique en Belgique. *Soc. naturalistes Belges Bull.* Nos. 7, 8, pp. 1–6.

_____. 1948. *Végétaux Dévonien Superieur de la Belgique*. Mém. Mus. Roy, Hist. Nat. No. 110, 85 p.

_____. 1968. *Végétaux Mésodévoniens récoltés aux confins du Massif du Brabant (Belgique)*. Mém. Institute Roy. Sci. Nat. de Belg. Mem. No. 159, 49 p.

Streel, M. 1964. Une association de spores du Givétien inférieur de la Vesdre, a Goé (Belgique). *Ann. Soc. Géol. Belg.* 87: 1–30.

_____. 1967. Associations de spores du Dévonien inférieur belge et leur signification stratigraphique. *Ann. Soc. Geol. Belg.* 90: 11–54.

_____. 1972. Dispersed spores associated with *Leclercqia complexa* Banks, Bonamo and Grierson from the late Middle Devonian of eastern New York State (U.S.A.), *Rev. Palaeobot. Palynol.* 14: 205–15.

Strother, P. K. and Traverse, A. 1979. Plant microfossils from Llandoverian and Wenlockian rocks of Pennsylvania. *Palynology* 3: 1–21.

Stubblefield, S. and Banks, H. P. 1978. The cuticle of *Drepanophycus spinaeformis*, a long-ranging Devonian lycopod from New York and eastern Canada. *Amer. Journ. Bot.* 65: 110–18.

Stur, D. R. J. 1875–77. Beiträge zur Kenntnis der Flora der Vorwelt-Die Culm Flora. Teil I. Die Culm-flora des mahrisch-schlesischen Dachschiefers. *Abh. K. K. Geol. Reichsanst.* Wien 8, 1–106.

_____. 1882. Die Silur-Flora der Etage H-h in Böhmen. *Kais. Akad. Wiss. Wien Sitsungsber.* 84: 1–62.

Surange, K. R. 1952. The morphology of *Stauropteris burntislandica* P. Bertrand and its megasporangium *Bensonites fusiformis* R. Scott. *Phil. Trans. Roy. Soc. London* 237B: 73–91.

Sussex, I. M. 1966. The origins and development of heterospory in vascular plants. In *Trends in Plant Morphogenesis*. Cutter, E. C. ed. New York: John Wiley and Sons, pp. 140–52.

Tanner, W. R. 1982. A new species of *Gosslingia* (Zosterophyllophytina) from the Lower Devonian Beartooth Butte Formation of northern Wyoming. In *Third North American Paleontological Convention, Proceed.* 2: 541–46.

Taylor, T. N. 1981. *Paleobotany, an introduction to fossil plant biology*. New York: McGraw-Hill Book Co. 589 p.

Taylor, T. N., and Brauer, D. F. 1983. Untrastructural studies of *in situ* Devonian spores. *Barinophyton citrulliforme*. Amer. Journ. Bot. 70: 106–12.

Taylor, T. N., Maihle, N. J., and Hills, L. V. 1980. Morphological and ultrastructural features of *Nikitinsporites canadensis* Chaloner, a Devonian megaspore from the Frasnian of Canada. *Rev. Palaeobot. Palynol.* 30: 89–99.

Thomas, D. E. 1935. A new species of *Calamopitys* from the American Devonian. *Bot. Gaz.* 97: 334–45.

Traité de Paléobotanique: see Boureau, É.

Tryon, A. F. 1964. *Platyzoma*—a Queensland fern with incipient heterospory. *Amer. Journ. Bot.* 51: 939–42.

———. 1967. *Platyzoma*: a new look at an old link in ferns. *Science* 156: 1109–10.

Tryon, R. M., and Tryon, A. F. 1982. *Ferns and allied plants*. New York: Springer-Verlag. 857 pp.

Unger, F. 1856. Beiträg zur Palaeontologie des Thuringer Waldes—Teil II, Schiefer und Sandstein Flora. *Kgl. Akad. Wiss. Denkschr.* 11: 139–86.

Vigran, J. O. 1964. Spores from Devonian deposits, Mimerdalen, Spitsbergen. *Norsk Polarinstitut Skr.* Nr. 132, 32 pp.

Wagner, W. H., Beitel, J. M., and Wagner, F. S. 1982. Complex venation patterns in the leaves of *Selaginella*: megaphyll-like leaves in lycophytes. *Science* 218: 793–94.

Walton, J. 1925. Carboniferous Bryophyta. I. Hepaticae. *Ann. Bot.* 39: 563–72.

———. 1928. Carboniferous Bryophyta. II. Hepaticae and Musci. *Ann. Bot.* 42: 707–16.

Walton, J. 1957. On *Protopitys* (Göppert): with a description of a fertile specimen "*Protopitys scotica*" sp. nov. from the Calciferous Sandstone Series of Dunbartonshire. *Trans. Roy. Soc. Edinburgh* 63: 333–40.

———. 1964. On the morphology of *Zosterophyllum* and other early Devonian plants. *Phytomorphology* 15: 155–60.

———. 1969. On the structure of a silicified stem of *Protopitys* and roots associated with it from the Carboniferous Limestone, Lower Carboniferous (Mississippian) of Yorkshire, England. *Amer. Journ. Bot.* 56: 808–13.

Weiss, C. 1889. Über *Drepanophycus spinaeformis* Goeppert. *Deutsch. Geol. Gesell. Zeitschr.* 41: 167–554.

White, D. 1903. Description of fossil algae from the Chemung of New York with remarks on the genus *Haliserites* Sternberg. *New York State Mus. Ann. Rep. 1901*, 55: 593–605.

White, D. in Smith, G. O. and White, D. 1905. The Geology of the Perry Basin in southeastern Maine. USGS Prof. Paper 35, 107 pp.

Witham, H. T. M. 1833. *The internal structure of fossil vegetables found in the Carboniferous and oolite deposits of Great Britain*. Edinburgh: Adam and Charles Black. 84 p.

Worsdell, W. C. 1904. The structure and morphology of the ovule: an historical sketch. *Ann. Bot.* 18: 57–86.

Zalessky, M. D. 1911. Étude sur l'anatomie du *Dadoxylon tchihatcheffi* Goeppert sp.: *Comité géol. Russie Mem. n. ser.* 68: 1–29.

Zdebska, D. 1972. *Sawdonia ornata* (= *Psilophyton princeps* var. *ornatum*) from Poland. *Acta Palaeobotanica* 13: 77–97.

_____. 1982. A new zosterophyll from the Lower Devonian of Poland. *Palaeontology* 25: 247–63.

Ziegler, W. 1979. *Historical subdivisions of the Devonian.* In: The Devonian System, House, M. R., Scrutton, C. T. and Bassett, M. G. Eds., *Spec. Papers in Palaeontology* 23: 23–47.

Zimmerman, W. 1930. *Die Phylogenie der Pflanzen.* Jena.

Index

Index*

Aberlemno, Scotland, quarry, 22, 79
Acanthotriletes famennensis, 166
Actinopodium nathorstii, 235
Actinoxylon, 231, 232, 235
   Actinoxylon banksii, 235–236, **236, 237**
Alaska, 169
Ambitisporites, possible spore of
   Cooksonia, 324
Amphidoxodendron, 158
Anatomical features, summary chart
   showing early appearance of,
   341
Ancyrospora, 317
Aneurophytales, 228, 237–254
Aneurophyton, 237–243
   Aneurophyton germanicum, 240–
      241, **240, 243**
   relationship with
      Eospermatopteris, 241
   Aneurophyton olnense, 245
Apiculiretusispora
   in *Psilophyton, Krithodeophyton,*
      Cooksonia, Zosterophyllum, 326
   in *Horneophyton*, 68
Arachnoxylon kopfii, 205–207, **206**
Arborescent habit
   in lycopods, 144
   in progymnosperms, 215, 223–224,
      257
   in seed plants, 285–288, 294
Archaeocalamites, 169–170, **170, 171**
Archaeoperisaccus, 311, 317
Archaeopteridales, 228–236
Archaeopteris, 214–221, **214, 221,**
   226–227
   Archaeopteris fimbriata, **219**

Archaeopteris fissilis, 226, 231
Archaeopteris halliana, 219–221,
   220
Archaeopteris hibernica, 215
Archaeopteris jacksoni, 225
Archaeopteris latifolia, 225
Archaeopteris macilenta, 215–216,
   216, 218; nature of the
   "frond" or modified branch
   system, 215–217; **217**
Archaeopteris obtusa, 226
   flora, 2, 11
Archaeosigillaria vanuxemii, 143–144,
   144, 145
Archaeosperma arnoldii, 264–267,
   265, 266, 267, 293
Archegonia (presumed), 106–110
Arctic studies, 5, 104, 226, 229; *see
   also* Bear Island, Spitsbergen
Arctophyton gracile, 192, 300
Arthrostigma gracile, 123
Asteropteris noveboracensis, 207–208,
   209
Asteroxylon mackiei, 68–69, 117–120,
   118, 119, 120
Assemblage zones, plant, 8–9, 11
Australia, see Victoria, Australia

Baragwanathia, 32, 129
   Baragwanathia abitibiensis, 130–31
   Baragwanathia longifolia, 117, 129–
      30, **130**, 156
Barinophyton, 300
   Barinophyton citrulliforme, 301–307,
      302, 303, 304, 305
   Barinophyton cf. *obscurum*, 307
   Barinophyton richardsoni, 304, 307

---

*Bold numbers indicate pages on which figures appear.

Bear Island, 5, 148–149, 166, 219, 339
Belgium, 5, 110, 134, 137, 140, 164, 166, 176, 189, 194, 196, 245, 335, 338
Bergeria mimerensis, 151
Beroun, Czechoslovakia, 20, 60
biostratigraphy
    megafossil, 9–11
    in New Brunswick and Gaspé, 336–337
    use of spores in, 318
Bothrodendron, 145, 149
    Bothrodendron kiltorkense, 146, 148
Brecon Beacons, Wales, quarry, 80, **81**
Bryophytes, Devonian occurrences, 107–112
Bucheria ovata, 103

Calamophyton, 162–163, 187, 189–191
    Calamophyton bicephalum, 189–192, **190, 191**
Callixylon, 214, 222–224, **223, 224**
    Callixylon arnoldii, 224, 226
    Callixylon newberryi, 223, **224**
    Callixylon whiteanum, 224
Cambial activity
    origins and evolution, 181–183
    unifacial, in: *Lepidodendron schizostelicum* and *Arthropitys communis*, 182
    in: *Aneurophyton*, 239, 243
    Callixylon, 183, 223
    Cladoxylon, 186
    Eddya, 232
    Phytokneme, 155
    Protopitys, 258
    Tetraxylopteris, 252
    Triloboxylon, 253
    Triradioxylon, 256
Chaleuria cirrosa, 298–300, **299, 300, 301**
China, 84, 74, 123, 133
Cladoxylales, 184–203
Cladoxylon, 185

Cladoxylon dawsonii, 187
Cladoxylon hueberi, 188, **188**
Cladoxylon scoparium, 185–187, **186**
Cladoxylopsida, 14, 184–203
Classification
    of Devonian plants, 11–14
    of early land vascular plants, 35–36
Coenopterids (Coenopteridopsida), 14, 175–181
Colpodexylon deatsii, 140–142, **142**
Colpodexylon trifurcatum, 142
Conostoma oblongum, 282
Continental positions in the Devonian, 329–330, **331**
Cooksonia, 19–25, **21**, 67
    Cooksonia caledonica, 22, **24**
    Cooksonia crassiparietalis, 23
    Cooksonia hemispherica, 20, **21, 25**
    Cooksonia pertoni, 20, **21**
cortical features
    in *Asteroxylon*, 118
    in *Gosslingia*, **87**
    in *Phytokneme*, 156
    in *Psilophyton*, 37
    in *Triloboxylon*, 253
Crenaticaulis verruculosus, 101–102, **102, 103**
Cyclostigma kiltorkense, 145–150, **147, 148, 149, 150**
Cystosporites devonicus, possible megaspore of *Archaeosperma*, 267, **268**, 317

Dawsonites, 60–61
    Dawsonites roskiliensis, 61
Deltasperma fouldenense, 281–282
Devonian landscapes
    Early, 332
    Middle, 335
    Upper, 335
Dibolisporites, in *Calamophyton*, 192
Dichotomous, definition, 34
Dispersed spores
    as indicators of environment, 317–318

as related to the megafossil record,
   316–317
compared to spores from
   megafossils, 320–323
significance in upper Silurian strata,
   316–317
significance of *in situ* spores, 319
variations in longevity, 318
Downtonian, 20
Drepanophycus, 121–126
   Drepanophycus gaspianus, 126, **127**
   Drepanophycus spinaeformis,
      124–126, **122, 124, 125**

Eddya sullivanensis, 232–234, **233,
   234**
Emergences, and evolution of, 35, 46,
   53, 94, 99, 101
Emphanisporites, 317
Enigmophyton superbum, 308–311,
   309, 310
Enigmophytospora, spores of
   Enigmophyton, **309**, 311
Eogaspesiea gracilis, 75–76, **75, 76**
Eospermatopteris erianus, 237–238,
   238, 242
   history of discovery, 242–244
Eorhynia, 25
Equisetum, 161–162
Estinnophyton, 134–138, 158
   Estinnophyton gracile, 134–136,
      135, 136
   Estinnophyton wahnbachense, 137,
      138
Eurystoma angulare, 277–279, **278**
Eurystoma trigona, 279, **279**
Eviostachya hoegii, 164–165, **164,
   165**

Fern
   classification of living ferns, 174
   use of term, 14
      as applied to Devonian fossils,
         173–175
      origins, 175, 212
Filicopsida, 14

Flabelliform leaves in Devonian
   plants, 235, 311
Floras, 332–340
Future studies in Devonian
   paleobotany, 340

Gametophytes of Devonian plants,
   106–110, 113
Gaspé, Canada, 26, 37, 39, 44, 46,
   57, **59**, 71, 75, 93, 95, 101,
   123, 126, 333, 334, 337
Genomosperma kidstoni, **274**,
   274–275
Genomosperma latens, 275, **274, 275**
Germany, 4, 81, 109, 121, 123, 132,
   135, 140, 154, 185, 239, 241,
   248, 335
Goé, Belgium, 5, 189, 193, 196, 245
Gosslingia breconensis, 85–89, **87, 88**

Hedeia corymbosa, 131
Hepaticites, 112
Heterospory, 295–314
   evidence from dispersed spores,
      312
   in Carboniferous plants
      Calamostachys americana, 314
      Calamostachys casheana, 313
      Selaginellites, 296
      Stauropteris burntislandica, 313
   in Devonian plants
      Archaeopteris halliana, 221
      Archaeopteris latifolia, 225
      Archaeopteris macilenta, 221
      Barinophyton, 300
      Chaleuria, 298
      Cyclostigma kiltorkense, 147, 297,
         313
      Enigmophyton, 310
      Kryshtofovichia, 311
      Lepidostrobus gallowayi, 153
      Protopitys, 259
   in living plants
      Isoetes, 296
      Platyzoma microphyllum, 296
      Selaginella, 295–6

Hicklingia edwardii, 89–93, **92, 93, 94**
Historical review, 3–5
  of literature, 2–3
Hornea lignieri, 62
Horneophyton, 68
  Horneophyton lignieri, **67**, 68, **69**
Hostinella, 59–60
Hsüa robusta, 74
Hydrasperma tenuis, 270–273, **272, 273**
Hyenia, 162–163, 192–196
  flora, 11
  Hyenia banksii, 196
  Hyenia elegans, 193–195, **193, 194**
  Hyenia sp., **195**
  Hyenia sphenophylloides, 192
  Hyenia vogtii, 195–196
Hyeniales, 192

Ibyka amphikoma, 163, 207, **208, 209**
In situ spores, 32, 319
Iridopteridales, 203–209, 211
Iridopteris eriensis, 204–205, **205**
Isoetes, 296
Isospore, definition, 316

Kaulangiophyton akantha, 126–128, **128**
Kiltorcan, Ireland, 146, 269
Koniora andrychoviensis, 105–107, **106, 107**
Kryshtofovichia africani, 311, 317

Leaf evolution, 34–35, 212, 217–218, 261–262
Leclercqia complexa, 137–140, **139, 141**, 158
Leiotriletes minutissimus, 95
Leiotriletes pullatus, 95
Lepidodendropsis, 153, 158
Lepidostrobus gallowayi, 153
Libya, 32, 37
Ligule, in *Leclercqia*, 138
Liverworts, in the Devonian, 112

Localities
  comments on those in Maine and southeast Canada, 57–58
  in other areas, 58–59, 332–335
Lycopodites oosensis, 153–154, **155**
Lycopodium, 115, 138, 158–159
Lycopodolica, 25
Lycopods
  classification, 117
  distinctive features, 115–116
  variation in leaf morphology and venation, 115, 138, 146, 152, 158
Lycospora svalbardiae, spores of *Svalbardia*, 230, **230**
Lyginorachis papilio, 288–291, **291**
Lyonophyton rhyniensis, 108–109, **109**
Lyrasperma scotica, 279–280, **280, 281**

Maine, USA, 46, 48, 52, 57, 126, 140, 300
Megafossils, earliest evidence of, 15–16
Megaphyllous leaf, origin of, 56, 212
  (*see also* leaf evolution)
Microfossils, as early evidence of land plants, 16–18, **18, 19**
Microphyllophyta, 115
Microphyllous leaf, origin of, 35, 56
  (*see also* leaf evolution)
Milleria, see discussion under *Rellimia*, 249
Miospore, definition, 316
Monopodial, definition, 34
Moresnetia zalesskyi, 269, **271**
Morphological features
  and development (vegetative), 342
  summary chart showing early appearance and development (reproductive), 343
  summary chart showing early appearance
Myreton Quarry, Scotland, 78

Nematophycus, 26

Nematophyton, 26
Nematothallus, 16, **17**
New Brunswick, Canada, 30, 38, 47,
    57, 81, 89, 98, 140, 298, 332,
    334, 335
New York State, USA
    Middle Devonian: 133, 142, 158,
        191, 196, 200, 204, 205, 207,
        209, 235, 335
    Silurian, 21
    Upper Devonian: 101, 140, 143,
        187, 188, 201, 207, 215, 225,
        232, 239, 249, 253, 301, 335,
        338
Nikitinsporites, possible spore of
    Kryshtofovichia, 311, 317, 326
Nomenclature, special problems with
    Devonian plants, 56
Norway, 60, 110, 123, 192
Nothia aphylla, 68–69, **70, 71**, 121

Ontario, Canada, 28, 37, 130
Oocampsa, 325
Oricilla bilinearis, 89, **90, 91**
Ovule, 263 (*see also* seed)

Pachytheca, 28–30, **29**
Paleobiogeography, 329–332
Pallavicinites, 112
Palynology (*see also* spores, *in situ*
        spores)
    contributions to Devonian
        megafossil studies, 315–327
    definition, 315
Parka, 30–31, **31**
Pectinophyton, 307
Periderm in Devonian plants, 256,
    257
Peripheral loop
    general discussion, 184–185, 212
    in
        Calamophyton bicephalum, 191
        Cladoxylon dawsonii, 187
        Iridopteris eriensis, 204
        Reimannia aldenense, 210
        Asteropteris, 208

Pertica quadrifaria, 52–53, **54**
Pertica varia, 53, **55**
Phloem, production of secondary, 183
    (*see also* cambial activity)
Phylogeny of late Silurian and
    Devonian vascular plants,
    chart, 344
Physostoma stellatum, 282
Phytokneme rhodona, 155–157, **157**
Pietzschia polyupsilon, 201
Pietzschia schulleri, 201
Pitus, 285–292, **292**; species
    distinction in, 287
    Pitus antiqua
        history and presumably related
            plant parts, 285
        great size attained in Lower
            Carboniferous, 286–288
    Pitus dayi, 287, 290, **289, 290**
    Pitus primaeva, **288**, 286
Platyspermic seeds
    Lower Carboniferous, 279
    Devonian, 269–270
Platyzoma microphyllum, 296
Podolia, USSR, 22, 24–25
Poland, 96, 105
Pollen-drop mechanism, 283–284
Prehepaticites, 25
Pre-lycopods, 117
Preservation of Devonian plants, 40
Primochara, 25
Progymnospermopsida, 14, 213–262
    classification and concept of the
        group, 213–215, 227–228,
        260–262
Prolepidodendron breviinternodum,
    116, 152–153, **154**
Protoarticulates, 162
Protobarinophyton obrutschevii, 308
Protobarinophyton pennsylvanicum,
    308, **308**
Protohyenia, 163
Protolepidodendrales, 131
Protolepidodendron
    leaf morphology, 158
    Protolepidodendron gilboense,
        133–134, **134**

Protolepidodendron scharianum,
131–133, **132**, 140
Protolepidodendron wahnbachense,
134–135, 137
review of described species, 131–
133
Protolepidodendropsis pulchra,
150–152, **151, 152**
Protopitys buchiana, 256–260, **258**
Protopteridium, see discussion under
Rellimia, 249
Prototaxites, 26–28, **27**
Prototaxites logani, 26
Prototaxites southworthii, 28
Pseudobornia ursina, 166–169, **167,
168**
Pseudomonopodial, definition, 34
Pseudosporochnus nodosus, 196–199,
197, 198, 199
Psilophytales, 12
Psilophyte, use of term, 3, 11–12
Psilophyton
"flora", 3, 11
history of generic concept, 43–45
Psilophyton charientos, 47, **49, 50**
Psilophyton crenulatum, 38–39, **41,
42**
Psilophyton dapsile, 47, **48**
Psilophyton dawsonii, 37–38, **38, 39**
Psilophyton forbesii, 48, **51**
Psilophyton princeps, 43–46, **44, 46,
47**
Psilophyton princeps var. *ornatum,*
44–45, 93
Psilophyton robustius, 43
Psilotum, 11, 110
Pteridosperm, 14, 264
Pyritic petrifactions, 88–89, 141

Rebuchia capitanea, 104
Rebuchia ovata, 103–104, **104**
Reimannia aldenense, 209–211, **210**
Rellimia thomsonii, 245–249, **246,
247, 248**
Renalia hueberi, 71–74, **72, 73**
Retispora, 318

Retusotriletes, possible spore of
Barinophyton, 303
Rhabdosporites
in *Rellimia thomsonii,* 247
in *Tetraxylopteris* and *Rellimia,* 318
Rhacophyton ceratangium, 176–180,
178, 179, 180, 181
Rhacophyton condrusorum, 180
Rhymokalon trichium, 201–203, **202,
203**
Rhynia, 61
Rhynia gwynne-vaughanii, 64–66,
64, 65
Rhynia major, 65, **66**
Rhynie, Scotland
flora, 4, 11
nature and discovery of the fossil
plant bed, 61–63
Rhyniophytina, 36, 61–76
Roots, presence in Early Devonian
plants, 35, 113

Sawdonia, 93
Sawdonia acanthotheca, 98–99, **98,
99**
Sawdonia ornata, 45, 93–97, **95,
96, 97**
Sciadophyton, problematical nature of,
109–110, **110**
Sclereids, in *Pseudosporochnus,* 199
Scotland, 4, 26, 30, 60, 61, 77, 90,
110, 123, 249, 255, 259, 270,
286
Secondary xylem (*see* cambial activity)
Selaginella, 295–296
Seed(s)
definition, 263
early lines of evolution in seed
plants, 264
evolutionary trends in, 282–285,
294
morphology, terms used, 271–273
oldest records, 264–269
significance of symmetry,
platyspermic or radiospermic,
283

significance of nucellar (salpinx) modifications, 283
theories of seed evolution, 282
Serrulacaulis furcatus, 99–101, **100**
Silurian non-vascular plants, 26
Spermolithus devonicus, 269–270, **270**
Sphenopsida, 161–172, 175
Sphenophyllum subtenerrimum, 166
Spitsbergen, 5, 30, 150, 151, 166, 192, 195, 229, 235
Spores
  as used in biostratigraphic correlations, 318
  characteristics of in major plant groups, 324–325
  from pre-Devonian strata, 17, 317
  in situ Devonian records, 319, 320–323
  large numbers per sporangium, 40, 48, 53, 65, 70–71, 72, 81, 89, 99, 258, 299, 303, 311
  ultrastructure studies, 326
Sporogonites exuberans, 111–13, **111**
Stamnostoma huttonense, 276, **276, 277**, 293
Steganotheca, 24, 74
  Steganotheca striata, 74–75
Stomata, 37, 65, **79**, 87, 96, 118, **119, 122**, 126, 341
  oldest record, 80
Stratigraphy, literature, 7, 10
Svalbardia
  Svalbardia avelinesiana, 232
  Svalbardia banksii, 231
  Svalbardia osmanica, 232
  Svalbardia polymorpha, 229–230, **230**
  Svalbardia scotica, 231
  validity of the genus, 229–232
Synangia, 68

Taeniocrada, 307, 340
  Taeniocrada dubia, 45
Terminology (morphological), 34
Tetraxylopteris schmidtii, 249–253, **250, 251, 252**
  relationship to *Rellimia*, 253

Time divisions of the Devonian: Lower, Middle, Upper vs. Early, Middle, Late, 10
Timetable, 7–11
Tracheidal anatomy, details of
  in *Callixylon*, 224
  in *Leclercqia*, 140
  in *Psilophyton*, 37, **39**
Trichotomous, definition, 34
Triloboxylon ashlandicum, 253–254, **254**
Trimerophytina, 35–61
  evolution in, 56
Trimerophyton robustius, 49–52, **52**
Triradioxylon primaevum, 254–256, **255**
Trochophyllum breviinternodum, 152
Tortilicaulis, 23, 21
Tubercles, 101

USSR, 22, 23, 89, 94, 222, 311, 335

Vallatisporites, 318
Vascular plants
  earliest evidence of, 16–17
  literature on origins of, 15–16
Victoria, Australia, 32, 129, 131

Wales, 19, 20, 23, 24, 26, 60, 74, 80, 81, 85, 124
West Virginia, USA, 176, 177, 215, 219, 268, 338
Wyoming, USA, 103

Xenocladia medullosina, 200, **200**

Yarravia oblonga, 131

Zostera, 78
Zosterophyllophytes, 76–106
Zosterophyllophytina, 36
Zosterophyllum, 77
  Zosterophyllum divaricatum, 81–84, **83, 84, 85**
  Zosterophyllum fertile, 80

Zosterophyllophytina *(continued)*
    Zosterophyllum llanoveranum, 80,
        82

Zosterophyllum myretonianum,
    77–79, **77, 78, 79**
Zosterophyllum, from China, 84, 86

About the Authors

Patricia G. Gensel received a BA degree in biology from Hope College, Holland, Michigan and then worked as a research assistant in palynology in London for a year. She then returned to the USA to earn her MS and PhD degrees in botany under the direction of H.N. Andrews at the University of Connecticut, specializing in the study of Devonian and Early Carboniferous plants. There followed three years of post-doctoral research in collaboration with Henry Andrews, mainly on Early Devonian plants of southeastern Canada, which resulted in several joint papers. In 1975 she joined the faculty at the University of North Carolina, Chapel Hill where she presently is associate professor, teaching courses in introductory biology, plant morphology, paleobotany, and palynology. Her research, funded since 1974 by the National Science Foundation and the UNC Research Council, has concentrated on working out the morphology, systematics and evolutionary significance of early vascular plants, especially those of Early Devonian age from eastern North America. She also continues to study Early Carboniferous plants. These studies have led to extensive collecting of fossil plants in southeastern Canada and eastern USA, and also in Great Britain and other parts of Europe. She has examined museum collections of comparable fossils in England, Scotland, France, Belgium, and Australia. She has presented numerous talks (seminars) and published many technical papers on early land plants, some in collaboration with H.N. Andrews.

Henry N. Andrews received his formal education at the Massachusetts Institute of Technology, Washington University (St. Louis) and Cambridge University and has served as member of the faculty at Washington University and the University of Connecticut. He has also taught at Poona University, India, and the University of Aarhus, Denmark; paleobotanical investigations have taken him to the Canadian Arctic, Sweden, and Belgium. His earlier research activities dealt chiefly with primitive seed plants of the Carboniferous (Coal Age), and later with early vascular land plants of the Devonian period. The latter has involved extensive explorations in northern Maine and along the coasts of New Brunswick and Quebec in collaboration with Patricia Gensel, resulting in numerous joint research papers describing previously unknown plants. Andrews, a member of the National Academy of Sciences, is also the author of several books including *Ancient Plants and the World They Lived In* (1947), *Studies in Paleobotany* (1961), and *The Fossil Hunters* (1980).